SCIENCE

IS GOD SHOUTING?

THE BIBLICAL FOUNDATION FOR SCIENCE

RAY MONDRAGON

TRUE GRACE BOOKS

Published by True Grace Books, Tacoma, WA.

ISBN 978-1-964184-40-1

Cover Design: Bailey Plemel
Graphics by Canva, licensed use

Printed in the United States of America

I dedicate this book to my mother, who died as Alyce Martinez Trujillo. She started my life off with a purpose when she sat me down on a little stool at age eight and with just a few words set the course of the rest of my life. She gave me early guidance, freedom, and the DNA that has resulted in many things, including this book. Thank you, Lord, for the powerful influence of mothers who can shape the course of kids, even unknowingly.

PRAISE FOR *Science: Is God Shouting?*

"*Science: Is God Shouting?* does a wonderful job of using Scripture and logic in presenting the biblical foundation for science. Ray Mondragon demolishes the purported bases of the fable known as evolution. The Bible is not a science textbook, which is a very good thing, because textbooks need constant revision as we discover more empirical data and rely less on assumptions. The author shows how the eternal unchanging truth in the Bible provides sufficient foundation to properly interpret the data we do have. If a person reads this book to the end, he or she will come to the conclusion that the only reasonable explanation for the existence of people and the entire universe is the truth given to us throughout the Bible."

> **JOSEPH KEZELE**, MD, President of Arizona Origin Science Association, Assistant Professor of Biology at Arizona Christian University, Logos Research Associate

"For too many years the educational and media establishments have not been forthright in distinguishing between operational science and historical science. The credibility of the former has illegitimately been passed to the latter to persuade Bible students to abandon a straightforward understanding of biblical texts. Since modern science became secularized over two centuries ago, Bible interpreters have naively believed that speculative pictures of cosmic and terrestrial origins share the same credibility as scientific disciplines like physics and chemistry. Thus, instead of taking foundational biblical texts like Genesis 1-11 as straightforward narrative as the Church did until the late 1700s, such interpreters began to invent ways of making such texts appear to say whatever the latest conjectures of the historical sciences claim. The resulting 'adjustments' sweep away God's Word as authoritative, historic, verbal revelation. This book by Ray Mondragon is an easy-to-follow, reasoned defense of the Bible's claim that God not only has spoken to humanity verbally, but He speaks—no, shouts—through the created environment surrounding every person."

> **CHARLES CLOUGH**, BS in Mathematics from MIT, MS in Atmospheric Science from Texas Tech, ThM in Hebrew and Old Testament from Dallas Theological Seminary

"Today, the word 'science' is bandied about quite a bit. Unfortunately, much of what is termed modern-day 'science' is actually a mere man-centered philosophy masquerading as 'science,' which, in turn, is nothing more than pseudoscience (1 Tim. 6:20; KJV). For something to be categorized as authentic science, not only must it conform to the scientific method, with its emphasis on testability, observability, and repeatability, but it also must be consistent with the immutable principles of the Creator-God, as revealed in His eternal Word (Isa. 40:8). For a comprehensive exploration of these vital subjects, I highly recommend the following work by Dr. Ray Mondragon, who is academically credentialed in both the areas of science and Theology. Personally, I am delighted that he is one of our faculty at Chafer Theological Seminary, where he has the opportunity to present a proper understanding of these topics to the future pastors of America and the world. For a proper balance in these areas, I highly recommend the following volume, which has been authored by Dr. Mondragon."

ANDY WOODS, PhD, law degree, Pastor-Teacher at Sugar Land Bible Church, President of Chafer Theological Seminary

"Whereas theoretical science, such as evolutionary biology, cosmology and historical geology, are unaccountable, engineering has real-world accountability since an error could lead to catastrophic failure. Ray Mondragon applies an engineer's analytical mind to the monolith of science, perceived by many today to be the source of all truth. He deconstructs this edifice into component parts, showing that all are ultimately accountable to the Word of God. Every true scientist should say, 'Amen' to that."

MARSHALL C. JORDAN, MD, FACS, MS in Biophysics, retired medical doctor of 40 years, Creation Research Society member

"This work by Dr. Ray Mondragon brings out the truths of both faith-based and scientific thought, showing them to be not just capable of coexistence but synergistic to the understanding of each, separately and together. And he does so without falling prey to the tendency to present dogmatic assertions addressing lessor issues which so often confuse and obscure the fundamental truths of observational science and Scriptural revelation. With his careful and insightful approach, Dr. Mondragon demonstrates the truth of Ps. 19:1, 'The heavens declare the glory of God, and the sky above proclaims His handiwork.' This book should be a part of any honest consideration of the issues that supposedly divide science and faith."

BILL MCCULLOCH, PhD, Retired Distinguished Member of Technical Staff of a U.S. National Laboratory, recipient of first PhD in Mechanical Engineering, Texas Tech University

"*Science: Is God Shouting?* by Dr. Ray Mondragon effectively argues that science has been commandeered by an anti-God and humanistic viewpoint resulting in immense harm along various fronts. Ironically, the anti-God worldview provides an inadequate basis for true and reliable science, the field that was indisputably built within a region and time dominated by Christian thought and launched with Christian thinkers at the helm. Those bold thinkers sought to explore and explain the wonders of creation, an endeavor that was an extension of their worship of the Creator. Through solid evidence and argumentation, *Science: Is God Shouting?* exposes many flaws and weaknesses of modern science in general, and evolutionary theory in particular. It also defends why God and His revelation to humankind in Scriptures is the only adequate basis for science and should be restored to their rightful place. This work is an important contribution to the ongoing battle to retake and right the ship of science. It is well worth the read. The reader will particularly appreciate the quotations of various experts, references to other sources for further study, and the author's dry wit."

NATHAN PURTZER, BS in Mechanical Engineering, Professor of Bible and Original Languages at Central American Theological Seminary

"When you think about a scientist, there are a variety of images which can come to mind. You might see someone in a lab coat, standing before an apparatus with bubbling chemicals. You might see someone plunged in thought paused before a whiteboard covered with equations. However, every scientist who ever lived was a human being like you or I. Science is the process by which human beings explore the world around them. What do we see when we examine the natural world? That depends on the mind of the observer. Consciously or unconsciously, we interpret the world around us through a framework of presuppositions. Ray Mondragon helps us examine those presuppositions. Let's pause and consider who we are and how we think about the amazing universe in which we live. Whom do we glorify?"

MELINDA HOYT, BS in Chemistry, Materials Science and Engineering Technologist at U.S. National Laboratory

"In these days it is all too easy to accept the evolutionary thinking that all the universe has been evolving from some kind of soup of Hydrogen atoms into the complex animal, vegetable, and mineral designs that we observe today. Such thinking is easy because it relieves the average person of the necessity of considering evidence which lies all about. I find that it takes greater faith to believe that nothing has created everything than it is to believe that everything has been uniquely created by the omniscient, omnipotent, omnipresent God who has revealed Himself to us in His Book, the Bible. Dr. Mondragon leads us to see how viewing creation through a Biblical worldview provides us with the key to understanding our place and purpose in the universe. His book asks the question, 'Is God Shouting?' Indeed, He is! The Heavens declare His glory as Psalm 19 tells us. As scientific advances reveal ever deeper glimpses into the universe that appears to extend to unimaginable limits, and as science probes ever deeper into the unseen world of the microscopic and sub-microscopic nature of life, one should be filled with awe and wonder of God's creation. A secular worldview cannot adequately account for any of this, but as Dr. Mondragon argues, we must look at these things through a worldview based on the Bible as God's absolute truth."

ROBERT A. MALSEED, Bachelor of Aerospace Engineering, Air-Launched Weapon System Engineer and Laser Weapon System Analyst for U.S. Air Force, Associate Fellow Emeritus of the American Institute of Aeronautics and Astronautics

"Coming from an engineering background with a BS Civil Engineering, an understanding of what it takes for 'things' to successfully come together, and a thorough grasp of the literal, historical, and grammatical interpretation of the Scriptures, Dr. Mondragon with a ThM from Dallas Theological Seminary, and a D.Min from Chafer Theological Seminary, brings forth a compelling argument for the historicity of the creation of the universe and the Genesis Flood according to the Genesis record."

ANTHONY GRIEGO, BS in Electrical Engineering, MDiv in Systematic Theology, retired engineer at Sandia National Laboratory

"Many things called facts in our day are, in reality, merely opinions or speculations. Dr. Mondragon weaves his way through many widely held so-called scientific facts which are attacks against Scripture and demonstrates that they do not conform to reality. In an easy-to-read style he clearly and concisely shows that the demonstrable facts of science are always in agreement with the Word of God. This book gives the searching Christian answers to many pronouncements of modern science which are in conflict with biblical truth."

JAMES F. MYERS, President of Word of God Bible College, Kyiv, Ukraine

"I have known Ray for several years and appreciate that he allowed me to preview his book and offer my feedback to him. I have been glad to be able to sit under his teaching and be encouraged by his God given wisdom. His knowledge of creation science is considerable as he has spent many years in its study, which shows in the depth of information contained in this book. I hope that you will find it to be a treasure trove of new and interesting information and sound analysis as I have."

TOM GOODWIN, AS Semiconductor and AS Radiographic technology fourteen years at Intel Corp.

CONTENTS

APPENDIX

ACKNOWLEDGMENTS

This book could not have been written without the help of key friends. Since I am not a writer, the following help was crucial. Jerry Coryell (computer support specialist, former president of Creations Science Fellowship of NM), Anthony Griego (electrical engineer, seminary professor, and Bible teacher), Jim MacGillivray (Chafer Theological Seminary board member and retired businessman), and Tom Goodwin (AS Semiconductor and AS Radiographic technology) read the entire manuscript and made helpful corrections and comments.

Dr. Joseph Kezele (medical doctor and university professor) and Nathan Purtzer (mechanical engineer and seminary professor) provided much technical, theological, and grammatical correction, advice, and help.

Dr. John Baumgardner (geophysicist, researcher, and university professor) added fascinating scientific details along with helpful technical corrections. You will especially enjoy his insights about his Flood model and the second law of thermodynamics.

Nancy Northcott (editor and co-author of two bestselling novels) was the most helpful in the actual writing of the manuscript. She offered not only grammatical corrections but re-wrote portions of my scratchings. She also made additions to the text from her own research and added several quotations from scientists. I consider her nearly a co-author of this book. She was also helpful in encouraging me to continue on to publication.

Certainly, this book could not have been published without the guidance, enlightenment, content, motivation, and enablement of our Lord Jesus Christ through the work of the Holy Spirit. Our Lord has worked throughout my life from even before my regeneration and providentially provided everything needed for this book. As I attempted to put ideas down on paper (digital "paper"), I often sensed providential content being provided at important moments.

As you will see, much of this book comes directly from His inspired Word. I hope every word in it glorifies His name and His Word.

FOREWORD

D. RUSSELL HUMPHREYS, PH.D. PHYSICS

Most educated people believe in deep time, the idea that the world is billions of years old and that living things evolved, gradually getting better and better, during that long time. I had been raised and educated to believe in that idea. Even after I accepted Jesus Christ as my Savior from sin, as a 27-year-old graduate student in physics, I continued to try to believe in deep time. But as I studied the Bible and tried to take it at face value, I found that it did not allow deep time. I was not happy with various theological systems which re-interpreted the words of Scripture to allow billions of years to be shoved into it.

I continued in this quandary for about one year. Then two things happened. First, a teacher mentioned that Noah's flood might indeed have been the worldwide cataclysm that Genesis chapters 7 and 8 record. Then the geologic layers containing the fossils would have been deposited within about one year about 5,000 years ago. That would erase half a billion years' worth of "geologic ages," the bulk of the alleged evidence for deep time.

Second, my pastor gave me a copy of a book by Dr. Henry M. Morris, *Biblical Cosmology and Modern Science*. It outlined much scientific evidence that the world is thousands, not billions, of years old. That evidence, plus over fifty years of personal scientific research since then, is what has made me a young-earth creationist.

I found out another interesting thing during my career as a physicist at a National Laboratory: most scientists do not believe in deep time because of any evidence. They believe it because they believe most other scientists believe in deep time. The idea was just "in the air" as they were educated, and they simply absorbed it without much thought, as I did. The irony is that, usually, each scientist thinks other scientists have done the homework that he himself has neglected. The idea just keeps rolling on because of sheer intellectual inertia! That is something both scientists and non-scientists need to know as they evaluate the ideas of deep time versus the young world of Scripture.

This book will help you do the evaluation. Ray Mondragon carefully lays out what we need: a perspective on science, the deficiencies of the evolutionist worldview, the reliability of Scripture, and a Biblical foundation for science. Along the way he discusses evidence, but his main goal is to equip us to sort out the evidence with clarity. The thoughtful reader will find this book well worth his time.

D. Russell Humphreys, Ph. D.
Fellow, Creation Research Society
Retired from Sandia National Laboratories, Albuquerque, NM
Author of *Starlight & Time*

FOREWORD

JOHN BAUMGARDNER, PH.D. GEOPHYSICS

With all the technological marvels most of us enjoy, such as social, weather, and GPS navigation aps on our phones, it is not surprising that most of us generally tend to regard science in a favorable way and even with a measure of awe. Many people today therefore find it difficult to grasp that the scientific enterprise with its many benefits could at the same time also be a vehicle of grave deception and error.

Remarkably, many of the pioneers of modern science such as Kepler, Pascal, Newton, Boyle, Faraday, Joule, Maxwell, and Kelvin were earnest and outspoken in their confidence in God and the Bible as an accurate account of the past. However, an ominous change in character of science occurred through an intellectual movement that came to prominence in the 1700s. This movement, known as the Enlightenment, led by such anti-God intellectuals as Voltaire, Diderot, Hume, and Spinoza, insisted that nature could be understood with no reference to God or the supernatural whatever.

Within a few decades, Enlightenment thinking with its concept of deep time came to dominate the interpretation of the rock record as the field of geology began to develop in the early 1800s. The view that slow and gradual change had produced the entire geological record was soon carried over to biology with the assertion that slow and gradual changes in the living world had produced the astonishing diversity of plants and animals we currently see around us.

Of course, these perspectives are radically at odds with the Bible's account of God's supernatural creation of the heavens, the earth, plants, fish, birds, land creatures, and people only a few thousand years ago. They are also radically at odds with the Bible's account of a subsequent world-destroying cataclysm sent by God because of the out-of-control wickedness of a fallen human race only some 1,600 years later.

Today these ideas of deep time and evolution are taught so pervasively and so persuasively that few human beings on earth have not been deeply influenced. Indeed, the vast majority of Christians today are largely unaware of how readily these deceptions collapse in light of honest and careful science combined with the revelation we have in God's word about the past. This gap in awareness has left many Christians with a severely compromised ability to trust the Bible in a wholehearted way.

My friend Ray Mondragon does a masterful job in this book of cutting through the maze of confusion that Enlightenment philosophy has produced relative to issues of science and the Bible. I highly recommend it to anyone who is unaware of how marvelously good science affirms a straightforward, according to the normal meaning of the words, understanding of the Biblical text.

John Baumgardner, PhD, Geophysics
Research Professor Emeritus, Liberty University
Retired from Los Alamos National Laboratory

PREFACE

I am not a writer. I chose to be an engineer not only because I like designing structures but also because I am weak in English and the humanities. My favorite toy as a kid was a Gilbert Erector Set, not Scrabble. Fortunately, I have an expert professional editor and author who has made not only extensive edits but valuable contributions to this book.[1]

Because of my engineering background, this book may not follow the conventional rules of writing and footnoting. I will try to at least spell words correctly (thanks to spell check I must admit). I will even attempt to throw in some dry humor to help keep you reading. Some of it you may miss unless you look for it. Ok, ok, that is a trick to get you to at least look for the humor. In the process, who knows, you may even get something helpful out of the reading.

I don't have any special area of expertise in science, except in its adjunct engineering, so this will not be a highly technical read. You will not need to be steeped in scientific jargon to follow the thoughts in this book. I will, however, attempt to accurately present scientific issues and their relationship to theology. Actually this book will give more attention to the truth found in Scripture than truth discovered by science.

1. Nancy Northcott was my editor.

I have been studying the area of creation science, off and on, for over thirty five years and have grown in my knowledge of it. My appreciation of God's creation has grown as well. I am eager for you to better understand the Creator's intricate plan and its relationship to science.

If I do have an area of expertise, it would be in interpreting Scripture. The fancy words for that are hermeneutics and exegesis. Though I teach it often in theological seminaries, I do not consider myself a scholar. I have never wanted to be thought of as a nerd though some would argue at my success.

I have not studied or read much about the philosophy of science, and I do not have any training at all in philosophy in general. It is that engineering thing again. I mention this because many areas of science have become riddled with philosophy, rather than with verifiable fact. When I address philosophical issues in science, my lack of educational background will be an advantage because my thinking is fresh, untainted by the current trend of science toward secular philosophy. My opinion of scientific philosophy is the same as Nobel prize winning physicist Richard Feynman, who wrote, "Philosophy of science is about as useful to scientists as ornithology is to birds."[2]

My purpose in writing this book is two-fold, to encourage you to seek truth and to help you learn to evaluate all truth claims. Most of the ideas that bombard you throughout every day come from the minds of fallible humans. Many will be false—if not outright lies. How can you know? More specifically, how can you know what is true and why it is true? This book is all about discovering, understanding, and even interacting with absolute truth, especially in the natural realm. Should that not also be the goal of science?

About the title of this book: I got the idea from James Nickel's book, *Mathematics: Is God Silent?*[3] In his book, Nickel makes a strong case that we could not have mathematics without the God of the Bible. If you read his book, after you finish, you will answer the question of his title with a resounding NO! God is definitely not silent. My hope is that after you read my book, and I hope you do—you will answer the question in my title, *Is God Shouting?*, with a resounding YES!!! I hope your response also will be "Holy, Holy, Holy, is the Lord of hosts. The whole earth is full

2. http://www.goodreads.com/author/quotes/1429989.Richard_P_Feynman?page=2.

3. Nickel, James, *Mathematics: Is God Silent*, Ross House Books, 2001.

of His glory."[4] I hope you will be able to add that "astrophysics is *shouting the glory of God* and astronomy is declaring the work of His hands. Day to day pours forth speech, and night to night reveals knowledge."[5]

My goal is to contribute to the enhancement of science, one based on God's shouts to us through His creation. Above all, I hope this book will bring glory to the Creator of all things, our Lord Jesus Christ.

> For by Him all things were created, both in the heavens and on earth, visible and invisible, whether thrones or dominions or rulers or authorities—all things have been created through Him and for Him. He is before all things, and in Him all things hold together.
>
> Colossians 1:16-17

> One way to learn the mind of the Creator is to study His creation. We must pay God the compliment of studying His work of art and this should apply to all realms of human thought. A refusal to use our intelligence honestly is an act of contempt for Him who gave us that intelligence.
>
> **ERNEST WALTON, 1951 NOBEL PRIZE IN PHYSICS**

4. Isaiah 6:4 and Revelation 4:8, "*HOLY, HOLY, HOLY is THE LORD GOD, THE ALMIGHTY, WHO WAS AND WHO IS AND WHO IS TO COME.*"
5. This is my paraphrase of Psalm 19:1-2 with synonymous parallelism.

PART I

INTRODUCTION

In the beginning
God created the heavens and the earth....
Then God said, "Let there be light;"
and there was light.

GENESIS 1:1,3

YAHWEH, THE CREATOR

CHAPTER 1

A PERSPECTIVE ON SCIENCE

Recently, I watched a series of videos on the solar system.[1] I was struck by the fact that even scientists know little about the heavens immediately over their heads. My attention was drawn to a comment that recent observations with more sophisticated equipment require us to revise our thinking about the formation of our solar system.

For example, the narrator said planet Mercury is "the enigma of the solar system's evolution." The density of a planet relates to the planet's size, except for Mercury. Because Mercury is much denser than expected for its small size, most astrophysicists assume something strange happened during what they believe to have been its long, evolutionary formation. The existence of its magnetic field surprises them also, especially given its slow rotation of once every fifty-seven earth days. These observations do not agree with standard evolutionary theories. Shouldn't we at least consider the possibility that something in the underlying evolutionary model might be flawed?

Venus, the planet closest to Earth, also puzzles secular scientists who have thought for years that Venus and Earth had similar beginnings but followed different paths in their evolution. The video said that, unlike Earth, Venus is the only planet that takes longer to rotate on its axis

1. *The New Frontier - The Inner World*, video series, season 1, episode 3.

than it takes to orbit the sun. Unlike Earth, it has no magnetic field. Its dense atmosphere is composed of 97% carbon dioxide. It has a surface temperature of 840 degrees F, a high atmospheric pressure, gentle surface winds, but winds of 186 miles per hour at 62 miles above the surface. Doesn't sound much like Earth, does it? If evolution is true, why are these similarly sized and situated planets so different? Could evolution be wrong, or is our perception of the planets wrong? We need to look at the facts so we can decide intelligently.

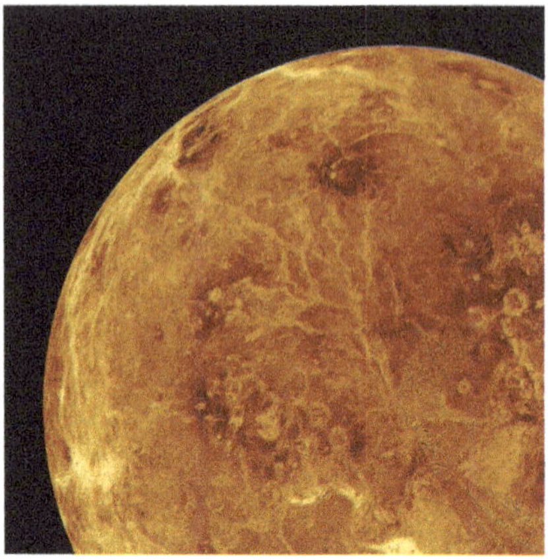

FIGURE 1 VENUS

In other episodes, the video series discussed each planet and its own peculiarities. Throughout, evolutionary assumptions were made for a fourteen-billion-year-old universe, give or take a few billion years. Even though the series revealed numerous evolutionary contradictions, it never questioned the validity of its basic evolutionary assumptions. Is it out of the question to ask for proof of these fundamental assumptions? Of course not. Any true scientist would.

Some scientists are waking up and beginning to question these long-held beliefs. Today's news is full of admiration for the James Webb Space Telescope and its amazing discoveries of galaxies far beyond our previous observations. *World* magazine in its "Quotables" page reported a typical response from astrophysicists: "University of Kansas astronomer Allison Kirkpatrick describes to *Nature* how new images from the James Webb Space Telescope have upset theories of galaxy evolution. 'Right now I find myself lying awake at 3 in the morning wondering if everything I've ever done is wrong.'"[2] Isn't she suggesting evolutionary ideas may have defects? Wouldn't she be right, even scientific to want to examine them?

2. "World Magazine," Volume 37, Number 16, August 27, 2022, 21.

> That God existed before there were human beings on Earth, that He holds the entire world, believers and non-believers, in His omnipotent hand for eternity, and that He will remain enthroned on a level inaccessible to human comprehension long after the Earth and everything that is on it has gone to ruins; those who profess this faith and who, inspired by it, in veneration and complete confidence, feel secure from the dangers of life under protection of the Almighty, only those may number themselves among the truly religious.
>
> **DR. MAX PLANCK, 1918 NOBEL PRIZE IN PHYSICS**

THE TAKE AWAY: Similar mysteries like those related to our solar system and astrophysics are found in every area of science. In fact, all areas of science challenge our finite minds. The more we learn, the more we realize the limitations of our understanding.

The concept of gravity, which is found everywhere in the universe and is included in the property of all matter, should be well understood by now, shouldn't it? But even gravity is hard for scientists to quantify.[3]

Sir Isaac Newton, the first to describe gravity, was baffled by it. Michael Guillen, who earned PhDs in physics, mathematics, and astronomy, wrote of Newton, "Even Sir Isaac couldn't make heads or tails of what gravity is, really. 'Gravity must be caused by an agent, acting constantly according to certain laws,' he ventured, 'but whether this agent be material or immaterial, I have left to the consideration of my readers.' In other words, he was saying, 'Hey, your guess is as good as mine.'"[4]

Albert Einstein's equations on general relativity and gravity improved our understanding of Newton's theories, but they also introduced more unknowns. More recently, astrophysicist Karl Schwarzschild discovered that Einstein's theory predicted the existence of massive objects that cannot be seen.[5] We call them black holes.

> Gravity explains the motions of the planets, but it cannot explain who sets the planets in motion.
>
> **ISAAC NEWTON**

---

3. Guillen, Michael, *Amazing Truths–How Science and the Bible Agree*, Zondervan, 2016, 55.
4. Ibid, 55.
5. Ibid, 57-58.

Black holes are not the only huge mysteries related to gravity and the universe. Other discoveries indicate that galaxies and galaxy clusters must contain some sort of strange material that is invisible to even our most powerful telescopes. This material has been dubbed dark matter. Guillen wrote, "They haven't yet identified what it is exactly—or even established that it truly exists—but it is not for lack of trying."[6]

Not only are black holes and dark matter full of mystery but add to that dark energy, which is the only known way of explaining why galaxies are accelerating instead of decelerating as they move apart from one another in contradiction to evolutionary theory.[7] Again, we see contradictions. Such contradictions afflict every area influenced by the concept of evolution.

> The big bang model assumes the existence of so-called dark matter and dark energy. Neither of these have ever been observed in the laboratory. However, they are desperately needed to uphold the model and to avoid contradiction with the observational data. The big bang also has no reasonable explanation for the virtual non-existence of antimatter. This is an enemy of normal matter: when they meet, they annihilate each other with intense release of energy. But according to the model, an equal amount of antimatter and matter should have been generated.
>
> **DR. MARKUS BLIETZ, ASTROPHYSICIST**

DOES SCIENCE GIVE US THE WHOLE PICTURE?

As our understanding of gravity has grown, so has our realization that we know little about it. We continue to be puzzled by what should be elementary scientific knowledge. What could be more elementary than why we remain on earth rather than flying off into space? If the evolutionary model doesn't answer such basic questions, could the whole model be based on faulty assumptions?

The point I am making is that science, as marvelous a pursuit as it is, does not give us all the answers. In fact, it often leaves us with more questions than answers. Guillen understood the problems caused by our lack of knowledge: "All told, astronomers have concluded that dark energy

6. Ibid, 59.
7. Ibid, 59.

comprises some 68 percent of the total universe and dark matter, about 27 percent. That means that only five percent of the entire universe is visible to us. That astounding revelation bears emphasizing. Everything we call scientific knowledge is based on but a pittance of what there is to know about our world. *Ninety five percent* of it is hidden from us."[8]

FIGURE 2 ANDROMEDA GALAXY

THE TAKE AWAY: Think about that for a moment. According to Guillen, 95% of scientific data is still unknown to us!

Despite science's limitations and its ignorance about most of creation, many in our culture have elevated it to almost divine status. Commentator Dennis Prager wrote, "Modern men and women have substituted 'experts' for prophets and priests. Science is the secular religion, and 'experts' are its prophets and priests."[9]

> An expert is a person who has made all the mistakes that can be made in a very narrow field.
>
> **NIELS BOHR, 1922 NOBEL PRIZE IN PHYSICS**

The pandemic that began in 2020 revealed the many problems that develop when we view secular scientists as godlike and depend on them for solutions to our problems. During the pandemic, many insisted we "follow the science" while at the same time ignoring the abundant data that contradicted their assertions and accompanying demands. The consequences of their inaccurate "scientific" pronouncements about Covid-19 have been staggering, not only medically, including the unnecessary deaths of many, but also economically and geopolitically.[10] Many of America's small businesses were forced to close, shelves in

8. Ibid, 60.

9. http://daily-journal.com/opinion/columnists/dennis-prager-has-the-lockdown-worked/article_50c61c8e-7e29-11ea-b9eb-4f6ca587cbe4.html.

10. http://patriotpost.us/alexander/83599-fact-v-fiction-the-chicom-virus-vaccines-infections-and-deaths-2021-10-20.

remaining large retailers were almost empty, and masses of people were without jobs. Even more dangerous, our government used false scientistic information to curtail freedom of speech, assembly, and religion.

Government's vaccine mandates—for vaccines that even their own scientists admitted are only 3% effective in preventing the virus—usurped individual responsibility in the name of the good of many.[11] This misapplied science attacked the Judeo-Christian foundation that had made America a bastion of liberty.

THE TAKE AWAY: Scientists are not gods. They are not even reasonable facsimiles. We must carefully examine their pronouncements before accepting them as true.

Many today believe scientific truth comes as close to absolute truth as possible. As we have seen briefly, even in areas that seem absolute, such as the solar system and gravity, much, if not most of the knowable is still not understood. Man's knowledge, even of science, is finite. We need a better approach to understanding the natural realm than science currently supplies.

> When I tell a young person: Look, there is a new star, a galaxy, a neutron star 100 million light-years away, yet the protons, electrons; neutrons and mesons which are found there are identical with those which are found in this microphone... Identity excludes probability. That which is identical is not probable...Therefore there is a cause, outside of space, outside of time, the master of being, which made being to be in this way. And this is God..." "The being–I am speaking scientifically–which has caused things to be identical at a distance of billions of light-years, exists. And the number of identical particles in the universe is 10 raised to the 85th power...Do we wish then to take in the song of the Galaxies? If I were Francis of Assisi I would say: O Galaxies of the immense heavens, give praise to my Lord, for He is omnipotent and good. O atoms, O protons, O electrons, O bird-songs, O blowing of the leaves and of the air, in the hands of man as a prayer, sing out the hymn which returns to God!
>
> **ENRICO MEDI, CHAIR OF GEOPHYSICS, UNIVERSITY OF PALERMO**

11. http://www.informationliberation.com/?id=62617.

WHY SHOULD YOU READ THIS BOOK?

Many, maybe even you—I know I did at one time, believe science is clear cut and free from interpretation. They assume it to be objective and unprejudiced, that science is science and can only give us real facts with no underlying biases. Anyone in a white lab coat is presumed to have examined all available information and come to right conclusions. After all, science has gained, often rightfully so, a high degree of respect. We have science to thank for the many conveniences and gadgets we use every day. We live longer and enjoy more luxuries because of scientific advances. Entire professions, like engineering and medicine, depend on the findings and principles developed by science.

How dare I then question the main assertions of science? But that's exactly what I am going to do. Modern secular science needs to be challenged by everyone who loves God. I hope this book will give you facts that will make you a powerful and knowledgeable advocate for truth in science.

Even though the culture contradicts itself by believing that relative truth is the only truth, it also believes scientific truth comes as close to absolute truth as possible. I believe God has given us a better approach for understanding the natural realm, including all fields of science, than is currently understood in the secular world with its relative truth. God is able to do so because He is absolute truth.

THE TAKE AWAY: Scripture gives an inspired and, therefore, accurate understanding of the natural realm. Without its foundation, science cannot achieve maximum success. We limit our potential when we fail to examine all things scientific, in fact all things in life, from a biblical worldview.

What question should you be asking yourself about now? That's right. What does God have to say about science? Buckle your seatbelt because you are about to find out.

> I want to know how God created this world. I am not interested in this or that phenomenon, in the spectrum of this or that element. I want to know his thoughts; the rest are details.
>
> ALBERT EINSTEIN, 1921 NOBEL PRIZE IN PHYSICS

"Where were you when I laid the foundation of the earth?
Tell Me, if you have understanding.
Who set its measurements? Since you know.
Or who stretched the line on it?
On what were its bases sunk?
Or who laid its cornerstone,
when the morning stars sang together
and all the sons of God shouted for joy?"

JOB 38:4-7

ELOHIM, THE CREATOR

CHAPTER 2

THE NATURE OF SCIENCE

So that we are on the same page when we talk about science, we'll begin by defining it. If you go to a standard dictionary, you will find a definition similar to this one in the Cambridge Dictionary. Science is knowledge from "the careful study of the structure and behavior of the physical world, especially by watching, measuring, and doing experiments, and the development of theories to describe the results of these activities."[1] A similar but better definition comes from the Britannica Dictionary. Science is "any system of knowledge that is concerned with the physical world and its phenomena and that entails unbiased observations and systematic experimentation. In general, a science involves a pursuit of knowledge covering general truths or the operations of fundamental laws."[2]

> Physics filled me with awe, put me in touch with a sense of original causes. Physics brought me closer to God. That feeling stayed with me throughout my years in science. Whenever one of my students came to me with a scientific project, I asked only one question, "Will it bring you nearer to God?"
>
> **DR. ISIDOR ISAAC RABI, 1944 NOBEL PRIZE IN PHYSICS**

1. https://dictionary.cambridge.org/dictionary/english/science.
2. https://www.britannica.com/science/science.

Why do I prefer the latter definition? Why do I say it is better? Because the first limits scientific knowledge to that obtained from the study of the physical world. If you continue reading my book, you will soon realize that science needs to include other sources of knowledge, not simply that from the physical realm. I prefer to describe science more simply as the search for knowledge from any reliable source that accurately describes the physical realm. I will show that the Creator of all things revealed such indispensable knowledge about His creation. He has revealed insights in His Word that cannot be discovered by experimental methods. Scientists will lose their way if they don't include His revelations in their examinations.

Most people are unaware that science has two distinct branches, observational science and historical science. You have probably heard of observational science, though, if you are like me, you just called it science. Historical science is a newcomer to the science field, one you need to view with caution.

Each branch has a different purpose, operating principles, and underlying assumptions. If we don't understand their differences, we cannot properly evaluate their scientific statements. We may even start doubting God. Yes, understanding the underlying assumptions of both branches of science is that important.

WHAT IS OBSERVATIONAL SCIENCE?

Observational science, which is also referred to as operational or empirical science, includes any scientific endeavor that can be performed in present time. In general, observational scientists observe the natural realm directly, anything from the smallest quark to the greatest galaxy. They formulate explanations about them called hypotheses and then perform experiments or formulate scientific models to validate or disprove their hypotheses. They repeat their observations and experiments over and over, thus confirming or disproving their conclusions. After sufficient testing by many skilled experimenters, the hypotheses that pass the experimentation

OBSERVATIONAL SCIENCE
→ Observations in Present

HISTORICAL SCIENCE
→ Traces of Past

tests progress from theories into scientific laws. Finally, observational scientific principles are utilized in developing the gadgets, technologies, medicines, structures, weapons, and tools, the multitude of things we are glad to be able to use every day.

The broad category of observational science is generally divided into two sub-categories, the physical sciences and the life sciences. Examples of physical sciences within observational science are physics, chemistry, geophysics, astrophysics, and hydrology. Examples from the life sciences include biology, genetics, zoology, and botany.

Some would add to these two the social sciences, which focus on an individual's behavior in society. Others debate whether they should be considered sciences at all. I don't believe they should. They certainly can't be repeated experimentally with any exactitude.

> Science is a game...In the presentation of a scientific problem, the other player is the good Lord. He has not only set the problem but also has devised the rules of the game—but they are not completely known, half of them are left for you to discover or to deduce. The uncertainty is how many of the rules God himself has permanently ordained, and how many apparently are caused by your own mental inertia, while the solution generally becomes possible only through freedom from its limitations. This is perhaps the most exciting thing in the game.
>
> **ERWIN SCHRODINGER, 1933 NOBEL PRIZE IN PHYSICS**

WHAT IS HISTORICAL SCIENCE?

DATA
= Traces left by an event

HISTORICAL FACT
= Data + Interpretation

Historical science is a relatively new, and currently popular branch of secular science. Its practitioners study past events and conditions rather than the physical phenomena that fascinate observational scientists. Because all events are unique, they are non-repeatable. Because they happened in the past, they are non-observable. Because they are outside the reach of observational science, few conclusions of historical science are truly conclusive. All are open to debate.

Historical science depends on data from the past, referred to as traces, to make their deductions, but many events leave no traces and, therefore, cannot be known. Your great-great-great-grandmother's fifty-fourth birthday, for example, may lack trace data. Now, consider your great-great forefather who lived in Christ's time or even farther back in the time of King Nebuchadnezzar of Babylon, and you see the problems faced by historical scientists. Instead of observing the events themselves, they can only observe the traces left by those events, and most events are woefully lacking in traces.

Traces of most historical events have been lost or destroyed over time. Even more critical, few if any participants remain to explain or interpret with certainty whatever meager data has survived.

THE TAKE AWAY: Because of limited data and limited verification, conclusions drawn about events are dramatically more speculative than those derived from observational science. Their accuracy depends on the quality and quantity of the traces and the expertise and lack of bias (preconceived ideas) of the scientist, not on observable, repeatable phenomena. Subjective assumptions are an unfortunate but inevitable by-product of historical science.

Ian T. Taylor (1945-2001), a metallurgical engineer, understood this problem. When he was well into adulthood, Taylor believed in Christ as his Savior. After his conversion, he wrote, "Presuppositions can not only make us see what does not exist but also prevent us from seeing what does."[3] Historical science produces too many followers who see what they want to believe and ignore what contradicts those beliefs.

To keep their research as free from presuppositions as possible, legitimate historical scientists search for eyewitness records, the most accurate traces of an event. Going back to your great-great-great-grandmother, you know how little information about her life you, a direct descendant, have, unless she left behind a diary or journal. Even then, only a small portion of her life is revealed and even that is from her perspective alone.

Written accounts from ancient times are even rarer. Even so, conclusions from the few examples that remain are generally more reliable than scientific guesswork related to other trace data. We are reminded

---

3. Nickel, 6.

again of the danger that faces science because of man's tendency to see only that which supports his preconceived ideas.

The secular science that examines ancient events totally discounts the most important and only inspired document written about them, the Bible. They set aside a wealth of accurate information and unsurprisingly come to erroneous conclusions. Subjectivism based on presuppositions not supported by Scripture threatens the accuracy of most of historical science's conclusions.

WHAT ARE SOME SPECIFIC AREAS STUDIED BY HISTORICAL SCIENCE?

Archeology, criminology, and historical geology are branches of historical science. In each area, the accuracy of perceived theories depends on the quality and quantity of the traces. It also depends on the subjective interpretive skills of the scientist. A wide range of viable reconstructions and possible conclusions can come from any set of data traces, depending on the scientist's biases.

Archaeologists study artifacts from ancient and not so ancient cities, sites, or civilizations in an attempt to reconstruct life at the time the site was active. Artifacts, the traces, can include pottery, tools, bones, weapons, coins, walls, stone foundations, architectural features, anything that has survived the ravages of time. As we've seen, the most helpful artifacts are inscriptions and written records, including those chiseled on tablets and pillars.

While archeology studies long past events, criminology uses similar techniques to reconstruct more recent events. Forensic scientists study crime scenes to determine the truth about an event. They analyze artifacts (traces) like bullets, knives, blood spatter, fingerprints, as well as the items left behind by the victim and the perpetrator. Eyewitness accounts help in determining the meaning of the traces. Often the results of a forensic scientist's analysis are used as evidence in a court of law, where, again, the quality of the evidence depends on the quality and quantity of the traces. It also depends on the forensic scientist's assumptions and biases. We know poor investigative skills can lead to the arrest and even conviction of the innocent. All interpretations are flawed by man's innate imperfections and limited knowledge. Think again about that ninety-five percent information void.

> Consider finding a dead body in the park. Did the person die from natural causes, or was some other factor involved? If you find a knife in the back, then it is logical to assume that some outside intelligence was involved. However, if you start by assuming that the death occurred from natural causes, then you can never arrive at the correct conclusion. It is not a question of science, but the foundational assumptions you take to the data.
>
> **DR. GEOFF DOWNES, WOOD PHYSIOLOGIST**

Historical geology is a branch of historical science that studies the geological record, including the geological layers and the fossils contained in them. It attempts to reconstruct ancient history apart from any written records. Its assumptions about the origin of life cannot be verified. We will discuss historical geology in greater detail in later chapters. A not surprising summary of that information: The interpretation of geological traces is affected by the biases and preconceptions of the scientist examining them.

THE TAKE AWAY: I cannot stress enough that personal, subjective interpretation is involved in every branch of historical science. Subjectivism exists not only in which traces are studied and the interpretation of them but also in its assumptions, many of which cannot be verified by the trace data. This interpretative element is at the heart of the conflict between secular scientists and those with a biblical worldview. The disagreement is not over the physical data, the traces, but over the interpretation, the explanation, of that data.

> The gift of mental power comes from God, Divine Being, and if we concentrate our minds on that truth, we become in tune with this great power.
>
> **NIKOLA TESLA, PHYSICS**

We must be careful to understand the limitations of all branches of historical science. Their conclusions are often used to try to discredit God's Word and even God Himself. We must choose, instead, to approach every area of science as did Gottfried Leibniz (1646–

1716), who has been called the last "universal genius."[4] He said, "It is especially in sciences…that we see the wonders of God, His power, wisdom, and goodness…that is why, since my youth, I have given myself to the sciences that I love."[5]

WHAT CAN WE CONCLUDE ABOUT SCIENCE?

While observational science is verifiable and reproducible and can yield conclusions that can be accepted as fact, its deficiency is that much of physical reality remains unknown, ninety-five percent as we have learned. New data uncovered with new, more precise equipment leads to new and different conclusions.

In the case of historical science, not only are most of its details unknown, but its conclusions are largely interpretive rather than verifiable and reproducible. When it comes to comprehensive knowledge, both observational and historical science leave more unknown than they uncover.

By realizing the profound differences between verifiable scientific information (observational science) and unprovable subjective interpretations (historical science), we can avoid being deceived by unproven but highly promoted scientific claims. For example, no one has ever observed one species of any animal that has evolved into another species. We will recognize that much of settled science is not settled at all but merely subjective conjecture. If we are alert, we will be able to discern scientific flaws and the prejudicial foundations that undergird them.

As we continue our study, keep reminding yourself that even the most brilliant scientist has access to no more than five percent, if even that, of all that comprises the physical universe. Understanding that deficiency is essential if we are to understand the limitations of modern secular science.

When we make scientific inquiries, can we find another source Who knows all the knowable, created everything ever created, and was an eyewitness to all events in human history? As you have probably

4. Stanford Encyclopedia of Philosophy, https://plato.stanford.edu/entries/leibniz.
5. Nickel, 128.

guessed, God is that perfect source. His Word is the reliable written record of ancient history, including the history of the natural realm.

THE TAKE AWAY: God was present at creation because He is the Creator. He understands every atom of creation because He created each one. Of course, we should go to His revelation to gain an accurate understanding of our world! Any conscientious scientist could do no less. Comparing His Word to the visible natural realm will lead us to logical, accurate, and satisfying scientific conclusions. And we will enjoy doing it because we won't have to fudge our results to make them fit any preconceived ideas.

On August 20, 1897, Sir Ronald Ross made his landmark discovery that malaria is transmitted to people by Anopheles mosquitoes. On that day of discovery he wrote the following poetic words in his Journal:

"This day relenting God
Hath placed within my hand
A wondrous thing; and
God be praised.
At His command,
Seeking His secret deeds
With tears and toiling breth,
I find thy cunning seeds,
O million-murdering Death.
I know this little thing
A myriad men will save.
O Death, where is they sting?
Thy victory, O Grave?"

SIR RONALD ROSS, 1902 NOBEL PRIZE IN MEDICINE

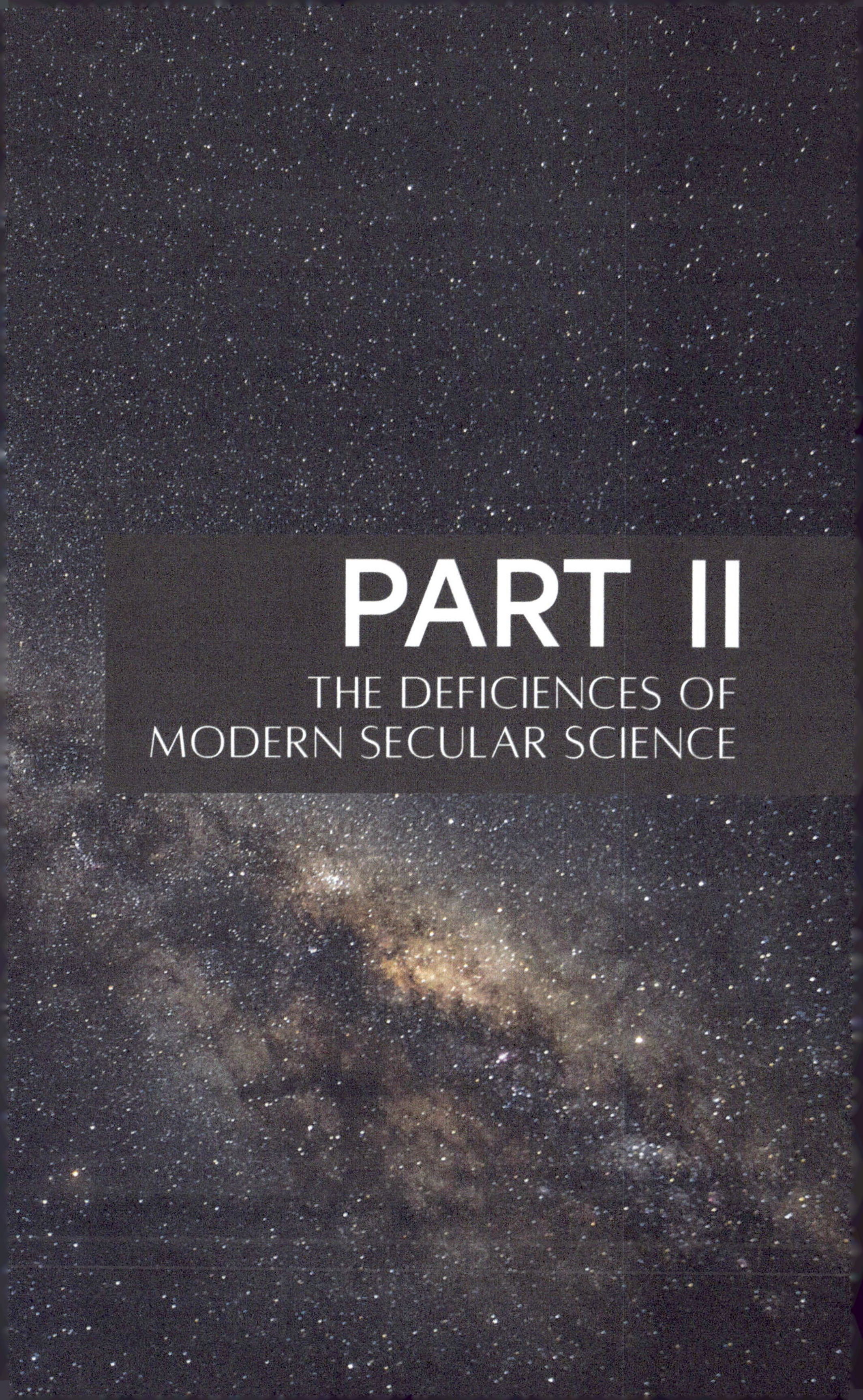

PART II

THE DEFICIENCES OF MODERN SECULAR SCIENCE

But now, thus says the Lord, your Creator, O Jacob,
and He who formed you, O Israel,
"Do not fear, for I have redeemed you;
I have called you by name; you are Mine!"

ISAIAH 43:1

Thus says the Lord, your Redeemer,
and the one who formed you from the womb,
"I, the Lord, am the maker of all things,
stretching out the heavens by Myself
and spreading out the earth all alone..."

ISAIAH 44:24

YAHWEH, THE CREATOR

CHAPTER 3

SCIENCE AND WORLDVIEWS

Today, science is, for the most part, secular and, therefore, riddled with problems and inaccuracies. To understand its inadequacies, we must understand the early development of modern science, which began about AD 1500, little more than five-hundred years ago. Join me on a short little trip into our scientific past.

HOW DID MODERN SCIENCE BEGIN?

Many today would be surprised to know modern science had its origin in men who were committed to the Scriptures and motivated to study the natural realm because of their biblical understanding. Isaac Newton (1642–1726), the famous English mathematician, physicist, astronomer, and discoverer of gravity, in his spare time wrote commentaries on Daniel, 2 Kings, and Revelation.[1] Although he was considered the premiere scientist of his day and possibly of all time,[2]

1. http://www.newtonproject.ox.ac.uk/texts/newtons-works/religious.

2. "In a 2005 survey of members of Britain's Royal Society (formerly headed by Newton) asking who had the greater effect on the history of science, Newton or Albert Einstein, the members deemed Newton to have made the greater overall contribution. In 1999, an opinion poll of 100 of the day's leading physicists voted Einstein the "greatest physicist ever," with Newton the runner-up, while a parallel survey of rank-and-file physicists by the site PhysicsWeb gave the top spot to Newton. New Scientist called Newton 'the supreme genius and most enigmatic character in the history of science." https://en.wikipedia.org/wiki/Isaac_Newton Einstein kept a picture of Newton on his study wall alongside ones of Michael Faraday and James Clerk Maxwell." https://en.wikipedia.org/wiki/Isaac_Newton.

because of these books and his numerous theological treatises, he was also praised as a profound theologian. The intertwining of theology and science in his life impacted his scientific advancements. He wrote, "All my discoveries have been made in answer to prayer." And, "I have a fundamental belief in the Bible as the Word of God, written by men who were inspired. I study the Bible daily."[3]

Louis Pasteur (1822-1895), a pioneer of bacteriology, also understood the essential link between God and scientific understanding. "The more I study nature, the more I stand amazed at the work of the Creator," he wrote. "Science brings men nearer to God." [4]

Robert Boyle (1627–1691), considered the founder of modern chemistry, argued that the study of science should lead to the worship and adoration of God.[5] "When with bold telescopes I survey the old and newly discovered stars and planets, when with excellent microscopes I discern the inimitable subtlety of nature's curious workmanship; and when, in a word, by the help of anatomical knives, and the light of chemical furnaces, I study the book of nature, I find myself oftentimes reduced to exclaim with the Psalmist, How manifold are Thy works, O Lord! In wisdom hast Thou made them all!"[6]

Francis Bacon (1561-1626), who was instrumental in developing the scientific method, wrote, "There are two books laid before us to study, to prevent our falling into error: first, the volume of the Scriptures, which reveal the will of God; then the volume of the Creatures, which express His power."[7] We would be wise to study both books, using Scripture to better understand the creation. Otherwise, as said Bacon, we will fall into error.

> We live in a Newtonian world of einsteinian physics ruled by Frankenstein logic.
>
> **DR. DAVID RUSSELL, CHEMISTRY**

3. https://quotefancy.com/isaac-newton-quotes.
4. Lamont, Ann, *21 Great Scientists Who Believed the Bible,* Creation Science Foundation, 1995, 153.
5. https://truthvanguard.wordpress.com/category/god-and-his-scientists.
6. http://hyperphysics.phy-astr.gsu.edu/Nave-html/Faithpathh/boyle.html.
7. https://www.biblia.work/sermons/baconsir-francis.

Newton, Boyle, and other founders of modern science operated within a biblical worldview. They believed that since God created the natural realm, the study of nature, of creation, fails when it doesn't depend on Him and His revelation in Scripture.

> Opposite to godliness is atheism in profession, and idolatry in practice. Atheism is so senseless and odious to mankind, that it never had many professors. Can it be by accident that all birds, beasts, and men have their right side and left side alike shaped, (except their bowels); and just two eyes, and no more, on either side of the face; and just two ears on either side of the head; and a nose with two holes; and either two forelegs, or two wings, or two arms on the shoulders, and two legs on the hips, and no more? Whence arises this uniformity in all their outward shapes but from the counsel and contrivance of an Author?
>
> **ISAAC NEWTON**

Even unbelieving evolutionist, Loren Eiseley, reluctantly admitted science flourished because of the Christian environment of Western Europe. "The most curious aspect of the scientific world we live in...is that it exists at all. Westerners often unconsciously assume a doctrine of Inexorable Progress, as though the mere passage of time leads inevitably to increased knowledge as surely as an acorn becomes an oak. Yet the archaeologist would be forced to tell us," said Eiseley, "that several great civilizations have arisen and vanished without the benefit of a scientific philosophy. The type of thinking known as scientific, with its emphasis upon experimentation and mathematical formulation, arose in one culture—Western Europe—and in no other."[8]

Eiseley identified Christianity as the impetus behind modern science. "In one of those strange permutations of which history yields occasional rare examples," he said, "it is the Christian world which finally gave birth in a clear, articulate fashion to the experimental method of science itself."

Eiseley is not the only secularist who credited Christianity as the originator of modern science. "Today a wide range of scholars recognize that Christianity provided both intellectual presuppositions and moral sanction for the development of modern science."[9]

---

8. Pearcey, Nancy R. and Thaxton, Charles B., *The Soul of Science—Christian Faith and Natural Philosophy*, Crossway Books, 1994, 17-18.
9. Ibid, 18.

Renowned Christian apologist and hydraulic engineer, Henry M. Morris wrote, "The basic compatibility of science with Christian theism is even more obvious when it is realized that modern science actually grew in large measure out of the seeds of Christian theism. It is absurd to claim, as modern evolutionists often do, that one cannot be a true scientist if he believes in creation."[10]

Morris quoted Stanley Beck, an anti-creation entomologist, concerning the basis of scientific premises. "The first of the unprovable premises on which science has been based is the belief that the world is real and the human mind is capable of knowing its real nature... The second and best-known postulate underlying the structure of scientific knowledge is that of cause and effect...The third basic scientific premise is that nature is unified... It should be pointed out, however, that each of these postulates had its origin, or was consistent with, Christian theology."[11]

Many Christian scientists have been credited with founding the different branches of science. All were committed to Scripture. All believed in a literal six-day creation. Isaac Newton founded dynamics; Johannes Kepler, astronomy; Robert Boyle, chemistry; Lord William Thomson Kelvin, thermodynamics; Louis Pasteur, bacteriology; Matthew Maury, oceanography; Michael Faraday, electromagnetism; James Clerk Maxwell, electromagnetic wave theory; John Ray, biology; Carolus Linnaeus, taxonomy; Gottfried Leibniz, comparative anatomy; Charles Babbage, computer science; John Eccles, electrophysiology; and William Harvey, physiology.[12]

> Many of our great scientists have actually been men of profound religious convictions and life: Sir Isaac Newton, Michael Faraday, James Clerk Maxwell, Louis Pasteur. All these men were not only religious men, but they were also faithful members of their communions. For the most important thing in the world is a belief in moral and spiritual values—a belief that there is a significance and a meaning to existence—a belief that we are going somewhere! These men could scarcely have been so great had they been lacking in this belief.
>
> **DR. ROBERT MILLIKAN, 1923 NOBEL PRIZE IN PHYSICS**

---

10. Morris, Henry, *The Biblical Basis for Modern Science*, Baker Book House, 1984, 30.
11. Ibid, 30-31.
12. Ibid., 30.

Morris gave brief biographies of one-hundred-and-one prominent founders of science who were also believers in the divine inspiration of the Scriptures.[13] He included well-known Christian scientists, doctors, and engineers, such as Leonardo da Vinci, Samuel F. B. Morse, Louis Agassiz, and Joseph Lister.

In explaining this phenomenon of scientists making breakthrough discoveries because of their understanding of the Creator, Stanley L. Jaki, Distinguished University Professor at Seton Hall University, wrote, "All great cultures that witnessed a stillbirth of science within their ambience have one major feature in common. They all were dominated by a pantheistic concept of the universe going through eternal cycles. By contrast, the only viable birth of science took place in a culture for which the world was a created, contingent entity."[14]

THE TAKE AWAY: In other words, only Christian nations have produced great scientists because only Christians recognize they cannot understand the creation unless they first understand its Creator. Regions that did not produce great scientists did not investigate creation to increase their understanding of the Creator. God truly is the Master Scientist on whom all beneficial scientific breakthroughs depend.

> I think all of science, in a sense, comes from belief in order in the universe. That's part of scientific faith, that there is order and reliability, and so on, and that's part of the Judeo-Christian tradition, that there is one God.
>
> CHARLES TOWNES, 1964 NOBEL PRIZE IN PHYSICS

WITH SUCH A PRODUCTIVE BEGINNING, WHY HAS SCIENCE BECOME SECULAR AND INCREASINGLY STAGNANT, EVEN DANGEROUS?

Gradually over the last two centuries, science has moved away from its biblical underpinnings. Pearcey traced the beginnings of that departure to a little over a century ago. "In late nineteenth-century England, several small groups of scientists and scholars organized under the

13. Morris, Henry M., *Men of Science, Men of God*, Master Books, 1988.
14. Nickel, 143.

leadership of Thomas H. Huxley to overthrow the cultural dominance of Christianity—particularly the intellectual dominance of Christianity of the Anglican Church. Their goal was to secularize society, replacing the Christian worldview with scientific naturalism, a worldview that recognizes the existence of nature alone. Though secularist, they understood very well that they were replacing one religion by another, for they described their goal as the establishment of the 'church scientific.'"[15]

> The evolutionary record leaks like a sieve...There are so many flaws in Darwinism that one can wonder why it swept so completely through the scientific world, and why it is still endemic today... The biggest thing going for Darwinism was that it finally broke the tyranny in which Christianity had held the minds of men for so many centuries.
>
> **SIR FRED HOYLE, ASTRONOMER, AND PROFESSOR C. WICKRAMASINGHE, MATHEMATICIAN**

Today, the scientific establishment almost exclusively practices methodological naturalism, which limits scientific research to the study of natural causes alone. The Merriam-Webster Dictionary describes naturalism as "a theory denying that an event or object has a supernatural significance, *specifically*: the doctrine that scientific laws are adequate to account for all phenomena."[16]

Ian T. Taylor summarized the modern definition of science, which is secular to the core: "Knowledge of the world comes to us either directly or indirectly through our five senses. Man has systematized that knowledge in order to gain an understanding of nature; the exercise is called science and the motivation is usefulness."[17]

Jaki posited that "science is now in possession of such a vast interconnectedness of data, laws, and instruments as to continue its progress even if no attention is paid any longer to that faith which played an indispensable role in its rise."[18] Is he right? Certainly, science can continue to progress simply on the momentum of its scientific foundation, but humans can develop nothing that doesn't tend toward

15. Pearcy, 19. For more details on this departure of science from its biblical groundings, I recommend this excellent book, which also includes discussions of some misconceptions of the conflicts between science and the church.

16. https://www.merriam-webster.com/dictionary/naturalism.

17. Nickel, 6.

18. Jaki, Stanley L., *Cosmos and Creator*, Scottish Academic Press, 1980, 139.

degeneration. Won't science itself eventually collapse when its departure from the absolute truth of God as Creator is complete?

Secular science is prone to massive failures. One of the most obvious is the SETI project (Search for Extra-Terrestrial Intelligence), which, after the expenditure of billions of taxpayer dollars, produced no evidence of alien life. Zero evidence. The project was initiated, not because of scientific evidence, but on the assumption the universe is so vast that surely life must exist somewhere else in space. Even SETI scientists admit they failed to find any hint of extraterrestrials. "The Search for Extraterrestrial Intelligence (SETI) is a field of science that's been around for over 60 years. While SETI tools and strategies have improved immensely over that time, we have yet to find definitive evidence for intelligent alien life in our universe."[19]

Wasn't the underlying, secular purpose of SETI to prove that God's account of creation is bogus and, therefore, God Himself is a figment of a demented imagination? An understanding of Scripture, particularly of Genesis 1 and 2, could have prevented the waste of considerable time and money. God created life on Earth. Period.

To find a current scientific approach that explicitly utilizes insights from a biblical worldview is rare. If only we would return to the biblical roots of science! Consider the possible advances that might be made in biology, medicine, agriculture, and even physics.

Reliance on the Creator inevitably leads to breakthroughs in scientific thought. What innovations might have been conceived had science remained grounded in biblical reality? The deleterious effect of secular thinking on scientific innovation is incalculable.

> To find the metaphysical beliefs...governing scientific research... it would have been enough to speak of one belief, the belief in a personal rational Creator. It was this belief, as cultivated especially within a Christian matrix, which supported the [scientific] view for which the world was an objective and orderly entity investigable by the mind because the mind too was an orderly and objective product of the same rational, that is, perfectly consistent Creator.
>
> **DR. STANLEY JAKI, TEMPLETON PRIZE WINNER, PHYSICS**

19. https://techcrunch.com/2016/02/04/seti-scientist-explains-why-we-havent-found-aliens-yet/.

WHAT DIFFERENCE DO WORLDVIEWS MAKE?

Naturalism and secularism are worldviews. So are Christianity, pantheism, hedonism, and globalism. Everyone has his own worldview though few of us are aware of it. In fact, most of the time, we are not even aware of our worldview's influence on us and on others. We are like fish that don't recognize the water they rely on until it is gone.[20]

Our worldview influences the way we process all information. It determines our response to every aspect of the world, including the friends we choose, the causes we support, and the food we eat. It silently and sometimes violently affects every aspect of our lives. It even affects whether a husband buys his wife flowers or a blender for her birthday!

Worldviews set the course of science because they fundamentally affect the interpretation of scientific data, especially in the area of historical science. They also influence which data is examined and which is ignored.

> Religion and science demand for their foundation faith in God. For the former (religion), God stands foremost; for the latter (science), at the end of all thought, For religion He represents a basis; for science, a crowning solution towards a world view.
>
> **MAX PLANCK, 1918 NOBEL PRIZE IN PHYSICS**

WHAT IS A WORLDVIEW?

Simply put, a worldview is how we view the world around us. Maybe that is why it's called a worldview. Easy, isn't it?

Eyeglasses are often used to illustrate the concept of worldviews. When we put on glasses with pink lenses, everything we see has a pinkish tint. We assume everything, even that which we can't see, has that same pinkish tint. Unconsciously, we think, "I am seeing things with my own two eyes. I can be sure everything I see is pink and predictable."

What happens if we put on a different set of glasses with different lenses, maybe with no tint at all? Our perceptions change. When we study

---

20. An illustration from a friend, Nate Purtzer.

science, we need to understand how worldviews with their different lenses influence all scientific ideas as well as our understanding of them.

A worldview is a set of beliefs about the trivial to the most important issues of life. It is how we perceive reality.[21] Author and lecturer James Sire wrote that "a worldview is a set of presuppositions (assumptions which may be true, partially true or entirely false) which we hold (consciously or subconsciously, consistently or inconsistently) about the basic make-up of our world."[22]

> [A worldview is]...a conceptual system by which we consciously or unconsciously place or fit everything we believe and by which we interpret and judge reality.
>
> R.H. NASH

The assumptions and beliefs we have already accepted as true are faith-based and foundational to our worldview. "A worldview is composed of a number of basic presuppositions, more or less consistent with each other, more or less consciously held, more or less true. They are generally unquestioned by each of us; rarely, if ever, mentioned by our friends; and only brought to mind when we are challenged by a foreigner from another ideological universe."[23]

Another good description is that "a worldview is a way of viewing or interpreting all reality. It is an interpretive framework thru which or by which one makes sense of the data of life and the world."[24]

In an excellent text describing several worldviews, author and social scientist David Noebel suggested our worldviews are based on our answers to such essential questions as "Why are we here? What is the meaning and purpose of life? Is there a difference between right and wrong? Is there a God? Are humans merely evolved animals?"[25] Sire added other questions. "What is prime reality—the really real? What is the nature of external reality, that is the world around us? What is a

21. Beckwith, Francis J., Craig, William Lane, and Moreland, J.P., Eds., *To Everyone an Answer: A Case for the Christian Worldview*, InterVarsity Press, 2004, Chapter 12 by Ronald H. Nash, 203.4
22. Sire, James W., *The Universe Next Door*, InterVarsity Press, 1988, 17.
23. Ibid, 17-18.
24. Geisler, Norman L., & Watkins, William D., *Worlds Apart*, Baker Book House, 1989, 11.
25. Noebel, David A., *Understanding the Times: The Collision of Today's Competing Worldviews*, Summit Press, 2006, 15.

human being? What happens to a person at death? Why is it possible to know anything at all?"[26]

Our answers and, therefore, our worldview are continually changing and refining as our knowledge, experiences, and opinions change. As you would imagine, everyone has a different worldview, and every worldview is in constant flux.

THE TAKE AWAY: Though everyone has a worldview, not all worldviews are compatible or even consistent with reality. Even more interesting, we can intellectually adopt one worldview but live inconsistently with it. For example, we may agree with the worldview, the platform, of the Republican party but vote for Democrats because our dear Granny always did. Our worldview affects every aspect of our lives from the inconsequential (what we will wear or eat today) to the more important (how, or even if, we will vote).

If we understand the tint of our worldview glasses, we are better able to understand life and come to more realistic decisions about it. In this study of science, understanding worldviews will help us evaluate the validity of scientific truth-claims. We will see that the main issue is not science versus religion but worldview versus worldview. The secular worldview puts a high value on science. We know they do because they constantly tell us. Though not as evident, the biblical worldview finds scientific study equally valuable. We must decide for ourselves which worldview is closest to reality.

> All men seek happiness. This is without exception. Whatever different means they employ, they all tend to this end. The cause of some going to war, and of others avoiding it, is the same desire in both, attended with different views. The will never takes the least step but to this object. This is the motive of every action of every man, even of those who hang themselves.
>
> **BLAISE PASCAL, MATHEMATICIAN**

26. Sire, 18.

WHAT ARE THE BASIC ELEMENTS OF A WORLDVIEW?

Now that we understand that our worldview reflects our answers to life's foundational as well as trivial questions, let's consider the basic elements that make up all worldviews. Nash suggested five categories: "God, ultimate reality, knowledge, ethics, and humankind."[27] I would add another important one: our eternal destiny.

These six categories involve the philosophies of theology, metaphysics, epistemology, ethics, anthropology, and eschatology.

- *Theology* is the systematic study of the nature of God. It answers the questions, does a supreme being exist and, if so, what can I know about that deity?
- *Metaphysics* "describes what is beyond physics—the nature and origin of reality itself, the immortal soul, and the existence of a supreme being."[28] Metaphysics attempts to answers the question, what is the nature of reality, material and/or spiritual?
- *Epistemology* is the science of determining if an idea or thought is valid. It answers the questions, why do I believe what I believe, and am I right?
- *Ethics* determines right and wrong. It answers the question, how do I know what I should do?
- *Anthropology* studies man and his culture. It answers the questions, what is man, and what is his purpose?
- *Eschatology* studies the future, including issues of death, judgment, and the final destiny of the soul, mankind, and the creation. It answers the question, what can I expect in the future?

ELEMENTS OF A WORLDVIEW

- GOD → Theology
- ULTIMATE REALITY → Metaphysics
- KNOWLEDGE → Epistemology
- ETHICS → Morality
- HUMANKIND → Anthropology
- DESTINY → Eschatology

27. Nash, 205.
28. https://www.merriam-webster.com/dictionary/metaphysics.

The most important element in every worldview is the first, its view of God, because our understanding of the divine influences every other view. Logically, then, our starting point in understanding and analyzing any worldview must be to investigate God and His impact on reality. In science, a scientist's view of God ultimately influences his view of science as well as his scientific conclusions.

Every worldview assumes certain ideas centered on these six categories. By studying a worldview's component beliefs, we can determine if it conforms to the real world. We can ascertain its strengths and weaknesses and adjust our own worldview and our response to other worldviews accordingly.

WHAT IS THE UNBELIEVING WORLDVIEW, THE WORLDVIEW OF SCIENCE TODAY?

All worldviews are unique and even conflicting because each of us looks at the world through unique glasses that have been influenced by our differing experiences and levels of knowledge. Even so, worldviews logically fall into two main categories, the biblical or believing worldview and the unbiblical or unbelieving worldview, unbelieving in the sense of not agreeing with the biblical worldview or, in the case of some, in direct opposition to it. Without exception, all worldviews not based on God's revelation in the Bible are unbelieving.

By using the foundational categories found in all worldviews, we can learn the main beliefs of the unbelieving worldview. Hint: If we took the truths of the Bible and said the opposite, we'd have many components of all unbiblical worldviews! Even so, many unbelieving worldviews have biblical truths imbedded in them. Everyone truly has a unique worldview!

1. *God:* In the unbelieving worldview, at worst, God does not exist. At best, gods other than

the Almighty God of the Bible have been and are now at work in human history. That god who is not God may be power or money rather than a being. Even an atheistic worldview believes something about God. Having rejected God, it assumes His non-existence.

2. *Ultimate Reality:* The unbelieving, naturalistic worldview denies the spiritual and unseen realm of God and angelic/demonic creatures. Only the material, natural realm of the universe is real to them. Even so, they often promote imagined gods or higher powers, like fate, luck, or the force. Contradictions abound in unbelieving worldviews.

3. *Knowledge:* In the unbelieving worldview, humans can ascertain all knowledge through reason, experience, and/or mystical feelings. It rejects the revelation of Scripture, considering it a collection of myths as improbable as all other man-made myths. They believe we all have the right to make up and believe our own truth, no matter how inconsistent and logic-defying it may be. To them, truth is fluid and cannot be absolute.

ELEMENTS OF AN UNBELIEVING WORLDVIEW

- GOD → Atheist/Distorted
- ULTIMATE REALITY → Natural/Spirit
- KNOWLEDGE → Reason
- ETHICS → Relativism
- HUMANKIND → Basic Goodness
- DESTINY → Varies

4. *Ethics:* In the unbelieving worldview, the determination of right and wrong is subjective, dependent on the desires of that individual or his society. The doctrine of absolute right and wrong as revealed in Scripture is deemed intolerant and unkind. God's commands are judgmental and, therefore, not reasonable or acceptable. Sin is relative and self-determined, as is good. Social justice and equity, however defined, should be the aim of all individuals and governments. Morality is unimportant.

5. *Mankind:* Humans are born good and then corrupted by experiences beyond their control, such as their environment. They are their own authority and determine their own reality.

In our culture today, people are struggling to give an answer to the simple question, "What is a woman?"

6. *Destiny:* Any belief other than a real eternal state that includes only Heaven and Hell is accepted by unbelieving worldviews. Some even believe they will come back as cockroaches if they beat their wives! Think about that for a moment. Yes, worldviews are important!

> None of the inventors of atheism was a naturalist. All of them were very mediocre philosophers.
>
> **DR. ARTHUR EDDINGTON, ASTRONOMY, MATHEMATICS, ASTROPHYSICS**

THE TAKE AWAY: Even if you remember nothing else about worldviews, remember that all worldviews are belief systems based on faith in something. Those beliefs inspire and circumscribe every bias, thought, and action of the one holding that worldview.

> And when you are convinced, be not ashamed to profess the truth. For otherwise you may become a stumbling block to others, and inherit the lot of those Rulers of the Jews who believed in Christ, but yet were afraid to confess him lest they should be put out of the Synogogue. Wherefore, when you are convinced, be not ashamed of the truth, but profess it openly and endeavor to convince your Brother also that you may inhereit at the resurrection the promise made in Daniel 12:3, that "they who turn many to righteousness shall shine as the stars for ever and ever." And rejoice if you are counted worthy to suffer in your reputation or any other way for the sake of the Gospel, for then, "great is thy reward"!
>
> **ISAAC NEWTON**

WHAT IS THE BIBLICAL, BELIEVING WORLDVIEW?

The biblical, believing worldview is different from all unbelieving worldviews, which are legion, because it emanates from the revelation of God to mankind through His written Word, the Bible. Not all Christians agree on every aspect of the Bible, but those who have a high view of Scripture agree on the main elements of the biblical worldview.

1. *God:* In the biblical worldview, the one supreme, all-powerful Creator is personal and Trinitarian. Scripture describes some of His attributes, His perfections, including His eternal self-existence, transcendence yet immanence, holiness, sovereignty, omniscience, omnipotence, omnipresence, immutability, goodness, love, justice, righteousness, and grace. He is infinitely complete and perfect in every way.

> Finite man cannot begin to comprehend an omnipresent, omniscient, omnipotent, and infinite God...I find it best to accept God through faith, as an intelligent will, perfect in goodness and wisdom, revealing Himself through His creation.
>
> **DR. WERNER VON BRAUN, ROCKET SCIENCE**

2. *Ultimate Reality:* The triune God is the ultimate reality. A spiritual realm exists outside of and in addition to the natural, physical realm. Creator-God created all things, the seen and unseen, *ex nihilo* (from nothing) by simply speaking them into existence. The natural realm is temporal, finite, and dependent on the Creator. The Creator is infinite and dependent on no one and nothing. As would be expected, a radical difference and separateness exists between the Creator and His creation.

ELEMENTS OF A BIBLICAL WORLDVIEW

- GOD → Personal, Creator, Trinity
- ULTIMATE REALITY → Spiritual Realm
- KNOWLEDGE → Revelation
- ETHICS → Absolutes
- HUMANKIND → Image, Fallen
- DESTINY → Heaven/Hell

3. *Knowledge:* The biblical worldview believes objective truth exists and can be known[29] but only through God's revelation, which is contained only in the sixty-six books of the Bible. God is truth. All absolute truth is embodied in the Trinity (God the Father, Son, and Holy Spirit). By using scriptural truth as a springboard, we can gain a better understanding of the Creator and all aspects of His creation and universe.

29. Appendix A contains a biblical word study on truth and knowledge.

> Where no guiding ideals are left to point the way, the scale of values disappears and with it the meaning of our deeds and sufferings, and at the end can lie only negation and despair. Religion is therefore the foundation of ethics, and ethics the presupposition of life.
>
> **DR. WERNER HEISENBERG, 1932 NOBEL PRIZE IN PHYSICS**

4. *Ethics:* God built a moral order into His creation and created all humans with an inner sense of good and evil called conscience. In His Word, He reveals an absolute standard for all thoughts, words, and actions. That standard is based on His perfect nature. He holds everyone accountable to conform to His righteous standard.

5. *Mankind:* The biblical worldview holds that God originally created mankind, both male and female, in His own image. That image reflects on a finite level some of His perfections, including spirituality, personality, intellect, volition, moral agency, creativity, communication, and immortality. God created us to have a relationship with Him. The original man and woman rebelled against the Creator, which damaged God's image in them. That original sin separated all the descendants of the first man and woman from their Creator God. It also brought death, disease, predation, and other natural evils into an originally "very good" creation.

6. *Destiny:* In the biblical worldview, Creator God has provided for the restoration of mankind from the damaging effects of sin. The basis for that restoration is the death and resurrection of Jesus Christ. Restoration begins with the spiritual regeneration of individual believers and climaxes in the first stage of the Kingdom, the prophesied thousand-year reign of Jesus Christ on King David's throne. This eternal restoration will return the entire physical realm to near its original status before sin damaged both mankind and the creation. The present heavens and earth will be burned up. There exists a real Heaven. Believers in Jesus Christ will live forever with Him in the New Heavens and New Earth. A real, eternal, and terrifying Hell (the Lake of Fire) will house forever those who reject God's provision of salvation through belief in Jesus Christ. God designed history as an

> unending linear sequence of events that ever continue to unveil new dimensions of His glory and fulfill ever more of His glorious purposes, which include the restoration of mankind and creation to perfect harmony with Himself.

James Nickel aptly summarized the biblical worldview. "The good news of the gospel is that the *logos* of God is the way (ethics), the truth (epistemology), and the life (metaphysics). The bad news is that this way, truth, and life is exclusive—no one can come to the Father except through Christ. A worldview is not only a network of presuppositions concerning how to live, how to know, and what is real, it is an *antithetical* network of presuppositions. There is one way to a true worldview and that way is in the person of the Lord Jesus Christ and His redemption."[30]

> But again I looked and saw Jesus, my substitute, scourged in my stead and dying on the cross for me. I looked and cried and was forgiven. And it seems to be my duty to tell you of that Saviour, to see if you will not also look and live: How simple it all becomes when the Holy Spirit opens the eyes!
>
> **SIR JAMES YOUNG SIMPSON, PIONEER OF ANAESTHETICS**

The biblical worldview contradicts all other worldviews. We must ask ourselves which worldview, the believing or unbelieving, best reflects and explains reality. Which corresponds best to the evidence? Which allows us to live our lives with joyful anticipation of the future?

We should note that the natural realm is an important element in both the believing and unbelieving worldviews. Science is devoted to the study of that realm. Clearly, a scientist's worldview affects his approach to science and his success in understanding and ruling the creation. If his worldview glasses have polka dots or specks of mud all over them, he will miss aspects of the natural realm that would be obvious if he wore clear lenses unaffected by prejudices and misconceptions. The strength of his microscope or telescope can't reveal truth if his eyes are already blinded by a subjective, secular worldview. His worldview glasses must be in line with reality if he is to reach the most accurate conclusions about the area of science he is studying.

30. Nickel, 62.

HOW DO SECULAR SCIENCE AND CREATION SCIENCE DIFFER?

Science today is secular because its worldview restricts all explanations to naturalistic ones. In general, it functions from a worldview that assumes the natural realm is the full extent of reality and, therefore, is sufficient to account for everything that is real. Such an outlook explicitly excludes God and the supernatural.

Though based today on the naturalistic approach, observational science is less prone to viewpoint-induced mistakes because it is self-correcting. On-going observations and experimentations reveal new information that forces changes in the interpretation of the data and, therefore, in the conclusions. While this sounds reassuring, observational scientists are human and too often allow their unbelieving worldviews to limit their scope and corrupt their findings. Scientific facts that disprove popular ideas often take many years to find acceptance.

> If it disagrees with experiment, it's wrong. And that simple statement is the key to science. It doesn't make a difference how beautiful your guess is, it doesn't matter how smart you are, who made the guess, or what his name is. If it disagrees with experiment, it's wrong. That's all there is to it.
>
> RICHARD FEYNMAN, 1965 NOBEL PRIZE IN PHYSICS

Historical science provides no avenue for self-correction since its findings are inherently subjective. We must be aware of the risk of unquestioning allegiance to history-based, "scientific" pronouncements. They are opinions, not verifiable facts. Scientists today did not experience past events—at least I don't know any billion-year-old or even several hundred-year-old people!

The true scientific approach, the one embraced by early scientists, accepts and even seeks data from all valid sources of knowledge. Creation scientists turn first to the Creator, who, in His Word, gives abundant insights not only about Himself (theology) but also about the creation (natural theology). Passages in Scripture give insights and understanding that cannot be derived from observations of the natural realm alone. They are authoritative since they come from the Creator.

The biblical worldview assumes the Scriptures are fully authoritative and inerrant, that all scriptural assertions are true to reality and of immeasurable value. We could accurately assert creation cannot be understood without this revelation from its Creator. Scientists ignore divine and, therefore, immutable and permanent truths at their own peril. You will learn how in later chapters.

Famed scientist Galileo Galilei (1564–1642), the father of observational astronomy, explained this truth nicely:

> And to prohibit the whole science would be but to censure a hundred passages of Holy Scripture which teach us that the glory and greatness of Almighty God are marvelously discerned in all his works and divinely read in the open book of heaven. For let no one believe that reading the lofty concepts written in that book leads to nothing further than the mere seeing of the splendor of the sun and the stars and their rising and setting, which is as far as the eyes of brutes and of the vulgar can penetrate. Within its pages are couched mysteries so profound and concepts so sublime that the vigils, labors, and studies of hundreds upon hundreds of the most acute minds have still not pierced them, even after continual investigations for thousands of years.[31]

> The sun, with all the planets revolving around it, and depending on it, can still ripen a bunch of grapes as though it had nothing else in the universe to do.
>
> GALILEO GALILEI, ASTRONOMER, PHYSICIST, ENGINEER

WHAT CONCLUSIONS CAN WE DRAW?

First, we have seen that, over the last five hundred years, science has changed its approach. Early scientists like Galileo were committed to the Bible. They sought to learn truth by observing what they knew to be God's creation. Over time, science has moved away from those biblical roots. We would be accurate in describing science today as totally

31. "Letter to Madame Christina of Lorraine, Grand Duchess of Tuscany, Concerning the Use of Biblical Quotations in Matters of Science," https://bertie.ccsu.edu/naturesci/Cosmology/GalileoChristina.html.

secular, devoid of any spiritual, immaterial understanding. Would science be more fruitful if it returned to its biblical roots? Of course. The Creator is the only one with complete knowledge of His creation, its foundation, and its possibilities.

Second, scientists observe the created realm, but, too often, their observations are distorted by faulty worldviews, by subjective prejudices. Their worldviews directly influence not only their observations but also their interpretations of the data, a critical weakness especially of historical science.

Astronomer Sir Fred Hoyle (1915-2001), who formulated the theory of stellar nucleosynthesis, defined the problem worldviews pose for scientific research:

> Writers on scientific method usually tell us that scientific discoveries are made 'inferentially,' that is to say, from putting together many facts. But this is far from being correct. The facts by themselves are never sufficient to lead unequivocally to the really profound discoveries. Facts are always analyzed in terms of the prejudices of the investigator.[32]

To achieve greater accuracy, scientists need to approach all science, but especially historical science, from a biblical worldview. Our friend of gravity fame, Sir Isaac Newton, understood this necessity:

> This most beautiful system of the sun, planets, and comets could only proceed from the counsel and dominion of an intelligent being. And if the fixed Stars are the centers of other like systems, these, being formed by the like wise counsel, must all be subject to the dominion of One...This Being Governs all things, not as the soul of the world, but as Lord over all: And on account of his dominion he is wont to be called *Lord God*, or Universal Ruler.

THE TAKE AWAY: Only by viewing scientific study through the lens of its Universal Ruler can we come to a correct, satisfying, and rewarding understanding of His creation. You must decide if God knows His creation better than any finite man, even well-educated scientists. Your decision will impact your life eternally.

---

32. https://answersingenesis.org/theory-of-evolution/quotes/how-prejudice-affects-science.

> There can never be any real opposition between religion and science; for the one is the complement of the other. Every serious and reflective person realizes, I think, that the religious element in his nature must be recognized and cultivated if all the powers of the human soul are to act together in perfect balance and harmony. And indeed it was not by accident that the greatest thinkers of all ages were deeply religious souls.
>
> **MAX PLANCK, 1918 NOBEL PRIZE IN PHYSICS**

"I have made the earth,
the men and the beasts which are on the face of the earth
by My great power and by My outstretched arm,
and I will give it to the one who is pleasing in My sight."

JEREMIAH 27:5

"Do you not fear Me?" declares the Lord.
"Do you not tremble in My presence?
For I have placed the sand as a boundary for the sea,
an eternal decree, so it cannot cross over it.
Though the waves toss, yet they cannot prevail;
though they roar, yet they cannot cross over it."

JEREMIAH 5:22

YAHWEH, THE CREATOR

CHAPTER 4

THE NATURE OF MAN

Are scientists human, or are they some sort of special creatures different from the rest of mankind? Of course, they are human! It doesn't take a genius to know that. Maybe some scientists are smarter than most of us, at least about science, but even so they have the same flaws and virtues we do. So, we must ask, what human characteristics do scientists have that are innate in all mankind? And how do those characteristics affect their scientific research and conclusions?

To fully understand the nature of man, we would need an entire course in biblically based anthropology. Such a course would include creation science (the original creation of man), hamartiology (the doctrine of sin), soteriology (the study of man's regeneration), and eschatology (the purpose and destiny of mankind). I say this so you will be thoroughly impressed and believe me when I say the following discussion is not only brief but merely skims the surface of available knowledge.

DID GOD CREATE MAN TO BE UNIQUE?

The creation of the entire universe and all it contains was a series of historical events, real events that really occurred during a six-day period.[1] On the sixth day, God gave life to the progenitors of all mankind.

1. I will demonstrate this in later chapters.

> Then God said, "Let Us make man in Our image, according to Our likeness; and let them rule over the fish of the sea and over the birds of the sky and over the cattle and over all the earth, and over every creeping thing that creeps on the earth." God created man in His own image, in the image of God He created him; male and female He created them.
>
> Genesis 1:26-27

Note that God created man in His image. Humans are special indeed. We are not merely products of millions or billions of years of random evolutionary processes. We are a special creation from the hand of God. As such, we are fundamentally different from the rest of the creation, even from animals. In an awesome sign of divine love and design, something in man reflects the splendor of the Creator Himself.

> Science and religion are very much alike. Both are imaginative and creative aspects of the human mind. The appearance of a conflict is a result of ignorance. We come to exist through divine act. That divine guidance is a theme throughout our life; at our death the brain goes, but that divine guidance and love continues. Each of us is a unique, conscious being, a divine creation. It is the religious view. It is the only view consistent with all the evidence.
>
> **DR. JOHN ECCLES, 1963 NOBEL PRIZE IN MEDICINE**

Even so, Almighty God reigns supreme. He is the Creator while we are merely the products of His creative forces. We are the shadow, the image. He is God, perfect, all-knowing, and all-powerful.

> Almighty God, who has created man in Thine own image, and made him a living soul that he might seek after Thee, and have dominion over Thy creatures, teach us to study the works of Thy hands, that we may subdue the earth to our use, and strengthen the reason for Thy service; so to receive Thy blessed Word, that we may believe on Him whom Thou has sent, to give us the knowledge of salvation and the remission of sins. All of which we ask in the name of the same Jesus Christ, our Lord.
>
> **JAMES CLERK MAXWELL, MATHEMATICIAN, PHYSICIST**

God is incomprehensible. Our finite minds are incapable of fully comprehending an infinite God. We can know Him truly, yes, but we cannot know Him exhaustively. Even a genius can understand only the bare minimum of God's majesty, dominion, might, and grace, and that only after a lifetime of studying both the creation and Scripture. God is so different from man that no analogies from creation can adequately describe His transcendence.[2] Theologian A. W. Tozer (1897–1973) noted God "is not exactly like anything or anybody."[3] Indeed, not. He is perfection incarnate.

God wanted the great believer Job to appreciate His incomprehensible superiority to all mankind, and so He confounded Job's thinking through a series of questions:[4]

> "Have you ever in your life commanded the morning and caused the dawn to know its place?...Have you entered into the springs of the sea or walked in the recesses of the deep?... Have the gates of death been revealed to you, or have you seen the gates of deep darkness? Have you understood the expanse of the earth? Tell Me, if you know all this....Do you know the time the mountain goats give birth? Do you observe the calving of the deer?...Do you give the horse his might? Do you clothe his neck with a mane?...Is it by your understanding that the hawk soars, stretching his wings toward the south? Is it at your command that the eagle mounts up and makes his nest on high?"
>
> Job 38:12, 16-18; 39:1-2, 19, 26-27

God not only knows everything but has complete control of it. Job must have trembled as he realized his arrogance in questioning God's motives and plan. God not only knows everything but has complete control of it. We, too, should tremble any time we think of ourselves more highly than we ought to think.[5] As a sidebar, reread God's words to Job and marvel at the many scientific insights He brought to light in this short section alone.

> After the knowledge of, and obedience to, the will of God, the next aim must be to know something of his attributes of wisdom, power and goodness as evidenced by His handiwork.
>
> **JAMES JOULE, AUTHOR OF FIRST LAW OF THERMODYNAMICS**

---

2. Transcendence is the biblical doctrine of God being totally separate from, distinct from, and outside of the creation.
3. Tozer, A. W., *The Knowledge of the Holy*, Harper & Row Publishers, 1961, 14.
4. See also Job 37:5; Psalm 139:6, 17; Isaiah 40:18; Romans 11:33-36.
5. Romans 12:3.

Later, God reminded the prophet Isaiah, "My thoughts are not your thoughts, nor are your ways My ways…For as the heavens are higher than the earth, so are My ways higher than your ways and My thoughts than your thoughts."[6]

Even though God is incomprehensible, He is knowable.[7] He wants us to know Him and so has revealed facets of Himself in His Word. We can't begin to comprehend His magnificence, but we can know enough to worship, revere, and serve Him.

> God's nature is revealed most perfectly in the life and teaching of Jesus of Nazareth, as recorded in the New Testament of the Bible, who was sent by God to reveal the divine nature, summarized in "God is Love."
>
> **DR. GEORGE ELLIS, ASTROPHYSICS**

Through revelation, we learn all He wants us to know about Himself. Revelation of God first came through His direct interaction with individuals in history, through visible appearances, communications, manifestations, and visions. The clearest revelation came when the Son of God became the man Jesus through the incarnation.[8] John 1:14 explains that "the Word [God the Son] became flesh and dwelt among us, and we saw His glory, glory as of the only begotten from the Father, full of grace and truth." Through His actions and words as recorded in His Word, Jesus Christ, the eternal Son of God, gave us a glimpse of God's majesty, only a glimpse because our finite minds cannot comprehend more. As theologian William Shedd (1820-1894) noted, "Some of the characteristics of the divine nature cannot be known by a finite intelligence."[9]

> I believe in God, who can respond to prayers, to whom we can give trust and without whom life on this earth would be without meaning (a tale told by an idiot). I believe that God has revealed Himself to us in many ways and through many men and women, and that for us here in the West the clearest revelation is through Jesus and those that have followed him.
>
> **DR. NEVILL MOTT, 1977 NOBEL PRIZE IN PHYSICS**

---

6. Isaiah 55:8.
7. Deuteronomy 4:35, "To you it was shown that you might know that the Lord, He is God; there is no other besides Him." And John 1:18, "No man has seen God at any time; the only begotten God, who is in the bosom of the Father, He has explained Him."
8. John 1:1-14, especially verse 14.
9. Shedd, William G.T., *Dogmatic Theology*, Vol 1, Thomas Nelson Publishers, 1980.

God's means of revelation today is His written Word, both the Old and New Testaments, though creation continues to help us appreciate His power, creativity, and love. Without Scripture to enlighten our understanding, we cannot pierce through the darkness of ignorance or perceive the life-path He prepared for us. No observation or experimentation of science can uncover His plan. No system of rational thought, including philosophy, psychology, and higher learning, is powerful enough to untangle the Gordian knot of His brilliance. Only through studying His Word under the Holy Spirit's guidance can we glimpse the width and breadth of His perfect person, plan, and purpose.

For any understanding at all, we are utterly dependent on His revelation of Himself in His Word. Gaining that understanding should be our lifelong pursuit. Don't think that when you are ninety-eight you will have God figured out! We can never exhaust His incomprehensibility.

Not only do we not know God exhaustively (or even close to it), but we also do not know what we do not know. Even so, we can be certain what we do not know does not contradict what we do know. Think about that astounding truth for a moment! Divine truth is the totality of all knowledge. No contradictions can enter into it.

We know only that which God has been pleased to reveal to us about Himself and His plan for us. Does that not imply that our knowledge is totally dependent on the revelation God makes available to us in His Word? Does that not show the inadequacy of our reasoning powers to obtain accurate scientific knowledge? Is it not obvious that science studied apart from the biblical worldview is inherently deficient and prone to error?

> "We are fortunate to have the Bible, and especially the New Testament, which tells us so much about God in widely accessible human terms...I find a need for God in the universe and in my own life."
>
> **DR. ARTHUR SCHAWLOW, 1981 NOBEL PRIZE IN PHYSICS**

WHAT DOES IT MEAN TO BE MADE IN GOD'S IMAGE?

Yes, God is incomprehensible, but we must understand facets of His essence if we are to rightly understand His creation and plan. Theology has divided God's attributes into two categories, communicable and non-communicable. I prefer to describe His attributes as perfections since God is perfect in every way. His non-communicable perfections are those that belong to Him alone. He did not communicate them or build them into humans, His image bearers. These perfections are known only through the revelation of Scripture. For example, we are not self-existent (having no needs outside of ourselves),[10] omnipresent (everywhere at the same time),[11] or immutable (unchanging),[12] but God is.

The perfections God shares with us because He created us in His image are described as communicable. He communicated them or built them into us in finite measure on the sixth day of creation. In a limited way, we can understand something about God by observing some human attributes. Even so, our understanding of these divine perfections is limited in that God is infinite while we are observably finite.

Scripture tells us God has infinite intellect (omniscience) while we have limited rational abilities. We can know a little about a few things. God knows everything about everything.

God created man originally without sin; thus, the first man reflected, on a limited scale, God's absolute moral qualities of righteousness[13] and holiness.[14] As believers, we reflect those qualities after we confess our sins to God the Father, an ongoing necessity.

The Creator built into us limited creative ability. We can't create the heavens and earth or even a flower, but we can build skyscrapers, decorate our rooms, and paint pictures.

---

10. God has existence in Himself and has no needs outside Himself. He has no cause or beginning. Isaiah 40:13-17 and Acts 17:25.
11. God is wholly everywhere at all times. Psalm 139:7-16; Ephesians 1:23.
12. God is unchanging and unchangeable in His being, essence, and nature. Malachi 3:6; Hebrews 1:11-12.
13. Psalm 11:7, "For the Lord is righteous; He loves righteousness; the upright will behold His face." Psalm 129:4; 145:17; Isaiah 45:21.
14. Isaiah 6:3, "Holy, Holy, Holy, is the Lord of hosts. The whole earth is full of His glory." Psalm 111:9; Isaiah 57:15; Leviticus 11:44.

All-powerful (omnipotent)[15] God created us with limited powers. We can't command the rain, but we can open an umbrella and close windows.

Because God communicates (in fact, He spoke the universe into existence[16]), we also have the ability to communicate, not only with each other but also with God through prayer.

> I believe it would be a great tragedy if men were to lose their faith in prayer. Without the help of prayer I might perhaps have failed where I have succeeded. In allowing me to attain what I have done, God has made of me merely an instrument of His own will, for the revelation of His own Divine power.
>
> **DR. GUGLIELMO MARCONI, 1909 NOBEL PRIZE IN PHYSICS**

We are like God but totally dependent on Him. Since God created the heavens and earth for our benefit, He designed us with the ability to observe and study all aspects of it. That study is called science. After the creation week, God commanded the first man and, implicitly, his progeny, to "subdue…and rule over" all creation, an impossibility unless we first study and understand it.[17] Unraveling creation's intricacies so that we can benefit through its usage is the responsibility of true science.

> What nobler ambition can one have than to cooperate with his Maker in bringing about a better world in which we live? When we view men's actions in the light of science we are thus presented with a new hope. Loyalty to our Maker, who has given us the ability, opportunity, and responsibility to mold our lives and our world according to a more perfect pattern cannot but inspire us to work with him heart and soul toward this great end.
>
> **DR. ARTHUR COMPTON, 1927 NOBEL PRIZE IN PHYSICS**

15. Psalm 62:11, "Once God has spoken. Twice I have heard this: That power belongs to God." Psalm 115:3; Exodus 15:6; Matthew 19:26.
16. Psalm 33:6-9, "By the word of the Lord the heavens were made, and by the breath of His mouth all their host….Let all the inhabitants of the world stand in awe of Him. For He spoke, and it was done; He commanded, and it stood fast."
17. Genesis 1:28, "God blessed them; and God said to them, 'Be fruitful and multiply, and fill the earth, and subdue it; and rule over the fish of the sea and over the birds of the sky and over every living thing that moves on the earth.'"

Adam, the first man, was the first scientist. One of the first tasks God gave him was to name the animals.[18] He had make biological observations to recognize the differing characteristics of each animal and come up with distinguishing names. Does that not reflect part of the role of biologists and zoologists, who even today name newly discovered parts of His creation?

One of the characteristics that distinguishes us from the rest of creation is our ability to communicate, not just about the physical realm but also about the unseen, about ideas and beliefs. Communication involves the ability to formulate ideas, process information, and communicate that information. Science reflects those abilities by gathering information, analyzing data, formulating statements to describe the data, and communicating the acquired knowledge. Because we are created in God's image, science should communicate both the seen and the unseen, weaving them together as a cohesive whole in the same way the Creator does.

Since God created us in His image, we have the potential to do good. Science can help us use creation for the benefit of all; however, because we inherited a sin nature from Adam, we can just as easily use scientific advances for evil purposes. Recently, scientists have utilized their knowledge of the creation to make massive strides in technology, but much of that which promised tremendous good, the computer, telephone, and television, for example, have led us further into the darkness. Do we really want pornography and brutality brought into our homes with the push of a button? Do we really want our children and grandchildren to be bombarded by anti-God propaganda? Because we are made in God's image, we have the ethical responsibility to limit scientific advancements to those that affect the world for good.

> It is not enough that you should understand about applied science in order that your work may increase man's blessings. Concern for the man himself and his fate must always form the chief interest of all technical endeavors; concern for the great unsolved problems of the organization of labor and the distribution of goods in order that the creations of our mind shall be a blessing and not a curse to mankind. Never forget this in the midst of your diagrams and equations.
>
> **ALBERT EINSTEIN, 1921 NOBEL PRIZE IN PHYSICS**

18. Genesis 2:19-20, "Out of the ground the Lord God formed every beast of the field and every bird of the sky and brought them to the man to see what he would call them; and whatever the man called a living creature, that was its name. The man gave names to all the cattle, and to the birds of the sky, and to every beast of the field, but for Adam there was not found a helper suitable for him."

WHAT EFFECT DOES HUMAN FALLIBILITY HAVE ON SCIENCE?

God did not create us to be robotic but to make choices and decisions. From the beginning, He gave us the right to choose to obey Him or not, to follow His plan or devise our own. Science today has chosen against God, a good reason to beware of any so-called "settled science" trumpeted by any secular scientist!

God also created us as moral creatures, reflecting His own moral attributes. He desires us to make moral judgments and freely choose to follow or reject Him. The first man exercised his volition by rebelling against his Creator.[19] His decision to walk in the darkness bereft of God's guidance brought catastrophic effects not only on all his descendants but also on the entire universe. Did you catch that? One event in history had a tragic effect not only on every descendant of Adam and Eve, including you and me, but also on the entire creation. Every particle of the universe was affected; therefore, all science was changed.

Adam brought death and destruction upon all the world by taking a bite of a piece of fruit. The fruit wasn't the problem. His rebellion against God was. Genesis 3 describes the consequences of that rebellion.

When sin entered the world through Adam's original sin, every aspect of mankind's nature changed. Adam and Eve lost their sinless and *very good* nature. They did not lose God's image entirely, but it was and continues to be severely damaged and obscured.[20] That first sin meant they no longer had a close relationship with Creator God. Of great significance to us, Adam genetically passed his sinful nature to all his descendants; therefore, we are all born sinful and separated from God.[21]

Theologians call this change in humanity's nature the doctrine of depravity. Paul described the unbeliever (which we all are at birth) as spiritually "dead" and under negative spiritual influences.[22] The prophet Jeremiah described our hearts (our inner being, including

---

19. Genesis 2:16-17, "The LORD God commanded the man, saying, 'From any tree of the garden you may eat freely; but from the tree of the knowledge of good and evil you shall not eat, for in the day that you eat from it you will surely die.'"
20. James 3:9 explains that even blatantly sinful men are described as being created in God's image.
21. 1 Corinthians 15:22, "For as in Adam all die, so also in Christ all will be made alive."
22. Ephesians 2:1-3.

mind, spirit, and all immaterial aspects) as "deceitful and desperately sick."[23] Creator Jesus said:

> "That which proceeds out of the man, that is what defiles the man. For from within, out of the heart of men, proceed the evil thoughts, fornications, thefts, murders, adulteries, deeds of coveting and wickedness, as well as deceit, sensuality, envy, slander, pride and foolishness. All these evil things proceed from within and defile the man."
>
> Mark 7:20-23

Paul wrote to the Ephesian church that, because of that first sin, man's mind and his ability to know truth (including true science) is "futile" and "darkened."[24] He further stated that those who reject God are "futile in their speculations," in their ability to evaluate data.[25] James, Jesus' half-brother and an early church martyr, wrote that man's volition is "carried away and enticed by his own lust."[26] Even our intellect is corrupted by our inherent sin nature.

Our bodies are equally corrupted. Every time we go to the dentist, put on our glasses so we can see, stifle a huge yawn, or groan as we move, we realize our nature wasn't the only part damaged by sin. Our physical bodies are no longer "very good" but subject to disease, decay, and eventual death.

Are scientists exempt from these mental and physical weaknesses? When they put on their white lab coats, do they gain special powers like that reporter for the *Daily Planet*? You may remember that, during a crisis, the very average Clark Kent went into a phone booth, took off his thick glasses, and changed from his bland-colored suit and tie into a blue, skin-tight outfit with a big "S" in front and a cape in back. After emerging, he took a leap and flew without an airplane to the problem spot. He bent steel with his bare hands and looked through walls with his x-ray vision.

When scientists put on their lab coats, do they somehow get Superman-like abilities? Are they no longer average, sinful human beings like the rest of us? When they receive their PhDs, do they gain immunity from fallibility, subjectivity, and prejudice? Or is the reality that even the best scientist is limited, fallible, finite, sinful, biased, imperfect, and, to a greater

23. Jeremiah 17:9.
24. Ephesians 4:17-18.
25. Romans 1:21-22.
26. James 1:13-15.

or lesser degree, self-seeking? I will let you put on your own blue, skin-tight suit to use your perfect, untarnished objectivity to answer those questions.

> Science means constantly walking a tightrope between blind faith and curiosity; between expertise and creativity; between bias and openness; between experience and epiphany; between ambition and passion; and between arrogance and conviction–in short, between an old today and a new tomorrow.
>
> **HENRICH ROHRER, 1986 NOBEL PRIZE IN PHYSICS**

WHAT CAN WE CONCLUDE?

God created the universe with us as its focus. He gave us responsibility and sovereignty over every aspect of creation. He expects us to study it in detail and utilize it through science. Only because He created us in His image is science even possible.

When the first man rebelled against his Creator, his one sin brought cursing on the entire universe. Every part of creation now decays until death. Every one of us is hindered by our inherent limitations and fallibility. We need God's view of the creation to counteract our own sin-related deficiencies.

THE TAKE AWAY: Since God created us in His image, we have the potential to do good. Science can help us use creation for the benefit of all; however, because we inherited a sin nature from Adam, we can just as easily use scientific advances for evil purposes.

> I believe that a full understanding of this remarkable human capacity for scientific discovery ultimately requires the insight that our power in this respect is the gift of the universe's Creator who, in that ancient and powerful phrase, has made humanity in the image of God (Genesis I: 26-27). Through the exercise of this gift, those working in fundamental physics are able to discern a world of deep and beautiful order–a universe shot through with signs of a mind. I believe that it is indeed the Mind of that world's Creator that is perceived in this way. Science is possible because the universe is a divine creation.
>
> **DR. JOHN POLKINGHORNE, MATHEMATICAL PHYSICS**

For in six days the Lord made the heavens and the earth, the sea and all that is in them, and rested on the seventh day; therefore, the Lord blessed the sabbath day and made it holy.

EXODUS 20:11

Before the mountains were born or You gave birth to the earth and the world, even from everlasting to everlasting, You are God.

PSALM 90:2

MOSES, WHO SAW THE GLORY OF THE LORD

CHAPTER 5

DISTORTIONS IN SCIENCE

You may be like I was when I first started wondering why science and the Bible don't seem to agree. Questions kept bouncing around in my head. But, I asked myself, doesn't science deal with facts, and don't scientists make every effort to be objective? Don't they verify their conclusions with experimentation? Are not opinions, biases, and politics contrary to the practices of science? For some reason, science had been able to convince me that it was above the problems that plague everything else in Satan's world.

Purity in science might be expected in an ideal world. Anyone who thinks we live in an ideal world must live on a different planet than I do! This book does not deal with that planet or with Superman's equally imaginary one. Sadly, science today on this planet tries to function without God, thereby limiting and even damaging many of its conclusions.

We already saw that on planet Earth, all mankind, including scientists, are born in total depravity. Because each of us inherits Adam's sin nature at birth, we should expect the politicizing and monetizing of science. We shouldn't be surprised that people often use science to bolster their preconceived ideas, sooth their egos, further their agendas, and satisfy their lusts for power, money, or admiration—or all the above.

CAN SCIENCE BE CORRUPTED THROUGH POLITICIZING?

Over the years, science has developed a stellar reputation. Today, though, everything, including science, can and is politicized. Trying to prove personal opinions has replaced searching for reality. Too often, truth is sacrificed for personal, agenda-driven gains.

In his book about the politicizing of science, Tom Bethell, an American journalist and senior editor of *The American Spectator*, wrote, "It turned out that science could easily be politicized. The most important reason is this: Often, there is a great deal of uncertainty as to what the facts are. Preferences can then be substituted for facts, and that can happen inconspicuously."[1] He is describing the substitution of worldviews for facts, isn't he?

Patrick Wood, a long-time advocate for truth in science, summarized this dangerous trend:

> Once upon a time, there was something called science. It included the discovery of truth about nature, the elements, the universe, etc. It was practiced by honest and accountable practitioners called scientists and engineers. These scientists often invented cool new things as a result of their studies, but generally they had no primal urge to use their knowledge to dominate other people, groups or even entire societies. Then certain other scientists and engineers rose up and made a discovery of their own. If true science was ever-so-slightly skewed and engineering disciplines were applied to society at large, then they could indeed use their "knowledge" to dominate and control other people, groups, entire societies or even, heaven forbid, the entire planet. The first group pursued science. The second group pursued pseudo-science.[2]

Science can be and often is politicized and corrupted through money.

---

1. Bethell, Tom, *The Politically Incorrect Guide to Science*, Regnery Publishing, 2005, vii-viii. For more details, I recommend you read this book, which is devoted entirely to this subject.
2. https://technocracy.news/the-miserable-pseudo-science-behind-face-masks-social-distancing-and-contact-tracing, Patrick Wood, June 19, 2020.

> I strongly believe that fundamental science cannot be driven by instructional, industrial and government or military pressures. This was the reason why I decided, as far as possible, not to accept money from the government.
>
> **C.V. RAMAN, 1939 NOBEL PRIZE IN PHYSICS**

Scientists and science projects are dependent on funding. Today, most funding sources expect the scientists they pay to produce specified results. Contrary results end the funding. Sometimes, livelihoods and careers are at stake. The cries of hungry children and demanding wives become strong motivations to do whatever it takes to keep the funding agency and, therefore, the family happy.

I saw this first-hand in the 1980s when I worked as a research engineer for an engineering group during the Reagan presidency. This was during the cold war when President Reagan was building up our military defenses. The funding for my project was secure since we were developing products related to this defense work. We had no motivation to politicize our work or to alter outcomes to please those paying us. When our work was completed, our group did, as expected, cease to exist.

> Whoever is careless with the truth in small matters cannot be trusted with important matters.
>
> **ALBERT EINSTEIN, 1921 NOBEL PRIZE IN PHYSICS**

Other groups of scientists are not as segregated from temptation as we were. Their vulnerability to losing their jobs makes them more susceptible to politicizing. Many succumb to the lure of money and prestige. Bethell explained the problem:

> Historically, the competition of theories has been the driving force behind science progress. Isolated individuals and private companies have been the most fruitful sources of this adventure. And just as a competitive market system forces innovation into private enterprise, so the competition of theories drives science to investigate new approaches. When all research eggs are in one basket, it's a different world. Competition may stagnate, or be eliminated entirely, if that is what the government decrees (as happened with Communism). When any single source of

funding dominates, science will almost certainly become the handmaiden of politics.[3]

> So, is naturalism actually demanded by science? Or is it just conceivable that naturalism is a philosophy that is brought to science, more than something that is entailed by science? Could it even be, dare one ask, more like an expression of faith, akin to religious faith? One might at least be forgiven for thinking that from the way in which those who dare ask such questions are sometimes treated. Like religious heretics of a former age they may suffer a form of martyrdom by the cutting off of their grants.
>
> **DR. JOHN LENNOX, MATHEMATICS**

HAS CLIMATOLOGY BEEN POLITICIZED?

Global warming (now called climate change) is perhaps the clearest, modern example of interested parties exerting pressure to get verifying "scientific" results that support their radical theories. They use fabricated scientific theories as well as subjective computer models to precipitate agenda-driven crises. They warn that if we do not act right now using "the current science" (their bought and paid for science), we will soon be faced with a crisis of immense proportions, one that science will be unable to solve. Horrors! Too many respond as these enemies of liberty desire, with fear rather than a thoughtful examination of their "facts."

We need to heed Bethell's warning: "All science based on dire warnings about the future should be suspect, and all such science is almost by definition politicized—if only because democracy as presently constituted responds with undue haste to any and all claims of crisis."[4]

Ironically, those adhering to this global warming pseudo-science had to change their crisis' name to climate change when the earth went through a period of cooling. Its leaders couldn't change the date or venue of a planned press conference even though one of the worst snowstorms in modern history was pummeling them! Thus, the term global warming was ditched, and climate change was born. As you can see, God, who keeps snow in His storehouses for just the right occasion, has perfect timing.[5] We might even say He laughs at our attempts to be Him.[6]

3. Bethell, xii.
4. Ibid, ix.
5. Job 28:22.
6. Psalm 2:4.

FIGURE 3 HURRICANE MATTHEW, 2016

The science of climatology has been distorted brazenly to promote a liberal-socialist-communist agenda aimed at stripping us of our freedoms and destroying the American way of life. We need to examine the issue rationally through a clear lens and ask two questions. First, does a damaging climate change even exist? A child could tell you such changes are called seasons and should be expected. Further, we can't stop natural events such as hurricanes or unusually warm or cold weather because God alone controls them. What arrogance to imagine otherwise!

The second question involves climate change advocates' declaration that man is the chief cause of supposedly disastrous climate fluctuations. They argue man is too stupid (though they use more politically correct words) to control his dangerous behavior and, therefore, needs to be put under their control. They advocate the removal of all cars and planes—except their own, of course. They even want to get rid of all cows, which they assure us produce too much carbon dioxide or methane or some other absurdity to be allowed to take up space on Earth! How does getting rid of cows line up with another of their pet causes, saving endangered species, I wonder? I also wonder how much warming the billions of American bison in our not so distant past caused. More contradictions.

FIGURE 4 MICHAEL RAMIREZ, 2014, USED WITH PERMISSION

As you have probably noticed if you watch television or read the news, the national media, which is predominately liberal, promotes only the data that supports climate change. Observational scientific data that challenges their liberal, subjective data is suppressed and ignored. The media rewards those whose views promote their preferred agenda while trying to silence and destroy those who scientifically bring their agenda into question. Truth has been sacrificed for the promotion of a communist agenda whose goal is a Godless society under their control.

> It does not seem to occur to people that once they can be persuaded that they are just smart monkeys produced by chance through a struggle for survival, they can be pushed around by dictators much more easily than someone who believes he was created with an eternal soul.
>
> **JAROSLAV JOSEPH HAVEL, CZECHOSLOVAKIAN-BORN BOTANIST, SILVICULTURIST, ECOPHYSIOLOGIST**

Any time we ignore God's truth about creation, foolishness is the by-product. Even so, too many, like sheep, are fooled. You don't want to be one of them. You need to be able to objectively evaluate scientific information, so anti-God pseudo-science won't lead you astray.

HAS THE USE OF BENEFICIAL SCIENTIFIC DISCOVERIES BEEN POLITICIZED AS WELL?

Climate change is only one of many areas in which politicization has corrupted scientific accuracy. In another example from the recent past, false science precipitated the removal of DDT (dichloro-diphenyl-trichloroethane) from the marketplace. Bethell briefly traced the development and history of DDT from its discovery by an Austrian scientist in 1874 to its ban in 1972.[7] This insecticide was successfully used for a century to prevent malaria and save millions of lives. By 1967, malaria had been virtually eradicated.

In 1970, the National Academy of Science reported, "To only a few chemicals does man owe as great a debt as DDT....In little more than two decades, DDT has prevented 500 million human deaths, due to malaria, that otherwise would have been inevitable."[8]

Unfortunately, due largely to an ecological book based on liberal subjectivism and pseudo-science, attitudes toward the use of DDT changed radically. Political authoritarianism precipitated the removal of DDT from the marketplace, thereby causing a dramatic increase in the number of disease-carrying insects, especially in third world countries. This politicization of science led directly to the deaths of millions. The Zika virus, a mosquito-borne illness, began causing severe medical problems even in America. Years later, even having proof of the devastating effects of the irrational misuse of science, DDT is still not available.

THE TAKE AWAY: All ideologies that deny God's existence use science corruptly to achieve their desired ends regardless of the cost to others. The deaths of the most vulnerable is a steep cost indeed.

As I write, the Covid-19 virus has spread around the world. Millions have died, mostly from bad medical practices, we now have learned, rather than from the disease itself. In the United States, disturbing lies about the pseudo-pandemic were used to help sway a presidential election. Biased reporters used inaccurate statistics to promote

7. Ibid, 73-85.
8. "The Life Sciences," National Academy of Sciences, NAS Press, 1970, quoted by Bethell, 75.

their political agendas, even twisting accurate data.[9] They denied the efficacy of several medicines that would have prevented many if not most deaths. They used the same tactics to promote dangerous vaccines that kill many more than did Covid. The buzz phrase today is "follow the science," but the science being followed is politicized pseudo-science. Those who swallowed the lies suffered the most.

These three examples illustrate the corrupting effect politics can and does have on science. A few others we don't have time to study are nuclear energy bans, stem cell research, gender fluidity, endangered species, biological weapons, population control, indiscriminate vaccine usage, weather control, and pollution. We could also add the present state of science education with its exclusive teaching of evolution. Public schools are not permitted to discuss alternatives to evolution because the dominant political ideology wants to abolish any mention of God and any attempt at freedom of speech.

Would true science cover up the weaknesses of any scientific idea? Would it omit viable alternatives to a weak theory? Would it falsify scientific data? No, but an anti-God, unbelieving worldview would.

> Another of the qualities of science is that it teaches the value of rational thought, as well as the importance of freedom of thought; the positive results that come from doubting that all the lessons are true....Learn from science that you must doubt the experts. As a matter of fact, I can also define science another way: Science is the belief in the ignorance of experts.
>
> **RICHARD FEYNMAN, 1965 NOBEL PRIZE IN PHYSICS**

DOES SCIENCE HAVE LIMITATIONS?

Though some, maybe most, in western culture have elevated science to an almost God-like status, it should come as no surprise that science has limitations. By stating the obvious, I'm not degrading science but explaining reality. All scientists are aware of these limitations and subscribe to nearly all of them. Remember: Even Superman had limited power when kryptonite was near.

---

9. Many facilities in Florida wrongly reported 100% positive tests but later data proved them wrong by a factor of 10. An example of some of the reporting can be found at https://www.usatoday.com/story/news/health/2020/07/15/florida-coronavir.

LIMITATION 1: THE ABSENCE OF ABSOLUTE TRUTH

What is absolute truth? Being absolute means being free from imperfection and, therefore, perfect, settled, unchanging, and unchangeable. No matter how many agenda-driven scientists declare otherwise, scientific truth is not absolute. It cannot be because change is the essence of science. Most scientists would agree.

World-famous evolutionist George Gaylord Simpson wrote:

> The concept of 'truth' in science is thus quite special. It implies nothing eternal and absolute but only a high degree of confidence after adequate self-testing and self-correcting....Above the level of triviality there is hardly any scientific subject on which agreement is literally universal.... In any complex situation the data are rarely so complete that only one explanation can conceivably be correct.[10]

In general, scientific truth is accepted as truth only after investigation, testing, and consensus, not because it is considered absolute. "In science, consistency of data interpretation constitutes proof."[11] Data interpretation, not data, is the hallmark of secular truth.

THE TAKE AWAY: Every scientific truth eventually reveals deficiencies. Its source is fallible man and, therefore, reflects the imperfections of that source. Science may aim for objectivity, and it often does, but newly discovered data changes it. In fact, the history of science is one of advancement and change, not of absolute, unchanging convictions.

Even settled laws of science have had to be abandoned. A famous example is the law of biogenesis. In the 1600s, science taught that life spontaneously generated from garbage. Yes, garbage. All the scientific observations of the day pointed to the maggots and worms that came from decaying meat as the source of life. In essence, they believed that life came from non-life, from garbage. You may laugh, but scientists in that day were as deadly serious in their belief in biogenesis as scientists today are about evolution, climate change, some vaccines, and other agenda-driven opinions. You will be interested to know that the theory

10. Simpson, George Gaylord, *Notes on the Nature of Science*, Harcourt, Brace, and World, 1962, 11, 7. Quoted in Davidheiser, Bolton, *Evolution and Christian Faith*, The Presbyterian and Reformed Publishing Company, 1969, 152.
11. Dye, David L., *Faith and the Physical World: A Comprehensive View*, Eerdmans, 1966.

that life comes from non-life is the basis of modern evolutionary thought, but it takes millions of years.[12] More garbage?

Another famous example of the fallibility of scientific laws was the belief the earth was at the center of the universe and that the sun orbited it.[13] That supposed absolute truth changed when Nicolaus Copernicus (1473–1543), a Christian, showed that all planets, including Earth, orbit the sun. He was ridiculed by other scientists, but time proved the accuracy of his observations.

Before Louis Pasteur (1822–1895) discovered the existence of unseen bacterial life forms, medical science caused almost as many problems, often fatal ones, as it cured. Medicine changed dramatically after his radical discovery was believed and implemented.

Science is constantly changing as more, better, or new observations are made. The essence of science is progress and advancement, not eternality, immutability, and absolute truth. Those belong only to God.

> Science is founded on uncertainty. Each time we learn something new and surprising, the astonishment comes with the realization that we were wrong before... In truth, whenever we discover a new fact it involves the elimination of old ones.
>
> DR. LEWIS THOMAS, MEDICINE

LIMITATION 2: THE INADEQUACIES OF THE HUMAN INTELLECT

Since science is the product of the human intellect, it has all the limitations of the human nature. We already discussed humanity's fallibility, including those limitations related to the intellect and the acquiring of truth and knowledge. We saw that scientists are not automatically immune from those limitations.

Man is fallible and, thus, makes mistakes. Scientists spend more of their time figuring out what doesn't work than what does. They can and have come to very wrong scientific conclusions. Like all humans, they are innately sinful, self-seeking, and prone to introducing biases and, thereby, politicizing scientific data.

12. https://www.britannica.com/science/abiogenesis.

13. Geocentric model, https://en.wikipedia.org/wiki/Geocentric_model.

Boyce Rensberger, a science writer for more than forty years, sounded the alarm:

> At this point, it is necessary to reveal a little inside information about how scientists work, something the textbooks don't usually tell you. The fact is that scientists are not really as objective and dispassionate in their work as they would like you to think. Most scientists first get their ideas about how the world works not through rigorously logical processes but through hunches and wild guesses. As individuals, they often come to believe something to be true long before they assemble the hard evidence that will convince somebody else that it is. Motivated by faith in his own ideas and a desire for acceptance by his peers, a scientist will labor for years knowing in his heart that his theory is correct but devising experiment after experiment whose results he hopes will support his position.[14]

The finiteness of the human mind informs us of the limitations of the human intellect. Only the God of the Bible is omniscient, knowing all that can be known. Humans have minds with finite capacity, able to reason and make choices, but not even close to omniscient. We certainly can't know the future or even reconstruct the past with accuracy. As for knowing the repercussions had we made different decisions, we haven't a clue. When we add to that many changed decisions by every person who has ever lived, a headache is our only conclusion. We are foolish to equate ourselves with God, who knows all the knowable, even the might-have-beens and certainly-will-bes.

Today, the specializations within the many disciplines of science more than hint at the enormous amount of scientific data available to study. Consider the broad science of physics, which includes many sub-specializations within even the following diverse and broad specializations: astrophysics, astronomy, geophysics, nuclear physics, atomic physics, nanotechnology, acoustics, optics, electricity, electronics, electromagnetism, mechanics, statics, meteorology, thermodynamics, laser physics, quantum mechanics, and the list goes on. If we are not capable of acquiring more than 5% of all the data within even one sub-sub-specialty of

14. Rensberger, Boyce, *How the World Works*, William Morrow, 1986, 17-18, as quoted by Jonathan D. Sarfati, *Refuting Evolution: A Responding To The National Academy Of Sciences' Teaching About Evolution And The Nature Of Science*, Master Books, 1999, 16-17.

one scientific specialty, how can we presume to know everything about every aspect? Only arrogance makes us think we can.

Albert Einstein questioned even that five percent knowledge deficit, saying, "We still do not know one thousandth of one percent of what nature has revealed to us."[15] If Einstein, the unquestioned scientific genius of the last century, considered himself ignorant of the hidden wonders of creation, should we not humbly admit to our limitations, too?

> Nevertheless, Jesus knows our world. Wherever we are at any moment of day or night, Jesus is at our disposition. We can reach Him simply by turning toward Him our desire and our love. It is an easily observable fact that, even in the society created by science and technology, this need of God has persisted.
>
> **DR. ALEXIS CARREL, 1912 NOBEL PRIZE IN MEDICINE**

Even some evolutionists realize the limitations of man's intellect. These secularists may have full confidence in science because their materialistic worldview demands it, but even they don't believe in the infallibility of scientists and scientific processes.

Richard Lewontin, a genetics professor and Marxist, admitted his bias when he wrote:

> We take the side of science in spite of the patent absurdity of some of its constructs, in spite of its failure to fulfill many of its extravagant promises of health and life, in spite of the tolerance of the scientific community for unsubstantiated just-so stories, because we have a prior commitment, a commitment to materialism. It is not that the methods and institutions of science somehow compel us to accept a material explanation of the phenomenal world, but, on the contrary, that we are forced by our a priori adherence to material causes to create an apparatus of investigation and a set of concepts that produce material explanations, no matter how counter-intuitive, no matter how mystifying to the uninitiated. Moreover, that materialism is an absolute, for we cannot allow a Divine Foot in the door.[16]

15. https://www.quote.cc/science-quotes/.
16. "Billions and Billions of Demons," *The New York Review*, Jan 9, 1997, 31, as quoted by Sarfati, *Refuting Evolution*, 43.

Even this communist denier of God's existence admits man needs to exclude the "Divine Foot" from science. His bias, his worldview, works to exclude God from science because, innately, he knows God exists. Why would he warn against God if God is imaginary?

Creationist author Ken Ham, founder of the Answers in Genesis creation organization, the Creation Museum, and the Ark Encounter, wrote a wonderfully accurate summary of bias. "It is not a matter of whether one is biased or not. It is really a question of which bias is the best bias with which to be biased."[17]

My bias is that the biblical worldview gives us the best foundation to do good science because only it depends on the Creator for insights into His creation. I am clearly and definitely biased toward the Creator.

Scientists are finite and do not have the mental capacity to adequately understand the many aspects of creation. They are part of the creation, not the Creator of it. Only the omniscient Creator God understands it all.

> Great spirits have always encountered violent opposition from mediocre minds. The mediocre mind is incapable of understanding the man who refuses to bow blindly to conventional prejudices and chooses instead to express his opinions courageously and honestly.
>
> **ALBERT EINSTEIN, 1921 NOBEL PRIZE IN PHYSICS**

LIMITATION 3: TIME

Many believe scientific principles are unchanging and, therefore, not dependent on or affected by time. They believe that the scientific "truth" accepted today has always been true and always will be true. Are they right, or does time put limits on science?

Observational science, by definition, studies the creation and its processes in ongoing, present time and, therefore, is dependent on present circumstances for repeatable results. This in itself is a limitation in terms of time. How can we know with certainty that these present-time scientific processes and phenomena are unlimited in scope? How can we know they have always and will always react just as they do now? We

17. Ham, Ken, *The Lie: Evolution*, Master Books, 2012, 9.

can't. Chapter 16 will demonstrate that the natural realm has undergone radical changes in the past such that the laws and constants of science observed today were very different in prior eras.

The universe is finite, not infinite or eternal. Only God is infinite and eternal. That means man in his finiteness can only make finite observations about the finite universe. He cannot presume to make accurate predictions about the continuation of scientific laws into the past or the future because both the past and the future are hidden from his finite view.

We can see this time limitation most clearly in historical science because events occurred only once in the past. They are not ongoing and cannot be reproduced. They can only be partially known from the meager remaining traces. By its very nature, historical science is severely limited by and vulnerable to time.

Another time limitation is obvious. Science cannot, with accuracy, tell us the future. Just as no time machine can take us back in time, so none can take us forward into the future. We can't even be certain of tomorrow. I mention this because we do have a source Who explains the future. In Scripture, God reveals not only future events but also some future scientific changes He will institute. And, yes, God promises that when eternity begins, time will change in a way we can't yet understand. He also promises that many scientific laws will undergo radical changes before the end of time. Science cannot perform experiments to discover these scientific changes. Only divine revelation can raise the veil that shrouds the future. Peter B. Medawar, the 1960 Nobel Prize recipient in medicine, wrote:

> My contention in the discussion so far has been that it is logically outside the competence of science to answer questions to do with first and last things...Catastrophe apart, I believe it to be science's greatest glory that there is no limit upon the power of science to answer questions of the kind science *can* answer.[18]

Science cannot answer every question. It is impossible, in part because of the limitations time puts on it. The prophet Isaiah recorded that only the true God of the Bible can proclaim the past and the future

18. Medawar, Peter B., *The Limits of Science*, Harper Row, 1984.

with accuracy because only He is eternal and omniscient. He is the First and the Last. He declared the end from the beginning.[19]

Though science is limited by time, God's revelation is not. We can trust Scripture to reveal the past as well as the future—at least all events relevant to our understanding.

> "What [Dr. Russell] Humphreys has done," he says, "is show us another parameter of something that most people view as a constant, and that is *time* itself. Using Einstein's Theory of General Relativity, he has shown how time can vary depending on your position in space–it affects your viewpoint. Time is slowed by gravitational forces. A clock at sea level has been shown to run more slowly than one on top of a mountain, because the one at sea level is affected by more gravity. This is an effect known as *time dilation*, and has been experimentally demonstrated. Humphreys uses this to great effect in his model to deal with the distant starlight issue. His cosmology starts with the earth near the centre originally, then the universe rapidly expanding in a 'white hole' or black hole running in reverse. At the beginning, gravity would slow earth 'clocks' far more than clocks further away, especially at the edge of the universe. Therefore, 'billions of years' would be available (measured by clocks in those distant regions of space) for light to reach the earth, for stars to age, etc.–while less than one ordinary day is passing on earth (measured by earth clocks, on which biblical time is based)."
>
> **DR. JOHN G. HARTNETT, PHYSICIST**

LIMITATION 4: FAITH AS PART OF SCIENCE

The goal of science is to gain a complete understanding of the natural realm. The quest for this understanding relies on certain starting points, certain unprovable assumptions. Science can't function properly apart from them.

As we have seen, all worldviews are accepted by faith. The unprovable starting points upon which the scientific enterprise relies are likewise accepted by faith. Contrary to popular understanding, scientists indeed exercise faith, faith in these assumptions, every time they do their scientific work.

---

19. Isaiah 44:6-7, "Thus says the Lord, the King of Israel and his Redeemer, the Lord of hosts: 'I am the *first* and I am the *last*, and there is no God besides Me. And who is like Me? Let him proclaim and declare it; yes, let him recount it to Me in order, from the time that I established the ancient nation. And let them declare to them the things that are coming and the events that are going to take place." Also Isaiah 46:9-10, "Remember the former things long past, for I am God, and there is no other; I am God, and there is no one like Me, declaring the end from the *beginning* and from ancient times things which have not been done, saying, 'My purpose will be established, and I will accomplish all My good pleasure.'"

The first basic assumption of science is that a physical, material, natural realm exists and can be studied. Though this assumption seems obvious to every one of our five senses, we need to address it because some philosophers have come to the opposite conclusion, insisting our senses can mislead us. Can they?

Second, science assumes that we can objectively observe the natural realm, that we have the ability to do science. Again, though this assumption is not obvious to some philosophers, scientists accept it as true.

A third assumption is that predictability, constancy, and repeatability are inherent to the natural realm. We don't dispute that natural characteristics are observable and testable to a limited degree, but is it sensible to assume they have always behaved similarly and will always continue to do so?

A fourth assumption is that logic exists, that man has the capacity to utilize it, that the natural realm exhibits logical relationships, and, therefore, that we can expect consistency as we undertake scientific measurements and observations. You could well ask, Where does logic come from?

All scientists, mostly unknowingly, accept these basic assumptions because they seem to work. They are accepted religiously, by faith. No scientist attempts to prove them since they cannot be arrived at by the scientific method. All scientists are forced to rely on them by faith.

Other prominent scientific conclusions are also faith-based. The most famous is evolution. Although it is almost universally accepted and often referred to as fact, any honest scientist would have to admit that evolution does not even rise to the level of a tentative hypothesis because so little evidence supports it. Evolution is a philosophical idea, a dogma elevated to the status of scientific fact by its supporters. Its validity is accepted by faith.

> We must therefore recognize evolutionary theory for what it is, a philosophy (indeed, for some a religion) and not basically a scientific discipline at all....It is the writer's belief that the consistent and satisfying interpretation (to the observations of nature)...is to be found in the Christian Scriptures, which provide an account of creation, nature, consciousness and being (both on the material and spiritual planes) as magnificent as it is complete.
>
> **DR. E.H. ANDREWS, PHYSICIST**

Even evolutionists, at least honest ones, admit evolution is faith based. British botanist D.H. Scott wrote that "a new generation has grown up which knows not Darwin. Is even then evolution not a scientific ascertained fact? *No!* We must hold it as an act of faith because there is *no alternative*."[20]

British physicist H. S. Lipson admitted that "in fact, evolution became in a sense a scientific religion; almost all scientists have accepted it and many are prepared to bend their observations to fit in with it."[21]

Another evolutionist not only made the same admission of evolution's fraudulent claims but also revealed the underlying reason so many accept the concept anyway. In 1929, D.M.S. Watson, Professor of Zoology at London University, summarized the thinking of most evolutionists: "The theory of evolution itself is a theory universally accepted, not because it can be proved by logical coherent evidence to be true but because the only alternative, 'special creation,' is clearly incredible."[22]

Evolution enjoys widespread acceptance because it posits a world in which God is not necessary. Many want, maybe even need, to accept this idea because God's existence strikes terror in their hearts. They are aware of their violations of His righteous standards while, sadly, unaware of His astonishing goodness and mercy. They fear His judgment. Their fear hides His love.

> Nothing in life is to be feared, it is only to be understood. Now is the time to understand more, so that we may fear less.
>
> **MARIE CURIE, 1903 NOBEL PRIZE IN PHYSICS, 1911 IN CHEMISTRY**

Secular science requires faith not only for its assumptions but also for its conclusions. The issue is not science versus faith but the faith of secular science versus the faith of the biblical worldview. Do we believe the Creator, or do we try to come up with our own ideas and manipulate the scientific process to prove them? Is our faith

20. Scott, D.H., "Presidential Address," *Nature*, 108:2709:153, 9/29, 1921, as quoted in Davidheiser, 155.
21. Lipson, H.S., "A Physicist Looks at Evolution", Physics Bulletin, V. 31, 1980, as quoted in *Impact #107*, Henry M. Morris, Institute for Creation Research, 1982, ii.
22. Watson, D.M.S., "Adaptation," *Nature*, Vol. 123, 1929, 233, as quoted in Morris, Henry M., *Scientific Creationism*, Master Books, 1974, 8.

grounded in God or in self, science, intellect, power, money, or some other unworthy object?

> Religion, with its theological reflection, builds on faith. Science too builds on faith. How? For successful science of the type we know, we must have faith that the universe is governed by reliable laws and, further, that these laws can be discovered by human inquiry. The logic of human inquiry is trustworthy only if nature is itself logical. Science operates with the faith that human logic can in the long run understand nature's laws and that they are dependable. This is the faith of reason.
>
> DR. CHARLES TOWNES, 1964 NOBEL PRIZE IN PHYSICS

LIMITATION 5: NO FIRM CONCLUSIONS

Many assume science comes to firm and final conclusions that are no longer subject to uncertainty. This impression, however, betrays a misunderstanding of science.

Scientific hypotheses, theories, and laws are all subject to ongoing testing against new observations. New data is constantly being reported, which requires adjustments to existing hypotheses and theories. As we've seen, on occasion even settled laws of science are entirely abandoned when new information is uncovered. (Remember putrefied meat with maggots as the source of life?) In principle, all scientific conclusions are subject to change as scientists expand their knowledge. By contrast, the scientific content found in Scripture is unchanging. It is final, finished, and reliable because God, its Author, is immutable and eternal.

LIMITATION 6: NO ANSWERS TO IMPORTANT QUESTIONS

How well does science provide viable answers to the really big questions? For example, what insights does it offer about why we are here, who God is, or the existence of a spiritual realm? Can it explain death? Secular science cannot answer these questions.

How can science address spiritual questions when it can't even deal with simple questions relating to the natural realm, such as "How did everything begin?" or "Why don't we go flying off into space?" Those who reject God and a spiritual realm generally invoke evolution in their attempt to address these larger questions, but as a purely materialistic hypothesis, evolution has no ability to provide answers to these inherently transcendent questions.

Often, science can explain, at least in part, how things work in present time. It can analyze and evaluate some details about the creation around us, but no experiment or observation can give insights into the reason things exist. Science cannot show the purpose of the universe or anything in it. The meaning of life or of anything in the creation is beyond its ability. Even philosophy can only speculate concerning origins, meaning, and purpose. Because of the limitations of science and philosophy, we need to look to divine revelation to determine the answers to the most basic, life-changing questions.

> Science may provide the most useful way to organize empirical, reproducible data, but its power to do so is predicated on its inability to grasp the most central aspects of human life: hope, fear, love, hate, beauty, envy, honor, weakness, striving, suffering, virtue.
>
> PAUL KALANITHI, NEUROSURGEON

LIMITATION 7: THE CHANGEABILITY OF SCIENCE

One of secular science's unverifiable foundational assumptions is uniformitarianism, which claims the natural realm in general and the earth specifically operated the same in the past as it does today. Its mantra is "the present is the key to the past." Most today believe the earth has proceeded in inexorable, predictable fashion over billions of years with no catastrophic or sudden discontinuities affecting that progression. In other words, scientists who believe in uniformitarianism discount the literal six-day creation, the fall of mankind, the Flood of Noah, and the tower of Babel. The biblical record of those historical events indicates the natural realm experienced radical changes during them. They prove the present is not the key to understanding the world before the tower of Babel event.

Here is a good description of uniformitarianism from Britannica:

> Uniformitarianism, in geology, the doctrine suggesting that Earth's geologic processes acted in the same manner and with essentially the same intensity in the past as they do in the present and that such uniformity is sufficient to account for all geologic change. This principle is fundamental to geologic thinking and underlies the whole development of the science of geology. When William Whewell, a University of Cambridge scholar, introduced the term in 1832, the prevailing view (called catastrophism) was that Earth had originated through supernatural means and had been affected by a series of catastrophic events such as the biblical Flood. In contrast to catastrophism, uniformitarianism postulates that phenomena displayed in rocks may be entirely accounted for by geologic processes that continue to operate—in other words, the present is the key to the past.[23]

Science depends on constancy and predictability. Without doubt, such stability does currently exist. For as long as scientific observations have been made, this principle appears to have been confirmed.

The question is not whether uniformity exists now and in the recent past but whether creation has always been unchanged and will always continue unchanged. In other words, have scientific constants always had the values they have today, and will they continue to have those same values? Are constants and the laws of science eternal? Because science is severely restricted in its ability to make measurements except in the present, it has no ability to answer these questions definitively. Observing the ancient past and predicting the future, even tomorrow, are beyond science's capabilities.

I will attempt to show that although science can't answer these and other fundamental life questions, divine revelation can. These answers are so important I will devote entire chapters to explaining the data given in the revelation of Scripture. I will show that several historical events have radically changed the universe. The world has not always been as it is today, that ancient historical events radically changed the universe.

23. https://www.britannica.com/science/uniformitarianism.

Evidence shows that not only some scientific constants but also some laws of science have undergone changes in the past, in some cases extreme changes. One example I will discuss later involves radioactive decay rates, which have not always been the same as they are today. I will also show from Scripture that we can expect radical changes in the future. You'll enjoy learning God's take on science, His answers to our many questions.

Are not scientific constants forever fixed and unchanging? Science assumes the answer to be yes, but we know the answer is an absolute no. Keep reading because the Bible has a lot to say about this. You may even be surprised.

> If you thought that science was certain—well, that is just an error on your part.
>
> RICHARD FEYNMAN, 1965 NOBEL PRIZE IN PHYSICS

LIMITATION 8: CIRCUMSCRIBED POTENTIAL

Discoveries, particularly in molecular biology over the past seventy years, have revealed the staggering complexity of the natural realm. The more we discover in every area of science, the more we realize the enormity of all that remains to be uncovered. From a biblical perspective, every scientist, even the most profound genius, is finite in his being and in his ability to acquire and process knowledge. Though a wealth of information remains to be uncovered, can sin-plagued scientists see beyond the most obvious truths about creation?

LIMITATIONS OF SCIENCE

1. The Absence of Absolute Truth
2. The Inadequacies of Human Intellect
3. Time
4. Faith as a Part of Science
5. No Firm Conclusions
6. No Answers to Important Questions
7. The Changeability of Science
8. Circumscribed Potential

The prophet Jeremiah transcribed these divine words that imply God placed limits on obtaining knowledge (science):

> Thus says the Lord, "If the heavens above can be measured [which even science can't do], and the foundations of the earth searched out below [which even science can't do],

> then I will also cast off all the offspring of Israel for all that they have done [which He won't]," declares the Lord.
>
> Jeremiah 31:37

Science's potential is not unlimited but circumscribed by its practitioners' human frailties. Science may have other limitations, but aren't these eight enough to make us examine and question the absolute assertions of science?

WHAT CAN WE CONCLUDE?

Adam's fall into sin, which resulted in the darkening of his progenies' intellect, led to a stifling of accurate, comprehensive scientific study. Mistakes and distortions are inevitable. Biases and politicizations abound. Evolution, climate change, and the Covid pandemic are sobering examples of the danger of imposing biases against God onto scientific study. The question is not whether all scientists can be trusted, but why do we, like sheep, accept their most preposterous, God-attacking opinions with barely a squeak of opposition. Why, indeed?

THE TAKE AWAY: Because science is practiced by finite humans and because the universe and all that is in it are also finite, science has limitations.

> Science is incompetent to reason upon the creation of matter itself out of nothing. We have reached the utmost limit of our thinking faculties when we have admitted that because matter cannot be eternal and self-existent it must have been created.
>
> JAMES CLERK MAXWELL, DEVELOPER OF ELECTROMAGNETIC WAVE THEORY

To Him who made the heavens with skill,
For His lovingkindness is everlasting;
To Him who spread out the earth above the waters,
For His lovingkindness is everlasting;
To Him who made the great lights,
For His lovingkindness is everlasting:
The sun to rule by day,
For His lovingkindness is everlasting,
The moon and stars to rule by night,
for His lovingkindness is everlasting.

PSALM 136:5-9

Covering Yourself with light as with a cloak,
Stretching out heaven like a tent curtain.
He lays the beams of His upper chambers in the waters;
He makes the clouds His chariot;
He walks upon the wings of the wind;
He makes the winds His messengers,
Flaming fire His ministers.
He established the earth upon its foundations,
So that it will not totter forever and ever.
You covered it with the deep as with a garment;
The waters were standing above the mountains.

PSALM 104:2-6

WORSHIPPING PSALMISTS

CHAPTER 6

THE GREATEST FRAUD IN MODERN SCIENCE

The dominant view of historical science today is evolution, the idea that life developed from the "goo to you by way of the zoo"[1] over billions of years. Most schools and universities, from kindergarten to doctoral programs, present evolution as an undisputed fact. The public perception is that this settled science has been proven conclusively. Even most scientists never think to question its assertions. Any alternative theory is presented as religious fantasy and not permitted in schools or polite society. I will consider it a compliment if you accuse me of being impolite.

Despite its public acclaim, evolution is a philosophical idea that cannot be elevated through observational science to the status of even a theory, much less a scientific law. No aspect of evolution is repeatable or observable. You haven't seen any half-man-half-apes playing in your backyard, have you? What about one-fourth-bird-and-three-fourths-cat?

We could accurately describe evolution as a pagan philosophical or ideological opinion and a failed scientific hypothesis.[2] Sadly for us and

1. https://books.google.com/books/about/From_Goo_to_You_by_Way_of_the_Zoo.html?id=hBbnAAAACAAJ.
2. The Creation Research Society (creationresearch.org), The Institute for Creation Research (icr.org), Answers in Genesis (answersingenesis.org), and Creation Ministries International (creation.com) have massive amounts of scientific evidence against the idea of evolution.

for scientific advancement, even with its shaky credentials, evolution is the driving principle in secular science and permeates much that is done in the name of science today.

Before we go farther, we need to discuss evolution, today's dominant view on origins. We need to show that the worldview based on evolution is not only inadequate but wrong. Can evolution be true science? Let's put on clear glasses and examine the evidence.

Let us start with the so-called evidence that the evolutionist himself offers in support of his idea of evolution. As you read ask the question: is it repeatable and provable or only superficial evidence?

HOW DO EVOLUTIONISTS USE HOMOLOGY TO SUPPORT THEIR BELIEF?

The area in biology called comparative anatomy (homology) provides a popular, easy to understand example evolutionists say proves evolution. They theorize that similarities in the basic anatomy of different species indicate a common ancestry. Because of this common ancestry, they say, similar structures are passed on to later descendants down through the various evolutionary stages.

Grouping creatures by their common traits is the basis for the entire system of classification in biology. For example, all mammals have similar bone structures (spinal columns, four jointed limbs, rib cages, skulls, etc.), similar circulatory systems (veins, arteries, blood cells, etc.), similar reproductive systems (I want to keep this book PG-13 so no details), and on and on.

As the late Horatio Newman, a geneticist and zoologist, said years ago, this line of argument about common ancestry is a crucial element in the evolutionary idea:

> Now, a careful survey of the situation reveals the fact that the only postulate the evolutionist needs is no more or less than a logical extension of what the layman considers a truism or self-evident fact, namely, that fundamental structure resemblance signifies generic relationship; that, generally speaking, the degree of structural resemblance runs essentially parallel with closeness of kinship. Most

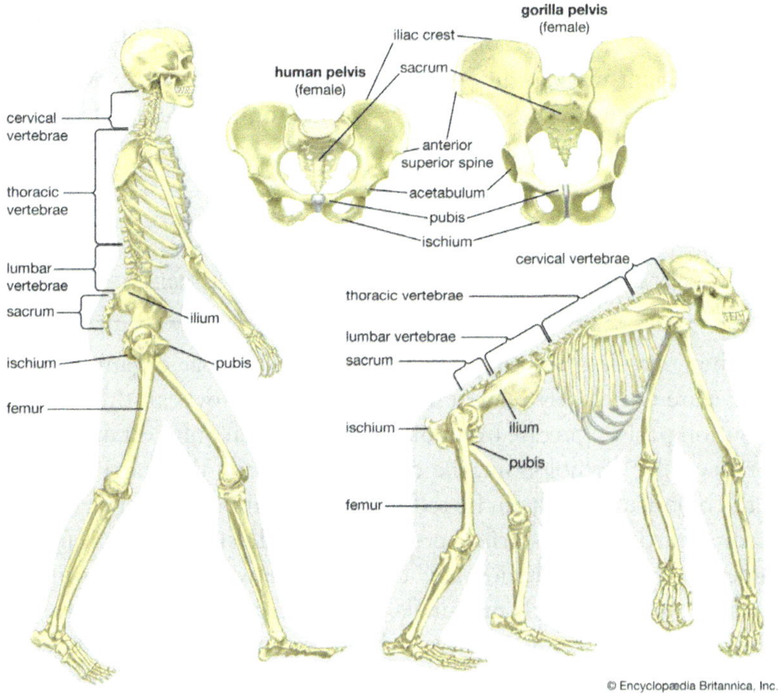

FIGURE 5 COMPARATIVE ANATOMY

> biologists say that this is not merely a postulate, but one of the basic established laws of life...If we cannot rely upon this postulate...we can make no sure progress in any attempt to establish the validity of the principle of evolution.[3]

More recently, Michael Denton, a biochemist, molecular biologist, and medical doctor, wrote that "the phenomenon of homology has remained the mainstay of the argument for evolution right down to the present day."[4] Why is that important? Because the evolutionary house of cards begins to collapse if homology doesn't provide sound evidence for evolution.

A few examples from their own discoveries will explain the inadequacies of homology as an indicator of evolution. Of particular note to evolutionists is the vertebrae, which is easily seen in many diverse species. Even you have a backbone, and aren't you glad you do! Though evolutionists

3. Newman, Horatio Hockett, *Evolution, Genetics, and Eugenics*, University of Chicago Press, 1932, 53, as quoted in Bolton Davidheiser, 232.
4. Denton, Michael. *Evolution: A Theory in Crisis*, Adler and Adler Publishers, 1985, 144.

attach much importance to the apparent homology of the vertebrae among many species, vertebrae actually form embryologically in significantly different ways in different groups of vertebrates. They even form from different embryonic sources. Despite the similar role vertebrae plays in all vertebrates, vertebrae in different species are not homologous.[5]

> Ultimately the Darwinian theory of evolution is no more nor less than the great cosmogenic myth of the twentieth century.
> **MICHAEL DENTON, BIOCHEMISTRY, MOLECULAR BIOLOGY, MEDICINE**

Another puzzling anomaly for evolutionists is that organisms they believe are not closely related have similarities. How can this be since evolution posits a precise order of change between only certain species? For example,[6] hemoglobin, the complex molecule that carries oxygen in the blood and results in its red color, is found in vertebrates but also in some earthworms, starfish, crustaceans, mollusks, and even in some bacteria. Evolution denies any closeness between earthworms and starfish, much less with mammals, so how can that be? Equally puzzling, an antigen receptor protein has the same unusual single chain structure in camels as in nurse sharks though neither has a common ancestor according to evolution.[7]

Some homologous structures, such as the body segments in fruit flies and locusts, depend on different genes, but many non-homologous structures, such as the eyes of octopuses and humans or the legs of mammals and insects, depend on similar genes. Again, how can that be since it contradicts basic evolutionary principles?

> Lots of genes are common to many creatures; bananas, for instance, share 50% of their genes with humans. But it's how the genes are regulated that makes a bat a bat and a cat a cat. We don't understand this very well at all in science. The so-called 'junk'? DNA is probably involved in that somehow. There are layers of additional complexity that we are only just discovering—codes within codes as it were.
> **DR. DUDLEY EIRICH, MOLECULAR BIOLOGY**

5. Dr. Alastair Noble, "Anomalous 'homology' refutes evolution," Centre for Intelligent Design, August 2024, www.c4id.org.uk/Articles/487033/Anomalous_homology_refutes.aspx
6. Jonathan Sarfati, Refuting Evolution 2—Chapter 6, Creation.com, Creation Ministries International, August 2024, creation.com/refuting-evolution-2-chapter-6-argument-common-design-points-to-common-ancestry.
7. Sarfati, 113.

In 1971, Gavin de Beer, a British evolutionary embryologist, wrote:

> It might be thought that genetics would provide the key to the problem of homology. This is where the worst shock of all is encountered"because "characters controlled by identical genes are not necessarily homologous...[and] homologous structures need not be controlled by identical genes.

De Beer concluded that "the inheritance of homologous structures from a common ancestor...cannot be ascribed to identity of genes."[8] He goes on to say:

> The evolutionary basis of homology is perhaps even more severely damaged by the discovery that apparently homologous structures are specified by quite different genes in different species....Over and over again, as knowledge of invertebrate zoology has increased over the past two centuries, structures of astonishing similarity which were first thought to be homologous were later found to be only analogous. In botany, too, homologous resemblance has often had to be later reclassified as convergence, or analogy.[9]

There is so much "slop" in both the data and its processing in the field of molecular homology that Colin Patterson, evolutionist and late senior paleontologist at the British Museum, spoke of it as "anti-knowledge" generating "anti-theory." Evolutionists allowed a false assessment of the facts to induce the false concept that evolutionary common ancestry offers some sort of explanation for the data (which, as he points out, has already been "massaged with evolutionary theory"). In fact, after explaining that the current evolutionary explanation for molecular homology was "precisely falsified," he went on to consider a creationist explanation for the data.[10]

We have already seen that evaluations of the same data can come to radically different conclusions because of differences in worldviews. Such is the case with homology. Similarities between species are better explained from a viewpoint other than evolution. Let's put on a different set of worldview glasses and take a look at the data. We

8. De Beer, Gavin. *Homology: An Unsolved Problem*. London: Oxford University Press, 1971, 15-16.
9. *Evolution: A Theory in Crisis*, 149-154.
10. https://www.icr.org/article/molecular-evolution.

must decide which conclusion fits all the scientific data and, therefore, is the best explanation.

I'll begin with a simple analogy in engineering, my area of study. Engineers design various structures. I might design a simple one-story office building in which I use a series of columns with adequate strength to support the entire roofing system. The columns might include reinforcing steel and high strength concrete. Later, I might use similar columns to support a bridge over a river or to elevate and support a large advertising sign. Though very different, these three structures utilize at least one similar structural element, the column. No evolutionary relationship exists between these structures. As an intelligent (Please don't snicker but assume it for argument's sake.) designer, I chose similar structural elements because they serve similar functional purposes.

Equally, similarities in creatures indicate a common Designer, not a common ancestry. That all vertebrates have vertebrae, their equivalent of a column, does not mean they are related. God merely used the same structural element in several different species.

Any honest scientist would be forced to note that varieties of creatures have far more differences than similarities. While these differences are often easy to see (cat, elephant, octopus, dragonfly), they are especially dramatic on the microbiological level. Clear boundaries between species prevent the transference of different genetic characteristics into subsequent generations. A cat can't become a dog. An ape can't become a child. God designed genetics, and He forbids it. He programmed every part of creation to reproduce "after its kind."[11] When examined fairly, rather than being a support homology proves evolution is scientifically impossible. Today there is an updated version of the homology argument from genetics called, "junk DNA." But current genetic research is showing it to be just as faulty an idea as homology.

> Another area is "junk DNA." Evolution predicts most DNA would be junk, and indeed, much of it doesn't code for protein. However, the ENCODE project [Encyclopedia of DNA Elements] shows most is used for other overlapping codes. E.g. they enable our bodies to make several times more

11. Genesis 1:21, 25.

proteins than genes, including those vital for brain function. So, the junk DNA concept belongs in the junk pile![12]

WHAT OTHER EVIDENCE DO EVOLUTIONISTS PROVIDE TO SUPPORT THEIR BELIEF?

Evolutionists supply other superficial and illogical points of evidence to support their evolutionary model. For example, they use what they call microevolution (evolutionary change within a species or small group of organisms, especially over a short period of time) to argue for what they call macroevolution (the evolution of whole groups of creatures like primates to humans over long periods of time). But this is a clever but deceptive use of language. It is more accurate and less deceiving to call microevolution "variation within a species." Variation within a species does exist but so-called macroevolution has never been observed.

Remember the finches that were the basis of Darwin's evolutionary opinion? While those finches, specifically their beaks, did change as they adapted to different environments (variation), they did not change into new species like squirrels or squid (macroevolution). To posit that minor changes within species indicate major changes into other species is nonsensical, a desperate attempt to exclude the divine Creator who made every species "after its own kind."

> No one has ever demonstrated macroevolutionary changes on a molecular level, yet many people readily speculate evolutionary links between bacteria, plants, animals and man. Are the gross structures not made up of individual cells with complex molecules? If macroevolution is unlikely on a molecular level, how can the whole be changed? Endless DNA sequence comparisons do not explain evolutionary development. Furthermore, the changes (mutations) observed on a molecular level, such as DNA, are predominantly disruptive, and always with loss of, not gain in, information and complexity. This led Lee Spetner to conclude 'Whoever thinks macroevolution can be made by mutations that lose information is like the merchant who lost a little money on every sale but thought he could make it up on volume.'
>
> **DR. JOHN K. G. KRAMER, BIOCHEMISTRY**

12. The marvelous molecule (creation.com).

Evolutionists also point to so-called vestigial organs, like the appendix and coccyx, as proof man has evolved. While they loudly proclaim such organs are useless and merely leftover bits from past evolutionary lifeforms, new scientific discoveries have largely discredited this view. Science has found that every vestigial organ is, in fact, important to optimal human health. Creation Magazine makes clear:

> Without the appendix, patients are four times more likely to get nasty Clostridium difficile (C-diff). The reason, as discovered only this millennium, is that the appendix is a "safe-house" for "good" bacteria that keep the "bad" ones away.[13]

Similarly, the coccyx is not a useless vestige of a tail as evolution posits but an attachment site for ligaments, tendons, and muscles as well as a support for the pelvis.

> The palmar grasp reflex occurs when the infant's fingers reflexively flex around an object placed in its palm such as a caregiver's finger. The response is the result of a spinal reflex controlled by the higher brain centres...The palmar grasp reflex is listed as one of the seven leading examples of vestigial organs by the Encyclopedia Britannica....Conclusion: The palmar grasp reflex is not vestigial, but has important functions including particularly as part of the developmental process leading up to the use of the hand for complex life in our human society. It is part of the design of the human body and involves many brain, nerve, muscle, tendon, and ligament systems. It also serves the important role of facilitating more effective bonding of the infant with its mother and other adults.
>
> **DR. JERRY BERGMAN, BIOLOGY**

THE TAKE AWAY: The human body is truly "fearfully and wonderfully made," just as God declares.[14] No part is superfluous or outdated.

Evolutionists also use arguments from embryology and cultural anthropology to support their ideas. Again, every new scientific discovery in these areas discredits evolution. For our purpose, we need not discuss discredited evolutionary ideas since many resources abundantly prove they are not scientific evidence but superficial opinions.[15]

---

13. Microbiome and virome (creation.com).
14. Psalm 139:14, "I will give thanks to You, for I am fearfully and wonderfully made. Wonderful are Your works, and my soul knows it very well."
15. A growing number of published works from both creationists and non-Christian authors

Remember that homology, misapplied microevolution, vestigial organs, and other discredited ideas are the best evidence evolutionists can provide to support their idea. Sadly, this evidence is presented as settled science in almost all schools. It is solemnly discussed as absolute truth by politicians, doctors, teachers, media talking heads, seemingly everyone in a position of power. We need to know the truth so human speculation won't hold us captive.[16]

> Creationists use these and similar examples to argue that much of God's creation was designed for human enjoyment and for God's own enjoyment, as He declared it 'good' several times before man was created...Evolutionists never have explained how and why so many structures could exist in humans (like the complex structures that enable music, song and dance) that confirm no survival advantage yet delight millions. Only creation can explain this observation. The clear conclusion is that the concept of evolutionary vestigial organs is useless, or largely speculative, and certainly is not good science.
>
> **DR. JERRY BERGMAN, BIOLOGY**

WHAT MAKES EVOLUTION WORK?

According to evolutionists, creatures evolve through mutations. They admit mutations are the only mechanism that can produce evolution.

Sadly for them, the sciences of microbiology and genetics have disproven their premise that evolution proceeds through mutations. These relatively young sciences have given us a great deal of new information about life, especially about DNA molecules. DNA, which stands for deoxyribonucleic acid, is inherited material in all humans and all other living organisms that allows them to pass their unique characteristics to the next generation. Biological chemist Dr. John P. Marcus explained the astounding complexity of the DNA in each cell of the body:

> If the DNA of one human cell were unraveled and held in a straight line, it would literally be almost one meter long and yet be so thin that it would be invisible to all but the most

demonstrate the failure of the theory of evolution. Many of these resources can be found at the websites of the Institute for Creation Research (icr.org), Answers in Genesis (answersingenesis.org), and Creation Ministries International (creation.com).

16. 2 Corinthians 10:5, "We are destroying speculations and every lofty thing raised up against the knowledge of God, and we are taking every thought captive to the obedience of Christ."

> powerful microscopes. Consider that this string of DNA must be packaged into a space that is much smaller than the head of a pin and that this tiny string of human DNA contains enough information to fill almost 1,000 books, each containing 1,000 pages of text. Human engineers would have a most difficult time trying to fit one such book into that amount of space; one thousand books in that amount of space boggles the mind! For compactness and information-carrying ability, no human invention has even come close to matching the design of this remarkable molecule.[17]

> If you take an entire DNA molecule, coding for all the proteins required for life, I think it's about 1 followed by 40,000 zeros (which is really no chance) that such coded DNA can arise by chance.
>
> **DR. ANDRÉ EGGEN, ANIMAL GENETICS**

Mutations cause changes in DNA's material composition that can change a creature's anatomy. Since these changes are in the DNA, they are passed on to later generations. Mutations are real. The question is, do they provide the mechanism that allows evolution to work?

All evolutionists recognize the importance of mutations as the only practical mechanism for evolution. Ernst Mayr, a Harvard University professor of taxonomy (the study of the classification of organisms), wrote, "It must not be forgotten that mutation is the ultimate source of all genetic variation found in natural populations and the only new material available for natural selection [evolution] to work."[18]

Theodosius Dobzhansky, another famous evolutionist and a geneticist from Columbia University, agreed: "The process of mutation is the only known source of new materials of genetic variability, and hence of evolution."[19]

A more recent evolutionist, paleontologist George Gaylord Simpson, agreed but emphasized a different twist: selection. "Only mutation supplies the materials for creation, but in the series of population

17. hppts://creation.com/john-p-marcus-biochemistry-in-six-days.
18. Mayr, Ernst, *Populations, Species and Evolution*, Harvard University Press, 1970, 103, as quoted in Morris, *Scientific Creationism*, 54.
19. "On Methods of Evolutionary Biology and Anthropology. Part I. Biology," *American Scientist*, 45:5:381, Theodosius Dobzhansky, 1957, 385, as quoted in Evolution and Christian Faith, 191-92.

genetics it is selection that is truly creative, building new organisms with these materials."[20]

Evolutionists are right that mutations cause changes in DNA's material composition that can change a creature's anatomy. Since thses changes are in the DNA, they are passed on to later generations. Mutations are real. The question is, do they provide the mechanism that allows evolution to work? To be an adequate mechanism for evolution, the mutations must occur frequently, abundantly, and beneficially. We can conclude that if they do not occur frequently, abundantly, and beneficially, evolution has no mechanism to make it work.

Since the time of Darwin, evolutionists have been motivated to demonstrate the viability of mutations; therefore, they have done most of the research in this area. In about one-hundred-sixty years[21] of research into the science of genetics, what have they found? Mutations provide no viable mechanism for evolutionary processes. That's right—none at all! Mutations can't cause evolution! If evolution doesn't have a mechanism to cause gradual but dramatic and positive changes, wouldn't we be wise to reject the whole idea? Let's examine their evidence.[22]

1. CONTRARY TO EVOLUTIONARY THEORY, MUTATIONS ARE RARE.

Evolution requires huge numbers of mutations to produce new species. Research shows the opposite, that mutations are relatively rare. Francisco J. Ayala, an evolutionary biologist, said a mutation occurs between 1 per 10,000 to 1 per 1,000,000 per gene per generation.[23] Try to use those numbers to figure out how many billions of years are required to evolve a new species composed of thousands of genes! No wonder the time needed for man to evolve has gone from millions of years to many billions!

---

20. Simpson, George Gaylord, *Tempo and Mode in Evolution*, Columbia University Press, 1949, 80, as quoted in Davidheiser, 192.

21. This time is from the publishing of Darwin's theory in the 1860s.

22. The 7-point list of results comes mainly from Zuck, Roy B., *Creation: Evidence from Scripture and Science*, Scripture Press Publications, 1975, 9-10, and *Scientific Creationism*, 55-57.

23. Ayala, Francisco J., "Teleological Explanations in Evolutionary Biology," Philosophy of Science, Vol.37, 1970, 3, as quoted by *Scientific Creationism*, 55.

In a more recent study based on newer scientific information, creation science biologist Gary Parker gave a rate of 1 mutation per 10 million (107) DNA duplications.[24] That is rare indeed!

Contrary to evolutionary claims, precise design is evident in even the most basic of genetic structures, the DNA molecule. In spite of the immense amount of information that must be reproduced in the creation of every individual cell, the DNA molecule does its job with amazing accuracy and stability. Because mutations are rare, isn't divine design a better explanation than random evolutionary processes?

> Although a biologist, I must confess that I do not understand how life came about....I consider that life only starts at the level of a functional cell. The most primitive cells may require at least several hundred different specific biological macro-molecules. How such already quite complex structures may have come together, remains a mystery to me. The possibility of the existence of a Creator, of God, represents to me a satisfactory solution to the problem.
>
> WERNER ARBER, 1978 NOBEL PRIZE IN MEDICINE AND PHYSIOLOGY

2. CONTRARY TO EVOLUTIONARY THEORY, MUTATIONS ARE UNPREDICTABLE.

Evolution requires that mutations be closely related to each other, serial in nature, and progressive. Instead, mutations occur unpredictably and often totally unrelated to each other. Mutations are mistakes in DNA reproduction, not positive advances. Again, research shows the opposite of evolutionary suppositions.

3. CONTRARY TO EVOLUTIONARY THEORY, MUTATIONS ARE MOSTLY HARMFUL.

Evolution requires an abundance of beneficial mutations, but the opposite is true. Not only are mutations rare, but few are beneficial. They are overwhelmingly harmful. Often, they cause the death of the organism. Even though an evolutionist, H. J. Muller, a world authority on mutations, was forced to admit that well over 99% of all

24. Parker, Gary, *Creation Facts of Life*, Master Books, 1994, 91.

mutations are bad. "Most mutations are bad. In fact, good ones are so rare we can consider them all bad."[25]

After studying mutations in fruit flies, evolutionist Theodosius Dobzhansky agreed:

> Most mutants which arise in any organism are more or less disadvantages to the possessor. The classical mutants claimed in drosophila [the scientific term for fruit flies] usually show deterioration, breakdown, and disappearance of some organs.[26]

Chemical engineer Linus Pauling (1901-1994) put a humorous spin on the problem with mutations:

> We can understand now why it is that mutations in these carefully selected organisms almost invariably are detrimental. The situation can be suggested by a statement by Dr. J.B.S. Haldane: "My clock is not keeping perfect time. It is conceivable that it will run better if I shoot a bullet through it; but it is much more probable that it will stop altogether."

Geneticist and 1958 Nobel Prize winner in medicine George Beadle asked while discussing mutations, "What is the chance that a typographical error would improve Hamlet?"[27] Yet, evolutionists still consider mutations the key that proves their theory.

Rather than supporting evolution, mutations support the biblical worldview that the fall of mankind into sin adversely affected the entire creation, including all plant and animal life. After the fall, the creation began the process of degeneration. We will expand on the fall of mankind later. Since ninety-nine percent of mutations are harmful, we should be glad they are rare. Science continues to validate the biblical worldview and refute all secular worldviews.

---

25. Muller, H.J., quoted in Zuck, 9.
26. Ibid., 9.
27. https://todayinsci.com/QuotationsCategories/M_Cat/Mutation-Quotations.htm.

> I study one of the fastest mutating viruses–HIV. While mutations and natural selection have a role in generating some diversity, the diversity is at the micro-level–modifying existing features, not generating any new functions. Furthermore, it is mostly detrimental to the virus's fitness. For evolution to explain the diversity of life would entail impossible chance events that would never happen–even with the secular ~14-billion-year timeline.
>
> **DR. SAMUEL GAN, BIOMEDICINE**

4. CONTRARY TO EVOLUTIONARY THEORY, MUTATIONS RESULT IN THE LOSS OF GENETIC INFORMATION.

The DNA molecule contains encyclopedic information, everything necessary to specify all the contents of all cells of every life form, both plant and animal. The vast amount of information in even the most primitive cells contradicts the evolutionary idea of progressively more information in the DNA molecule from supposed simple life forms to the more complex.

Unfortunately for evolution, mutations do not produce additional information. Instead of an increase of information, every mutation causes a loss of information. This loss is recognized by all geneticists today, including evolutionists. Biophysicist Lee Spetner wrote:

> In this chapter I'll bring several examples of evolution [i.e., instances alleged to be examples of evolution] particularly mutations, and show that information is not increased...But in all the reading I've done in the life-sciences literature, I've never found a mutation that added information. All mutations that have been studied on the molecular level turn out to reduce the genetic information and not to increase it.[28]

After years of research, Cornell University professor, a creationist, Dr. John Sanford came to the same conclusion. "Yet I am still not convinced there is a single, crystal-clear example of a known mutation which unambiguously *created* information."[29]

---

28. Spetner, L.M., *Not by Chance! Shattering the Modern Theory of Evolution*, Judaica Press, 1996, 7, as quoted by Sarfati, Jonathan, *By Design: Evidence for Nature's Intelligent Designer—The God of the Bible*, Creation Book Publishers, 2008, 11 and 18.
29. Dr. John C. Sanford, *Genetic Entropy*, FMS Publications, 2014, 17.

The ideas of evolutionary progress and increased complexity are imaginary. True science consistently demonstrates the opposite. Again, this degradation of life forms through mutations fits the biblical worldview perfectly. Jeremiah, a Jewish prophet who lived about 2,600 years ago, implied that mutations do not add new information to the genetic make-up of creatures. "Can the Ethiopian change his skin? Or the leopard his spots?"[30] The implied answer is an emphatic NO! Creatures have not and cannot evolve into new creatures.

5. CONTRARY TO EVOLUTIONARY THEORY, MUTATIONS IMPOSE AN INCREASINGLY HEAVY GENETIC BURDEN.

Geneticists tell us that the adverse effects of mutations increase over time through genetic heredity. Since mutations result in a loss of genetic information, as time passes, more and more beneficial information is lost. Geneticists call this phenomenon genetic burden or genetic entropy. Thus, our genetic burden continues to increase as we get farther from the perfection of the Garden of Eden. This means that ancient generations, especially before the Genesis Flood, had a purer genetic make-up than we do today.

Don't take my word for it. Believe Gary Parker, who has a PhD in biology. He wrote:

> Time, the usual 'hero of the plot' for evolutionists only makes the genetic burden worse. As time goes on, existing mutants build up to a complex equilibrium point, and new mutations are continually occurring.[31]

Parker explained genetic burden further by quoting a genetics textbook: "In her genetics textbook, Anna Pac makes it clear that the word 'load' is used intentionally to imply some sort of 'burden' that drags down the genetic quality of a species.[32] Over time these mutations accumulate in a detrimental way that makes every creature less viable rather than more complex.

30. Jeremiah 13:23.
31. Parker, Gary E., *Impact #89*, Institute for Creation Research, 1980, ii.
32. Pai, Anna, *Foundations of Genetics*, McGraw-Hill Books Co, 1974, 248-49, quoted in *Impact #89*, ii.

More recently, Dr. John Sanford's extensive research in genetics disproves, not proves, the idea of evolution, which he describes as the "Primary Axiom" in science today. His conclusions are both informed and emphatic:

> Arguably, I have studied the specific problem of genetic entropy more than any other scientist. My research has included careful analysis of how the processes of mutation and selection actually operate, combined with simple logic and mathematical formulation.[33]
>
> We have reviewed compelling evidence that, even when ignoring deleterious mutations, mutation/selection cannot create a single gene within the evolutionary timescale. When deleterious mutations are factored back in, we see that mutations/selection cannot create a single gene *ever*. This is overwhelming evidence against the Primary Axiom. In my opinion, this constitutes what is essentially a logical proof that the Primary Axiom is false.[34]

Once again, the process of mutational degradation is the opposite of evolution's assertions. Overwhelmingly, science proves evolution to be false! Mutations are better explained from the biblical worldview. Since the fall of man in the Garden of Eden, sin has caused all life forms to degenerate, not increase in complexity and purity as evolution falsely proclaims.

This increasing genetic burden answers questions raised by the early narratives of the book of Genesis. Where did the children of Adam and Eve get their spouses? The answer is they had to marry siblings not mentioned by name in Genesis.[35] What about the problems inherent in an incestuous relationship? The answer has to do with the progressive degradation caused by sin. The human genetic pool in the early generations after Adam was pure with possibly no mutations and virtually no genetic burden; therefore, the medical issues often found today in children of close relatives would not have been present so close to the perfection of the original creation.

---

33. Dr. John C. Sanford, *Genetic Entropy*, FMS Publications, 2014, 154.
34. Dr. John C. Sanford, *Genetic Entropy*, FMS Publications, 2014, 149.
35. Genesis 5:4.

The three sons of Noah had children. Those children had to have children by either first cousins or brothers and sisters. The human genetic pool would still have been mostly pure, and so the offspring would not have been genetically compromised. By the time of Moses, however, the genetic burden had increased appreciably and, therefore, in the Mosaic Law, God prohibited marriage between close relatives. The biblical worldview and true science always agree, even down to the smallest DNA molecule.

6. CONTRARY TO EVOLUTIONARY THEORY, MUTATIONS ARE PATHOLOGICAL.

The medical field treats mutations as pathological, meaning they cause diseases and medical problems. In fact, many fatal diseases can be traced to mutations, including hemophilia (bleeders), cystic fibrosis, sickle cell anemia, some cancers, and some brain malfunctions. Albinism (the loss of protective coloring), though not fatal, is an especially visible example of the damage caused by mutations.

> Mutations are pathological (disease-causing) and only "modify what pre-exists" as French zoologist Pierre-Paul Grasse says, so mutations have "no final evolutionary effect." Instead, mutations point back to the creation of the created order. There are 40+ variants of hemoglobin, for example. All are variants of hemoglobin; that points back to creation. All are less effective oxygen carriers than normal hemoglobin; that points back to a corruption of the created order.[36]

Mutations are pathological for the three reasons we have already described; they are mainly bad, they destroy genetic information, and they add to the genetic burden. Mutations are not only pathological for humans but for all creatures for the same reasons.

In 1980, Parker wrote, "The list of human mutational disorders, or genetic diseases, for example, has already passed 1500, and is continuing to grow."[37] Parker also wrote that mutations are responsible for over

36. Grasse, Pierre-Paul, *Evolution of Living Organisms*, Academic Press, 1977, as quoted in *Impact #89*, ii.
37. *Impact #89*, ii.

3,500 hereditary defects in humans.[38] Forty plus years later, according to the Stanford University Department of Genetics, "Scientists estimate that every one of us has between 5 and 10 potentially deadly mutations in our genes."[39]

Are you scared yet? Don't be. The advantage of the biblical worldview is that we know that the same Creator who covered the heavens with stars has a plan for each one of us. He promises to work all things together for good for those who follow Him—even when mutations cause damaging, even deadly, effects.[40] It makes sense to trust in Him even for those dangers we cannot see, even those growing inside us, doesn't it?

7. CONTRARY TO EVOLUTIONARY THEORY, MUTATIONS CAUSE CREATURES TO HAVE DIFFICULTY REPRODUCING.

We would expect that, the greater the mutation, the greater the problems produced by the mutation, including problems with reproduction. We would be right. Given enough time, mutations can cause the extinction of a creature, not its evolution into a new species.

Drs. Michelle Zorrilla and Alexander N. Yatsenko at the Pittsburgh School of Medicine wrote, "It has been estimated that nearly 50% of infertility cases are due to genetic defects [mutations]."[41]

According to Dr. L. C. Layman of the Institute of Molecular Medicine and Genetics at the Medical College of Georgia, "To date, most mutations characterized in humans cause pubertal failure and infertility, while a few result in normal puberty, but cause infertility."[42]

Dr. Ignatia B. Van den Veyver, professor of obstetrics and gynecology and molecular and human genetics at Baylor, wrote in a study:

> Researchers have found that women with mutations that lead to loss-of-function of some of the genes of the NLRP family can fail to reproduce for reasons that may include

38. *Creation Facts of Life*, 96.
39. https://genetics.thetech.org/about-genetics/mutations-and-disease, 2019.
40. Romans 8:28.
41. https://pubmed.ncbi.nlm.nih.gov/24416713/.
42. https://jmg.bmj.com/content/39/3/153.

MUTATIONS ARE...

- Rare
- Unpredictable
- Mostly harmful
- Result in the loss of genetic information
- Impose an increasingly heavy genetic burden
- Pathological
- Cause creatures to have difficulty reproducing

...NOT A MECHANISM FOR EVOLUTION

> recurrent loss of pregnancies with abnormally developing placentas, loss of the embryo before implantation, or, more rarely, having a baby with developmental disabilities. Women carrying these mutations are healthy in all other physical aspects, so they are unaware that they have these mutations that do not allow them to carry a pregnancy.[43]

As Gary Parker summarized so well:

> To make evolution happen—or even to make evolution a scientific theory—evolutionists need some kind of "genetic script writer" to increase the quantity and quality of genetic information. Mutations are just "typographical errors" that occur as genetic script is copied. Mutations have no ability to compose genetic sentences, and thus no ability to make evolution happen at all.[44]

Mutations are not a vehicle for evolution but a destroyer of it.

> It doesn't matter if one population breaks into several subgroups, even to the extent of not reproducing with each other anymore. In fact, you would expect that to happen after the Flood, so coyotes, wolves, dingoes, and so on might have had a common ancestor, but the key is that there's no new information–that natural processes don't create any new DNA information. I've observed 40 generations of selection of fruit flies. I've seen lots of defective flies because of mutations, but I've never seen new, additional genetic information appear which would give hope to evolutionists. The belief in amoeba-to-man evolution needs a huge amount of new genetic information.
>
> **DR. JIM ALLAN, GENETICS**

43. https://www.sciencedaily.com/releases/2017/03/170322092247.htm.
44. *Creation Facts of Life*, 104.

WHY IS SOME EVIDENCE NO LONGER EMPHASIZED BY EVOLUTIONISTS?

In the past, evolutionists have pointed to "scientific" discoveries that "proved" the validity of evolution. None have survived close scrutiny. All have been scrubbed from scientific discussions because of their obvious defects. We will study one of those, probably the one with which you are most familiar, the fossil record.

In the days of Darwin (1809–1882), the science of paleontology (the study of fossils) was relatively new. The discovery of fossil samples was in the early stages of the science. Many evolutionists were certain that further fossil discoveries would prove evolution. After over a century and a half, the opposite is true. Today, paleontologists have much greater knowledge of the fossil record. The problem is not a lack of fossil evidence but a lack of the kind that supports evolution.

So compelling is the fossil evidence that in his study of ornithology (birds), Douglas Dewar (1875–1957) came to reject evolution because of it. "The theory of evolution must stand or fail by the evidence of the fossils. They furnish the only direct evidence. If this be against the theory, no amount of indirect evidence can avail."[45]

WHERE ARE FOSSILS FOUND?

Geology divides the geologic column into two major rock formations, basement, igneous rock and the sedimentary layers that cover it. Basement rock, such as granite, supports a series of many sedimentary rock layers that extend to the surface of the earth. As the name indicates, sediment settling out of water formed the upper layers. The depth of the sediment layers varies from place to place around the planet. At the Grand Canyon in Arizona, the many exposed layers give a dramatic cross-sectional view, one that is nearly a mile deep. The major distinguishing characteristic of the sedimentary layers is the fossils they contain. Here, paleontologists find the data to do their work.

The geologic column categorizes the many layers of sedimentary rock by the fossils found in them. The lowest sedimentary layers, called the Cambrian layers, house a huge diversity of animal fossils

45. Dewar, Douglas, *Difficulties of the Evolution Theory*, E. Arnold & Co, 1931.

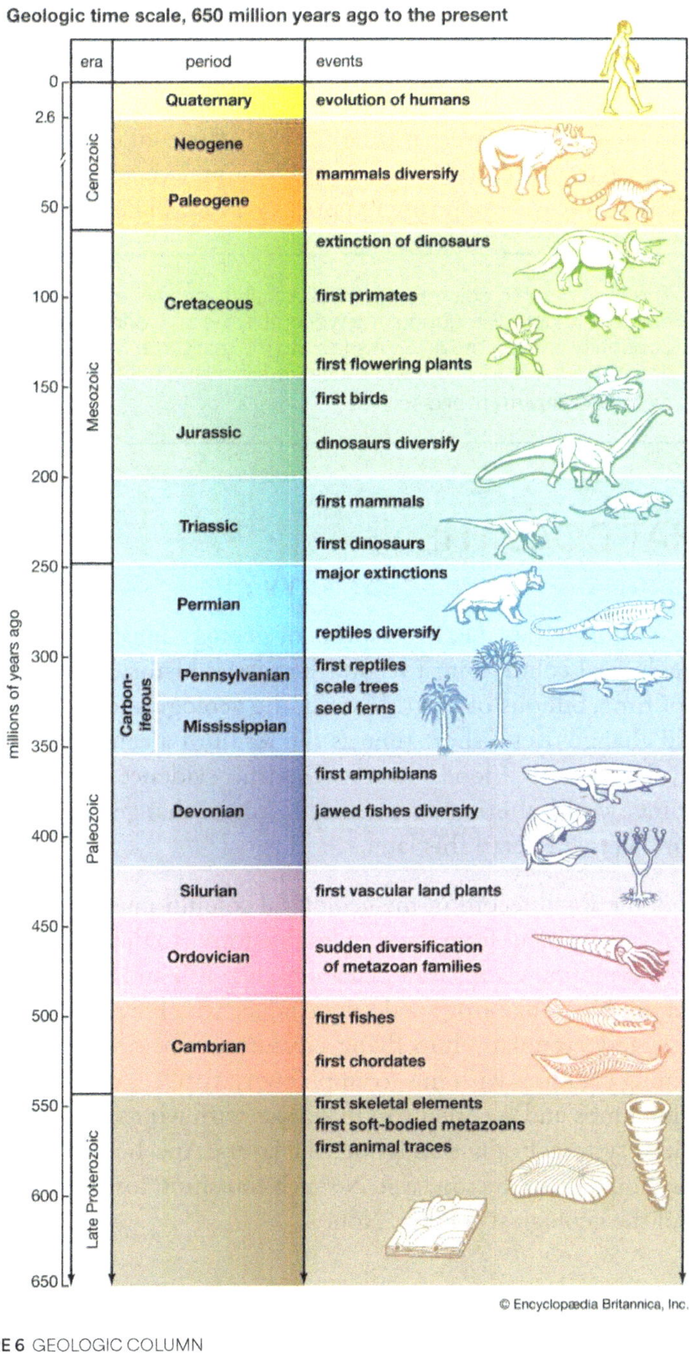

FIGURE 6 GEOLOGIC COLUMN

that abruptly appeared above the virtually fossil-less basement layer. For evolution to be true, the lower and older layers should contain the simplest life forms with the top layers becoming gradually more complex. Instead, basement rock has no fossils, but the early Cambrian sediment layers resting on top of it contain a multitude of fossils of almost every category of the animal kingdom, including vertebrates, the exact opposite required by evolution.

> The stratigraphic column, the 'stack' of all these layers, is essentially continuous throughout the world; there is no worldwide discontinuity or 'time break'. So it shows to me that there was indeed a worldwide Flood, and not just localized floods as many believe.
>
> **DR. ALAN GALBRAITH, HYDRO-SCIENCE**

WHAT DOES THE FOSSIL EVIDENCE SHOW?

For evolution to be true, two aspects of geology must be true. First, the geological column must record the earth's history, spanning vast ages of time, billions of years. Instead, the geological column reveals a rapid change over a short time as the result of a cataclysmic water event, the Genesis Flood. Strong scientific evidence points to this one-time, world-altering event, not to micro-changes over billions of years. We will detail this later.

Second, the fossil record in the geological column must show a step-by-step development from the simplest life forms to the most complex. Because evolution claims water creatures, like fish and octopi, evolved into amphibians, like frogs and salamanders, which evolved into land animals and eventually into flying creatures, transitional links from swimming creatures with fins to amphibians with legs to land animals with forelimbs and eventually to mammals with wings must be found fossilized systematically throughout the layers. Ape-human transitory fossils should be in the top layer. No such transition fossils exist in any layer of the geological column. None.

> The search for these transitional forms–"missing links"–has failed to produce any unequivocal examples. Some paleontologists suggest that they will never be found because they are not there. Evolution supposedly occurred so rapidly in the past that the evidence has vanished without trace. This explanation raises another problem. Evolution cannot be observed in the present because allegedly it is proceeding too slowly. Why should it have occurred so rapidly in the past?
>
> **DR. RALPH MATTHEWS, RADIATION CHEMISTRY**

Since the time of Darwin, paleontologists have come to three major conclusions:

1. LIFE APPEARED ABRUPTLY

The basement Pre-Cambrian layers contain virtually no fossils. In contrast, the lowest sedimentary layers, the Cambrian layers, have an abundance of fossils of many different kinds, mainly marine creatures. Evolutionary paleontologists call this rich layer of fossils the Cambrian explosion. Life seems to have appeared suddenly and abruptly, contradicting the evolutionary model of slow processes over long periods of time.

This problem is recognized by most evolutionists. George G. Simpson, perhaps the most famous paleontologist of the last century as well as an evolutionist, considered this Cambrian explosion the major mystery of the history of life. How sad that he clung to his false assumptions about evolution and didn't look for a true scientific explanation, one supported by the evidence. He said:

> It remains true, as every paleontologist knows, that most new species, genera and families, and that nearly all categories above the levels of family, appear in the record suddenly and are not led up to by known, gradual, completely continuous transitional sequences....[This] is the most puzzling feature of the whole fossil record but also is the greatest inadequacy.[46]

46 *Scientific Creationism*, 79-80.

Another evolutionist, Daniel I. Axelrod, a paleoecologist specializing in Tertiary Cordilleran floras, recognized the same problem:

> One of the major unsolved problems of geology and evolution is the occurrence of diversified, multicellular marine invertebrates in Lower Cambrian rocks on all the continents and their absence in rocks of greater age. These Early Cambrian fossils included porifera, coelenterates, brachiopods, mollusca, echinoids, and arthropods…However, when we turn to examine the Precambrian rocks for the forerunners of these Early Cambrian fossils, they are nowhere to be found.[47]

Evolution requires evidence that single cells evolved into more complex creatures, a theory disproved by the Cambrian explosion. Evidence of a gradual evolution to more complex forms of life does not exist. Instead, the fossil record shows that all life appeared abruptly.

2. COMPLEX LIFE IS FOUND IN THE CAMBRIAN LAYER

Evolution posits that the simplest lifeforms came first and slowly evolved into more complex creatures. The bottom fossil layer of sedimentary rock however, includes creatures like trilobites, corals, worms, jellyfish, and brachiopods. Some have both complex internal and external structures. Some, like trilobites, even have complex compound eyes. The quotes above from Simpson and Axelrod not only emphasize the sudden appearance of life but also the complexity of the creatures. Both profoundly weaken the possibility of evolution as an explanation for the origin of life.

3. NO TRANSITIONAL FORMS WERE FOSSILIZED

If those problems were not enough, the absence of transitional forms destroys the possibility of evolution as the origin of life. Evolutionists expected to find millions or even billions of transitional forms since changes were supposedly gradual, so gradual that billions of years were needed to achieve them. The fossil record does not reveal any

47. Axelrod, Daniel, "Early Cambrian Marine Fauna", *Science*, Vol. 128, 1958, 7, as quoted in Gish, Duane T., *Evolution: The Fossils Say NO!*, Creation-Life Publishers, 1973, 46.

transitional creatures. None. Zero. In fact, the few fossils that evolutionists have proposed as transitional are merely subjective speculations that can be interpreted differently. Some of the most publicized have been exposed as hoaxes perpetrated by those who couldn't bear to have evolution disproved.[48] Even Charles Darwin agreed that the fossil record reveals no transitional links:

> Why is not every geological formation and every stratum full of such intermediate links? Geology assuredly does not reveal any such finely graduated organic chain; and this is the most obvious and serious objection which can be urged against the theory.[49]

Darwin hoped more fossil discoveries would uncover these links, but after one-hundred-sixty years of searching, paleontologists have concluded that transitional fossils simply do not exist. This problem is lamented by most evolutionary paleontologists:

> Despite the bright promise that paleontology provides a means of "seeing" evolution, it has presented some nasty difficulties for evolutionists, the most notorious of which is the presence of "gaps" in the fossil record. Evolution requires intermediate forms between species and paleontology does not provide them.[50]

Evolutionist Stephen J Gould, who haunted our television screens for years as he lauded evolution and denigrated God, agreed this absence of transitional links is puzzling:

> The absence of fossil evidence for intermediary stages between major transitions in organic design, indeed our inability, even in our imagination, to construct functional intermediates in many cases, has been a persistent and nagging problem for gradualistic accounts of evolution.[51]

---

48. https://www.washingtonpost.com/national/health-science/five-of-the-most-famous-scientific-hoaxes/2015/03/02/a0b64f9e-b6cd-11e4-9423-f3d0a1ec335c_story.html.
49. Darwin, Charles, *Origin of Species*, 1859, 280, as quoted in http://darwin-online.org.uk/Variorum/1859/1859-280-c-1860.html.
50. Kitts, David B., "Paleontology & Evolutionary Thought," *Evolution*, volume 28, 467, as quoted in https://home.apu.edu/~jsimons/Bio101/quotes3.htm.
51. Smith, John Maynard, Ed., *Evolution Now: A Century After Darwin*, Macmillan Publishing Co., 1982, as quoted in *Refuting Evolution*, 48.

> I regard the failure to find a clear 'vector of progress' in life's history as the most puzzling fact of the fossil record.[52]

This lack of intermediary fossil evidence leaves evolution with no scientific evidence to uphold it. The fossil record, which was evolutionists' great hope, actually gives evidence against it. Any honest scientist who studied the facts would be forced to agree that no scientific data supports the idea of evolution.

> For close to a century, evolutionists taught that certain traits of coelacanths...then known only as fossils...meant they were the precursors of four-limbed animals (tetrapods) that walked on land...An enormous amount of research has been done on the coelacanth. Dozens of scientists have written scores of articles trying to demonstrate that it is a 'missing link' between fish and land animals. The coelacanth story has been long and extensive, but in the end has supported the conclusions of creationists–coelacanths may have varied a little, but have always been coelacanths.
>
> **DR. JERRY BERGMAN, BIOLOGY**

WHAT CAN WE CONCLUDE?

We could discuss much more evidence against evolution, but this discussion makes the point. When I speak to groups, I challenge them to let me know if I have missed any scientific evidence supporting evolution. I have yet to hear from anyone because such evidence does not exist. We've been brainwashed by pseudo-science and anti-God propaganda into accepting a discredited opinion as scientific fact.

> I admit that an awful lot of that has gotten into the textbooks as though it were true. For instance, the most famous example still on exhibit downstairs (in the American Museum) is the exhibit on horse evolution prepared perhaps 50 years ago. That has been presented as literal truth in textbook after textbook. Now I think that that is lamentable, particularly because the people who propose these kinds of stories themselves may be aware of the speculative nature of some of the stuff. But by the time it filters down to the textbooks, we've got science as truth and we've got a problem.
>
> **DR. NILES ELDREDGE,**
> **CURATOR AT THE AMERICAN MUSEUM OF NATURAL HISTORY**

52. "The Ediacaran Experiment," *Natural History*, 93(2):14-23, 1984, as quoted in *Refuting Evolution*, 48-49.

Science has demonstrated the failure of the theory of evolution. Even evolutionists have been forced to admit their pet theory lacks evidence. Further, the recently developed sciences of molecular biology and genetics confirm that evolution is impossible. No valid scientific evidence supports the "theory." Why do our schools continue to teach it as fact? Why do so many, including scientists, continue to believe it. Satan, who is ruler of this world and the father of lies, should come to mind.

Michael Denton (remember that the title of his book is *Evolution: A Theory in Crisis*) summarized the problems:

> By presenting a systematic critique of the current Darwinian model, ranging from paleontology to molecular biology, I have tried to show why I believe that the problems are too severe and too intractable to offer any hope of resolution in terms of the orthodox Darwinian framework, and that consequently the conservative view [conservative in the sense of trying to save evolution] is no longer tenable.[53]

Here is the revealing conclusion of his research and his book. This was written in 1985. More recent research has only made things worse for those advocating evolution:

> The influence of evolutionary theory on fields far removed from biology is one of the most spectacular examples in history of how a highly speculative idea for which there is no really hard scientific evidence can come to fashion the thinking of a whole society and dominate the outlook of an age. Considering its historic significance and the social and moral transformation it caused in western thought, one might have hoped that Darwinian theory was capable of a complete, comprehensive and entirely plausible explanation for all biological phenomena from the origin of life on through all its diverse manifestations up to, and including, the intellect of man. That it is neither fully plausible, nor comprehensive, is deeply troubling. One might have expected that a theory of such cardinal importance, a theory that literally changed the world, would have been something more than metaphysical,

53. *Evolution: A Theory in Crisis*, 16

> something more than a myth. Ultimately the Darwinian theory of evolution is no more nor less than the great cosmogenic myth of the twentieth century.[54]

Notice he was discussing a worldview, that of evolution. The heart of that unbelieving worldview is metaphysical, a philosophy, not a scientific truth. Science needs a better worldview.

You may be wondering why the idea of evolution continues to be held by many, if not most, scientists. Why does it remain the dominant explanation of origins in our culture today? The answer is simple but profound and has nothing to do with science. The only alternative to evolution is God as Creator. If God is accepted as Creator, all mankind stands accountable to Him. Most seek to avoid divine judgment at all costs. They have chosen to use evolution as their justification for rejecting their Creator.

> They (evolutionists) challenge science to prove the existence of God. But must we really light a candle to see the sun? They say they cannot visualize a Designer. Well, can a physicist visualize an electron? What strange rationale makes some physicists accept the inconceivable electron as real while refusing to accept the reality of a Designer on the grounds that they cannot conceive Him?
>
> **WERNER VON BRAUN, FATHER OF ROCKET SCIENCE**

WHAT EVILS ARE EVOLUTIONISTS UP TO TODAY?

For well over a century, those who hate God have worked to convince us we are no different from apes—or snails or ooze for that matter—just more complexly evolved. With recent advances in science that disprove evolution completely, evolutionists have been forced to change their direction. Instead of talking about our supposed past evolutions, they are now eager to introduce us to our future evolution, the culmination of all evolutions. They call it transhumanism, beyond human.[55]

54. *Evolution: A Theory in Crisis*, 358.

55. The following discussion on transhumanism was researched and supplied by my editor, Nancy Northcott.

Transhumanists assure us they can "use technology to artificially speed up the evolutionary process and bring on the next phase—merging humans with machines."[56] Yes, they proclaim, you can become better than human!

Wait a minute, you protest. Didn't God create us in His image? What could be better than that?

Becoming gods, they reply. And they are serious.

In the words of Yuval Noah Harari, prominent leader of the World Economic Forum (WEF) and the most powerful promoter of transhumanism (also called Humanity+): "History began when humans invented gods and will end when humans become god."[57] How is that possible? "Bioengineers will take the old Sapiens body, and intentionally rewrite its genetic code, rewire its brain circuits, alter its biochemical balance, and even grow entirely new limbs. They will thereby create new godlings, who might be as different from us Sapiens as we are different from *Homo erectus*."[58]

I'll bet Harari's words remind you of Satan's promise to Eve in the Garden of Eden: "The serpent said to the woman, 'You surely will not die! For God knows that in the day you eat from it your eyes will be opened, and you will be like God, knowing good and evil'" (Genesis 3:4-5). Satan's promise was not quite as it appeared. Instead of becoming perfect like God as she expected, Eve learned to know evil all too well, experientially.

Many today are as gullible as Eve and have been fooled into believing that transhumanism will help perfect them. And why wouldn't they when transhumanism's stated goal is to better mankind by taking away every disease, disability, suffering, and involuntary death?[59] Their grand promises hide an evil agenda.

Transhumanism is a religion.

> Where traditional theism promises personal salvation, the new faith [transhumanism] offers the prospect of

---

56. https://www.ucg.org/beyond-today/blogs/transhumanism-what-it-is-and-why-it-is-so-fundamentally-wrong.
57. https://www.ynharari.com/.
58. Harari Y.N., *Homo Deus: A brief history of tomorrow*, Harper Collins, Ch. 1, 2016, as quoted in https://creation.com/transhumanism-image-of-god.
59. https://www.humanityplus.org.

> *rescue* via radical life-extension attained by technological applications—a postmodern twist, if you will, on faith's promise of eternal life.[60]

According to transhumanist Harari:

> Religion is the most important *invention* [emphasis mine] of humans. As long as humans believed they relied more and more on these gods, they were controllable. With religion, it's easy to understand. You can't convince a chimpanzee to give you a banana with the promise it will get 20 more bananas in chimpanzee Heaven. It won't do it. But humans will. But what we see in the last few centuries is humans becoming more powerful, and they no longer need the crutches of the gods. Now we are saying, "We do not need God, just technology."[61]

Why do transhumanists cling so tightly to technology as their religion? They see death all around and are repulsed. They fear death, not that of others but their own. Rather than turn to God, who wants to save them from death, they turn to themselves. They truly believe they alone, the intellectual giants and future gods of our age, can save humanity. They can make death optional. Their plan? They will implant their own thoughts in machines. Yes, they truly put their life-and-death faith in machines. Human-made, destructible, deteriorating machines! Again, we see that logic plays no part in evolutionary thought.

> Today's transhumanists are, in one sense, nothing new. They merely seek to build a new Tower of Babel, "to reach unto heaven" and thus to "make a name" for themselves (Genesis 11:4). The Babylonians used the technology of their day, brick and mortar, whereas the new Babylonians use AI, genetic engineering and other modern technology. And just like the blind Babylonians of old, with desperate actions of self-preservation they seek life and unity outside themselves but not in God. They imagine god-like greatness in their demonically inspired arrogance

---

60. https://www.acton.org/religion-liberty/volume-28-number-4/transhumanism-religion-postmodern-times.
61. Ibid.

(Genesis 11:6), much of which will not happen and for which they will also be scattered and then shattered 'like a potter's vessel' (Psalm 2:9).[62]

> Helping a few couples condemned to childlessness towards getting a child may strike the obstetrical cytologist as such a laudable step, but we can see the beginning of human husbandry, of industrial breeding factories...Who can deny the scientific interest attaching to the production of chimaeras, to the study of human embryonic growth in an animal uterus?... What I see coming is a gigantic slaughterhouse, a molecular Auschwitz, in which valuable enzymes, hormones and so on will be extracted instead of gold teeth.
>
> **DR. ERWIN CHARGAFF, BIOCHEMISTRY**

We could legitimately say transhumanism began to pervade society in the 1960s "with reproductive technology. The Pill was the first transhumanist technology: it set out not to fix something that was wrong with 'normal' human physiology...It set out to interrupt normal in the interests of individual freedom."[63] Even more, the Pill suggested that man, not God, was the giver and determiner of life. From that seemingly innocent beginning, transhumanism has grown into a powerful organized force that works to destroy life as we know it.

Transhumanism's plan to radically change mankind includes three stages, the first two related to advanced biotechnology.

First, they continue their already invasive reproductive manipulation by using genetic engineering to breed their preferred humans. Through embryonic gene editing technology, they boast they can produce designer babies that have all genetic problems erased and any desired traits added. Not surprisingly, they are finding this more successful in theory than reality. Human mutations and desired traits both involve hundreds if not thousands of gene manipulations.[64]

For years, they have used gene editing technology to encourage the in-utero abortion of babies that have what they consider potential genetic problems. For example, approximately, 90% of babies suspected to have

62. https://www.orthodoxethos.com/post/transhumanism-the-end-of-evolution-apotheosis-of-carnal-man-and-preparation-path-to-antichrist
63. https://unherd.com/thepost/mary-harrington-transhumanism-is-already-here/.
64. https://www.nature.com/articles/d41586-019-01906-z.

Down syndrome are aborted each year in America.[65] More recently, the Covid vaccine has allowed transhumanists to induce infertility in both males and females, an important aspect of their population control plan.

In case you wondered, transhumanists are also the ones pushing the transgender obsession. They see it as a means

> ...to disintegrate the boundary between the sexes, meaning no youth or age, and no male or female. Transhumanism is boundary-less-ness—the transferring of reproduction to the tech sector. The first gender legislation got its start in the 1980s, laying the groundwork for the eventual disembodiment of humans into cyberspace, or virtual reality, or what has come to be known as the "metaverse."[66]

Truth distortion is a powerful manipulator. If they can trick us into accepting their lie that men can become women, we are but one small decision from relinquishing our God-image to their technologies.

Second, for over a decade, cybernetics has used biomechanical implants and prosthetics, whether necessary or not, to increase performance, both intellectual and physical.

> The engineering of organisms to develop unique purposes and abilities, which are not inherently available, is known as synthetic biology. It involves changing the organism's genetic code by infusing it with another creature's DNA, a radical step-up from genome editing. The WEF is a proponent of the methodology.[67]

Transhumanists "are openly and literally proclaiming their intention to...fuse microchips into people's brains that will be able to read and manipulate individuals' thoughts."[68] This seems so unbelievable but there is some potential evidence that they are at least attempting

---

65. https://en.wikipedia.org/wiki/Genetics_and_abortion#Legality_of_selective_abortion.
66. https://www.naturalnews.com/2023-04-18-transgender-movement-watch-jennifer-bilek-explains.html.
67. https://www.theepochtimes.com/those-who-master-artificial-intelligence-synthetic-biology-will-be-master-of-the-world-klaus-schwab_5064979.html?utm_source=Morningbrief&src_src=Morningbrief&utm_campaign=mb-2023-02-18&src_cmp=mb-2023-02-18&utm_medium=email&est=ghHUmVpzFm43wx%2F1APaFe6Za1gI6Sj721cFVrVsPjcXnRxqVAMpEtWk%3D
68. Alex Newman, *"Great Reset" Transhumanism: Merging Man & Machine*, December 2, 2020, @ https://thenewamerican.com/great-reset-transhumanism-merging-man-machine/.

to implant microscopic computer chips in us,[69] either surgically or through vaccines,[70] clothing,[71] and manipulated food.[72] Much more research is needed to verify these possibilities. But once in the body, transhumanists think they could program these chips to change our thoughts, behaviors, and physical abilities.[73]

In the words of WEF's Harari:

> Humans are now hackable animals that will be re-engineered…Previously, government surveillance was above the skin, now it's going under the skin…The free will to choose who we vote for or what we buy in the supermarket, human free will, that's over![74]

Is it possible that the Covid pseudo-pandemic allowed transhumanists to put self-assembling nano robots, synthetic biology, RFID (radio frequency identity) chips, metals, graphene oxide, parasites, and nano circuitry[75] into the human blood stream through Covid vaccines?[76] Through microscopic examinations of blood samples, many reputable scientists have found evidence of just such dangerous assaults on our genetic integrity. You may have seen vaccine victims collapse and die during sporting practices and events, on television, and in your neighborhoods and schools. Unexplained sudden deaths even in the young have become common among those who have taken the vaccine. Even with all this carnage, the worst may still be ahead. We are just beginning to see the damaging effect of these alien components on the billions of vaccine victims who have so far survived.

The Wall Street Journal suggested that, because the Covid virus made "our biological fragility more obvious than ever," it may be "just the kind of crisis needed to turbocharge efforts to achieve transhumanists'

---

69. https://www.brighteon.com/f4416239-9a57-4996-a16f-84dd34c40483.
70. https://www.naturalnews.com/2022-08-25-vaccinated-blood-is-tainted-blood-study-finds.html.
71. https://www.brighteon.com/b91cb526-4a3f-42c3-b74f-54cfe5c84e17.
72. https://www.brighteon.com/5eb3f505-20a7-4846-b6fe-52f300614f5b.
73. https://www.jonathanbrentner.com/https/jonathan-brentner-g8fgsquarespacecom/config/2021/5/25/transhumanism-why-should-i-care#:~:text=The%20globalists%20needed%20a%20worldwide%20atmosphere%20of%20fear,perhaps%20death.%20And%20that%20is%20what%20has%20happened.
74. https://harbingersdaily.com/transhumanism-the-evil-force-behind-the-new-world-order/.
75. https://rumble.com/v2d150e--pfizer-covid-vaccine-vial-contents-exposed-by-who-whistleblower-.html.
76. https://www.brighteon.com/5eb3f505-20a7-4846-b6fe-52f300614f5b.

goal of immortality."[77] Note their focus: The deaths of millions, even billions, means nothing if it helps achieve the next step in evolution, the immortality of the greatest among us. Themselves, of course.

Third, transhumanists have discarded biological exclusivity and instead are programming human thoughts into robots whose computer intellects and mechanical physiques, they assure us, will make them far superior to mere humans. They expect this merging of humans with machines (artificial intelligence, AI) to be completed by 2030, the same year the United Nations plans to have its one world government in place as detailed in its Agenda 2030.[78]

> No government has the right to decide on the truth of scientific principles, nor to prescribe in any way the character of the questions investigated. Neither may a government determine the aesthetic value of artistic creations, nor limit the forms of literacy or artistic expression. Nor should it pronounce on the validity of economic, historic, religious, or philosophical doctrines. Instead it has a duty to its citizens to maintain the freedom, to let those citizens contribute to the further adventure and the development of the human race.
>
> **RICHARD FEYNMAN, 1965 NOBEL PRIZE IN PHYSICS**

A theme plays throughout each of these stages. Really brilliant people, who consider themselves gods, will decide which parts of each of us should be kept and which destroyed. Those of us who do not fit any of their criteria will be eliminated.

Satan, who designed and promotes transhuman evolution, needs to kill all Jews and their Christian supporters if he is to keep God from fulfilling His promise to bring the Jews back into Israel in faith. Shouldn't we expect Satan to encourage a future holocaust of unparalleled brutality led by his transhumanist gods? He also needs to promote a one world government in which the antichrist can finally rule unimpeded. Interesting, isn't it, to see transhuman evolution working to fulfill both those satanic desires?

Transhumanists have given themselves authority to change the specific kinds that God created in those first six days of creation. Decades ago, they began genetically modifying our food sources, changing plants

77. https://www.wsj.com/articles/looking-forward-to-the-end-of-humanity-11592625661.

78. https://www.jonathanbrentner.com/https/jonathan-brentner-g8fgsquarespacecom/config/2021/6/1/the-path-to-the-mark-of-the-beast.

from the nutrient dense foods God designed into the contaminants (GMOs) they are today. They have begun vaccinating our protein supply with nanotechnology as well. Using geoengineering, they have sprayed millions of tons of toxic chemicals and synthetic biology (programmable matter) from planes, injuring, often killing, many types of wildlife, insects, and plants, and rendering the ground toxic. Already, lifespans in the United States have dropped, and illnesses have skyrocketed.[79] Now, they are focused on genetically modifying our bodies as well. What could possibly go wrong?

> Many people still do not believe that we live in the time of nanotechnological warfare, mind control and transhumanist synthetic biological assault on humanity–where our health, free will, soul and spirit are under attack. People do not believe we found nanotechnology in the C19 shots and have been inhaling nanotechnology and synthetic biology via geoengineering weather warfare for decades. They do not want to believe that carbon nanotubes aka graphene oxide can transmit all information about your thoughts via the spin state of subatomic particles in the synaptic cleft in the brain and modify this signal externally while sending the information to the cloud–hence controlling your mind. They do not want to believe that the C19 injected emit a MAC address via Blue tooth signal... I recommend against all forms of surveillance via any type of device, even if they claim it is for your wellness....It is all part of the transhumanist agenda.
>
> **ANA MARIA MIHALCEA, MD, PHD IN PATHOLOGY**

Through David, God prophesied our present time, the last days of the Church Age before the Tribulation judgment and Christ's second coming.

> Why are the nations in an uproar and the peoples devising a vain thing? The kings of the earth take their stand and the rulers take counsel together against the LORD and against His Anointed, saying, "Let us tear their fetters apart and cast away their cords from us!" He who sits in the heavens laughs. The Lord scoffs at them.
>
> Psalm 2:1-4

Is anything more vain than thinking a computer mind is superior to being made in God's image or that fallible humans can become gods?

---

79. https://www.cdc.gov/nchs/pressroom/nchs_press_releases/2022/20220831.htm.

Try to envision the robots transhumanists are designing to replace us, the last phase in their plan to replace God (whom they say they don't believe in). They alone will build the robots, including the computer "brains" that control them. They alone will determine every fact that enters their "brains." They alone will give them their strength and abilities. They truly will be gods, creating the greatest beings to ever exist—or so they believe.

Will their penultimate product be the Abomination of Desolation that will stand in the Jerusalem temple during the Tribulation?[80] Certainly, robots today can shoot fire and amass enough data to accurately guess an individual's thoughts. Is it too far-fetched to expect even greater computer sophistication soon, by the time of the Tribulation? If so, God will spectacularly destroy evolution once and for all when He uses His enemies to demolish the supposedly eternal statue and the defiled temple that houses it. And what about the mark of the beast? Are the "advances" of transhumanism paving the way for its implementation?

> According to Revelation 14:10-11, accepting the mark of the beast is a fateful decision with eternal consequences, rendering the recipient unredeemable...Could it be possible that transhumanism will modify humans genetically to such an extent... they are no longer created in God's image, making them beyond salvation?[81]

Did similar genetic destruction occur before the Flood? We cannot know, but both past and future possibilities are fascinating to consider. "There is nothing new under the sun" (Ecclesiastes 1:9).

Did you notice the glaring weakness in the transhumanists' plan? They believe they can use their fallible minds to create a being that will transform into infallible God. They completely ignore the problem of their own inherent evil. They misunderstand eternal life. They have everything backwards. Though no human can ever become God, God chose to become human. With His sacrificial death and resurrection, the God/man Jesus Christ solved the death problem. He died so that we might live forever with Him, not as corrupt machines but as His beloved children. We absolutely cannot become gods, but we absolutely

80. https://www.prophecynewswatch.com/article.cfm?recent_news_id=5869.
81. https://www.prophecynewswatch.com/article.cfm?recent_news_id=5916.

can spend eternity as the children of the only God if we accept Jesus' gift of eternal life through faith in Him.

Be assured that transhumanists' evil plans will fail. Not only will God not allow them to succeed, which is reason enough, but their plans have obvious limitations, which we will discuss in detail in later chapters.

1. Man is a creature and can never be God. Even Satan is a mere angel, though the greatest of all angels. His lack of omniscience dooms his plans from the start. Besides, he must get God's permission for everything he does, a sure sign of his eventual defeat. The Book of Job exposes Satan's subservience to God.[82]

2. God declares the human body fearfully and wonderfully made (Psalm 139:14), but transhumanists see our bodies as the greatest threat to our advancement as a species. Their ignorance of the human body's uniqueness has led to problems already. The coded implants transhumanists plan to use to control man's thoughts and movements have failed miserably in pets, whose bodies reject them. "Pets with chips develop fast growing lethal tumors. A tumor naturally encases foreign objects in the body as a way to protect the body from something unnatural."[83] Won't the human body be even more sensitive to invaders that try to change its basic makeup, its unique "kind" as God calls it?[84]

3. Science has shown that man can make nothing that does not disintegrate, including computers and robots. A robot's demise is as near as its next power failure, a rust-causing or byte-destroying drop of water, or a slippery spot of oil. The first time its programmer gets sick or dies, what can a computer do to help itself?

4. Robots cannot think. They are amazing at organizing information, but a human has to input all data. A robot's ability to function depends on the accuracy of the data it receives, which depends on the worldview of the one doing the

82. Job 1-2.
83. https://preventdisease.com/news/10/032610_pets_chipped_tumors.shtml.
84. Genesis 1:21, 24, 25; 6:20; 7:14.

inputting. Someone who believes he is a god or even going to become one has obvious holes in his thinking!

5. Morality cannot be programmed into a robot. "Human nature can't be changed through the augmentation of intelligence. Even a cursory view of history reminds us that sophisticated societies can still invent more horrific ways to destroy their neighbors and exalt themselves…It follows that technological advance does not inevitably lead to advance in human goodness."[85] Absolute moral values can come only from the omniscient moral Lawgiver. Only those with freewill can choose to obey Him. Robots are machines. They have no freewill. They are incapable of moral, righteous decisions.

6. Transhumanists, like all evolutionists, are materialists. They deny the spiritual, including man as God's image bearer. No computer can have compassion, humility, perseverance, lovingkindness, or any other fruit of the Holy Spirit. They can never be an improvement on those made in God's image.

The evolutionists' super-humans will inevitably deteriorate into the sub-human machines they are. How can a thing without the volition, sensibilities, and divine image God created in us be anything but an empty but frighteningly deceptive piece of machinery?

> The question of whether a computer can think is no more interesting than the question of whether a submarine can swim.
> **EDSGER W. DIJKSTRA, COMPUTER SCIENTIST**

Listen carefully and see if you can hear the transhumanists' siren song: "We alone can bring heaven to earth, banishing all suffering from life." But God says, I "will wipe away every tear from their eyes; and there will no longer be any death; there will no longer be any mourning, or crying, or pain; the first things have passed away" (Revelation 21:4). They say, "Don't listen to Him. You don't have to wait for heaven. We will end all suffering here on earth—and you won't have to die to enjoy it! Trust us! Submit to our will!"

85. Godde, Sandra, *Reaching for Immortality: Can Science Cheat Death?* Wipf and Stock, 2022, 40.

Many will be fooled by their promises. You don't want to be one of them. Remember who is behind transhuman evolution and be on your guard. Remember who God is and rejoice in His promise of an eternal life wonderful beyond comprehension. Rather than an eternal life promised by men who proclaim themselves gods, God sent His Son as the Man to die as our substitute. Will you believe the biblical worldview and trust the One who loved you so much He left Heaven to suffer as a man for your sake, or will you trust those who assure you they are gods and you can be one, too?

> Our world is under assault by Satan and his legions. Satanists rule this world. The technology used to conquer and destroy humanity is in the shots, in the air, in the food and water. Its purpose is to kill, corrupt and take away humanity from God. They will not succeed.
>
> **ANA MARIA MIHALCEA, MD, PHD IN PATHOLOGY**

As we have seen, evolution is a failed belief system that has distorted many Christians' understanding of reality. Evolutionists today are packaging their new vision, transhumanism, as the only true savior of mankind. Their promises will be hard to resist, but you can if you examine their assertions as any scientist should. Don't accept any claims, scientific or otherwise, without verification. Don't accept any pronouncements without examining the worldview behind them. Don't abandon reality because most of those you know have been fooled. If a transhumanist promise seems too good to be true, it is.

Trust God. Study and believe His Word. Pray. Be patient. At the right time, God will destroy evolution in all its evil manifestations.

> The fanatical atheists are like slaves who are still feeling the weight of their chains which they have thrown off after hard struggle. They are creatures who–in their grudge against traditional religion as the 'opium of the masses'–cannot hear the music of the spheres.
>
> **ALBERT EINSTEIN, 1921 NOBEL PRIZE IN PHYSICS**

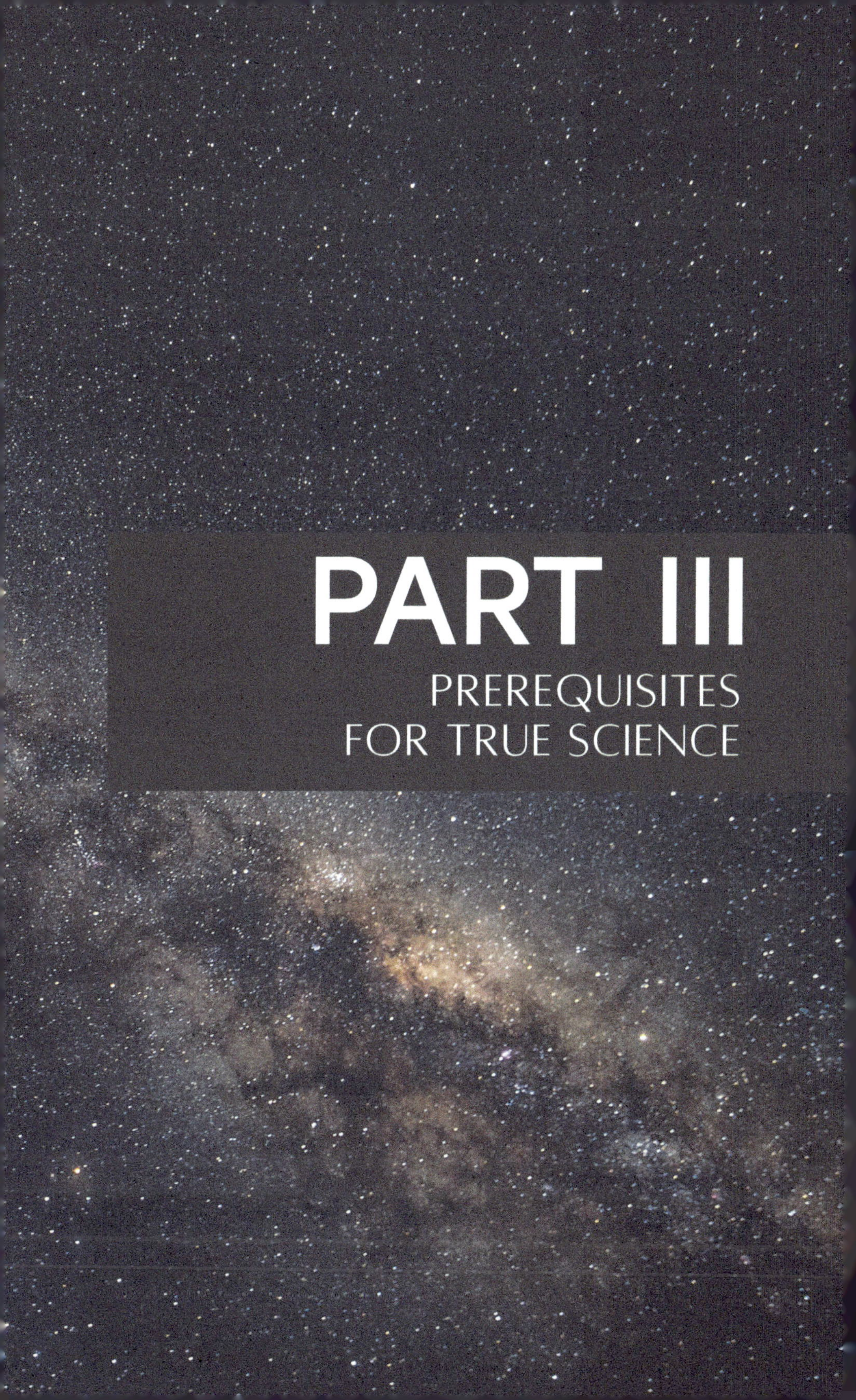

PART III
PREREQUISITES FOR TRUE SCIENCE

By the word of the Lord the heavens were made,
And by the breath of His mouth all their host.
He gathers the waters of the sea together as a heap;
He lays up the deeps in storehouses.
Let all the earth fear the Lord;
Let all the inhabitants of the world stand in awe of Him.
For He spoke, and it was done.

PSALM 33:6-9

For the LORD is a great God
And a great King above all gods
In whose hand are the depths of the earth,
The peaks of the mountains are also His.
The sea is His, for it was He who made it,
And His hands formed the dry land.
Come, let us worship and bow down.
Let us kneel before the Lord our Maker.
For He is our God.

PSALM 95:3-7

DAVID, WRITER AND KING

CHAPTER 7

THE RELIABILITY OF SCRIPTURE

Since I am advocating the Bible as the foundation for all scientific study, you have the right, indeed the responsibility, to demand I prove its reliability for all areas of life, especially for science. Can the Bible be trusted? That should be your question. I will show you that it can. Why? Because the Bible comes directly from the only One worthy of our worship and service. That's why we call it the Word of God.

Most people acknowledge that the Bible is not just any ancient book but unique in the realm of literature. I hear it is the bestselling book of all time, with an estimated six billion copies in existence.[1] By comparison, the Quran has sold approximately 800 million copies, the Book of Mormon, 120 million, and China's "Quotations from Chairman Mao Zedong," several hundred million copies.[2] Interestingly, in 2012, Amity, the only company then authorized to print Bibles in China, printed its 100th million Bible in the Chinese language.[3] Mao would not be happy!

---

1. https://wordsrated.com/bible-sales-statistics/ #:~:text=Bible%20sales%20revenue%20per%20year%20Bible%20sales%20revenue,Bible%20that%20generate%20at%20least%201%20sale%20annually.
2. https://sage-answers.com/how-many-different-bibles-have-been-written/.
3. https://www.christianpost.com/news/over-100-million-bibles-have-been-printed-in-china-84716/.

Not only would the Bible top the bestsellers list every year—if it weren't barred from the list, but it has been translated into over 2000 languages and dialects.[4] In the United States alone, more than 168,000 Bibles are sold or given to others every day.[5] The Gideon's International distributed 59,460,000 Bibles worldwide in one year alone.[6] That's more than 100 Bibles per minute. Today, the Bible can be read on many cell phones through apps. The YouVersion's Bible App alone was downloaded over 400 million times by the end of 2019. By 2020, its downloads totaled over 600 million.[7] More people are reading the Bible worldwide than ever before. The Bible is also the book most likely to be shoplifted.[8] Go figure that one! Obviously, a lot of people agree the Bible is worth reading—or at least owning.

As you would imagine, views concerning the nature of Scripture are as varied as the worldviews of those expressing them. The most theologically liberal classify the Bible as simply another human book of myths. At the other end of the spectrum, the theologically conservative attempt to apply sound interpretive hermeneutics to God's Word because they believe it truly is God-breathed and, therefore, the repository for invaluable information. I hold to that conservative view because I consider it the biblical view and the one most supported by the evidence.

WHAT IS THE NATURE OF SCRIPTURE?

Many skeptics attempt to discredit the Bible by pointing out it was written by fallible, imperfect humans. They are hypocrites, agreeing with the biblical description of the total depravity of man, but only when trying to discredit God's Word! It is essential we clarify how men with sin natures produced this book with its wholly reliable information.

> There is abundant evidence that the Bible, though written by men, is not the product of the human mind. By countless multitudes it has always been revered as a communication to us from the Creator of the Universe.
>
> SIR AMBROSE FLEMING, PIONEER IN ELECTRONICS

4. https://churchgists.com/how-many-copies-of-the-bible-have-been-sold.
5. https://brandongaille.com/27-good-bible-sales-statistics/.
6. Ibid.
7. Churchgists.com.
8. https://thebibleanswer.org/bibles-sold-each-year/.

Only God possesses truth and knowledge and only He can reveal them to us. "For who among men knows the thoughts of a man except the spirit of the man, which is in him? Even so the thoughts of God no one knows except the Spirit of God" (1 Corinthians 2:11).

Scripture encapsulates God's thinking. Its uniqueness is found in both its divine revelation and divine inspiration. First, Scripture claims its content came through revelation from God Himself. Second, it claims God inspired its authors to accurately record all He desired to reveal.

Theologian Henry Thiessen described the revelation of Scripture as "that act of God whereby He discloses Himself or communicates truth to the mind, whereby He makes manifest to His creatures that which could not be known in any other way."[9]

THE TAKE AWAY: Divine revelation includes ideas that cannot be derived by science through observation and experimentation or by the exercise of human intellect through philosophy or logic. Since absolute truth exists only in God and in Scripture, and God guarantees it does, aren't we foolish if we don't study the Bible to better understand all areas of life?

From Genesis to Revelation, the human authors of Scripture attributed their writings to revelation from God. Moses wrote that God gave him the information he recorded in the first five books of the Bible, including accurate accounts of events far before his time, back to the beginning of time. Moses' writings include entire chapters, even sequences of chapters, that are verbatim records of God's words to him.[10] Early in the Book of Genesis alone, he recorded God speaking directly to Adam and Eve (Genesis 2-3), Cain (4), Noah (6-9), Abraham (12-25), Isaac (26), and Jacob (28, 32, 35). Only God could have provided Moses with these conversations.

God commanded the Old Testament prophet Jeremiah, "Write all the words which I have spoken to you in a book" (Jeremiah 30:2). *All the words.* Through His communication with the human authors, God revealed everything necessary for us to serve Him with joy.

---

9. Thiessen, Henry C., *Lectures in Systematic Theology*, William B. Eerdmans Publishing Co., 1990, 7.
10. Deuteronomy 1:3, "Moses spoke to the children of Israel, according to all that the Lord had commanded him to give to them." Deuteronomy 6:1, "Now this is the commandment, the statutes and the judgments which the Lord your God has commanded me to teach you, that you might do them in the land where you are going over to possess it."

New Testament apostle Peter credited God for the words he wrote, many of which were the product of his study under Jesus Christ Himself.

> For we did not follow cleverly devised tales when we made known to you the power and coming of our Lord Jesus Christ, but we were eyewitnesses of His majesty.
>
> 2 Peter 1:16

Paul also declared God to be the source of his information: "The things which I write to you are the Lord's commandment" (1 Cor. 14:37).

> Yet we do speak wisdom among those who are mature; a wisdom, however, not of this age nor of the rulers of this age, who are passing away; but we speak God's wisdom in a mystery, the hidden wisdom which God predestined before the ages to our glory; the wisdom which none of the rulers of this age has understood; for if they had understood it they would not have crucified the Lord of glory; but just as it is written, "Things which eye has not seen and ear has not heard, and which have not entered the heart of man, all that God has prepared for those who love Him." For to us God revealed them through the Spirit; for the Spirit searches all things, even the depths of God.
>
> 1 Corinthians 2:6-10

Notice that Paul made the concept of the divine revelation of Scripture applicable to every writer of the New Testament when he referred to "we" (two times) and "us" (once). In principle, this divine revelation applied to all Old Testament authors as well.

In the last book of the Bible, the apostle John assured us his words had their source in God. "I heard behind me a loud voice [Jesus speaking] like the sound of a trumpet, saying, 'Write in a book what you see, and send it to the seven churches: to Ephesus…'" (Revelation 1:10-11). And so John obeyed exactly, and God's words to him are available to us today.

HOW DID GOD INSPIRE THE SCRIPTURES?

The authors of Scripture were not only recipients of God's revelation but also of His inspiration, the second stage of His communication to mankind. The doctrine of inspiration teaches that God controlled the writing process so that the human authors recorded His thoughts perfectly. In the words of theologian Charles Ryrie, "God superintended the human authors of the Bible so that they composed and recorded without error His message to mankind in the words of their original writings."[11]

For increased clarity, let's expand that definition. Inspiration means God the Holy Spirit controlled all the circumstances that resulted in the authorship of the individual books of the Bible (superintending). He chose certain individuals (the authors) to write messages with a purpose. He revealed the content of the books (the message) to the authors. He motivated (inspiration) the authors to put that content into writing, forming books (recordings). He controlled the composing of each book so the final product was exactly as He desired. This process shows that every book of the Bible had two authors, the human and the divine. We might say each had an author and an Author.

Inspiration also means that the Holy Spirit, the divine Author, overrode the human authors. This controlling process insured that the final product reflected His desires. We can describe the end-product as the Word of God in the sense that God Himself insured that the words in Scripture represent His thinking.

> No prophecy was ever made by an act of human will, but men moved by the Holy Spirit spoke from God.
>
> 2 Peter 1:21

Conservative scholars call this doctrine plenary inspiration, which means "all Scripture is inspired by God" (2 Timothy 3:16). God assures us every book in the canon of Scripture, all sixty-six of them, and every word in the original manuscripts of those books, from Genesis 1:1 to Revelation 22:21, carries His guarantee of authenticity.

---

11. Ryrie, Charles C., *Basic Theology*, Victor Books, 1986, 71.

Scripture itself supports the truth of its divinely guaranteed authenticity.

> The *sum* of Your word is truth, and every one of Your righteous ordinances is everlasting.
>
> Psalm 119:160

> But the Lord said to me, "Do not say, 'I am a youth,' Because everywhere I send you, you shall go, and *all* that I command you, you shall speak."
>
> Jeremiah 1:7

> All Scripture is inspired by God and profitable for teaching, for reproof, for correction, for training in righteousness.
>
> 2 Timothy 3:16

The inspiration of Scripture is also described as verbal inspiration, meaning that the words themselves are inspired, not just the ideas. Again, Scriptures in both the Old and New Testaments assure us of this truth.

> Then the Lord stretched out His hand and touched my mouth, and the Lord said to me, "Behold, I have put My words in your mouth."
>
> Jeremiah 1:9

> Then he [an angel] said to me [the apostle John], "Write, 'Blessed are those who are invited to the marriage supper of the Lamb.'" And he said to me, "These are *true words* of God."
>
> Revelation 19:9

Jeremiah 26:2 proclaims the importance God placed on the accurate transmission of His Word:

> Thus says the Lord, "Stand in the court of the Lord's house, and speak to all the cities of Judah who have come to worship in the Lord's house all the *words* that I have commanded you to speak to them. *Do not omit a word!*"

Every word of God's Word is essential to the unfolding of truth.

In His humanity, God incarnate, Jesus Christ, proclaimed the accuracy of His Word:

> "It is the Spirit who gives life; the flesh profits nothing; the words that I have spoken to you are spirit and are life."
>
> John 6:63

THE TAKE AWAY: As we read the Bible, we can easily discern that it reflects the distinct writing style and experiences of each author. Unlike every other book written by human authors, however, it is also a divine book, originating from the Creator of the universe. Because of God's revelation and inspiration, His book gives us access to absolute truth and knowledge. Because of His over-riding control of the process, we have, as an end-product, an inherently reliable book.

WHY IS THE INERRANCY OF SCRIPTURE IMPORTANT?

The Bible claims to be God's inspired revelation and, therefore, without error. This doctrine of inerrancy is directly related to the reliability, accuracy, and infallibility of its contents. Again, Charles Ryrie provided an excellent definition. "The Bible is inerrant in that it tells the truth, and it does so without error in all parts and with all its words."[12]

Most theologians who hold to inerrancy limit it to issues of theology and doctrine. Only the most conservative hold to its inerrancy when speaking about history and science. I agree with them. Scripture addresses every area of life and is inerrant in every comment it makes about the latter.

The clearest passage on the inerrancy of Scripture is Psalm 19:7-9. The psalms are poetic literature, so Psalm 19 has poetic features. The main ones are synonymous parallelism, repetition, and the use of synonyms for the word Scripture. Hebrew parallelism is composed of at least two lines of phrases or clauses, one following the other(s) in relationship. The first line states an idea and the following lines repeat the same or similar idea using different words.

In Psalm 19:7-9 we have six lines in synonymous parallelism. They encourage us to focus on the main idea, that no matter how we look at Scripture, its inerrancy is evident.

12. Ibid, 83.

> The law of the Lord is perfect, restoring the soul.
>
> Psalm 19:7a

David, the second king of Israel, used the word "law" to refer to Scripture. He assured us we can accept the validity of Scripture as readily as we would accept any legal document. Because of its perfection, it can fill our minds, our souls, with joy, peace, and contentment.

> The testimony of the Lord is sure, making wise the simple.
>
> Psalm 19:7b

We can believe God's Word because His "testimony," His legal witness, is sure, completely accurate. It is verifiably reliable because its source is reliable. When we study and believe it, we, the simple, those without omniscience, increase our wisdom.

> The precepts of the Lord are right, rejoicing the heart.
>
> Psalm 19:8a

The word "precept," another synonym for Scripture, has both legal and behavioral implications. Legally, a precept is a writ or warrant, a legal document based on accurate knowledge. Behaviorally, it is a general rule to promote correct, acceptable behavior. Taken together, God's precepts promote that which is right both legally and morally. David assured us God's precepts are free from error because they come from righteous God who personifies all that is right. His authorship guarantees no mistakes, distortions, or falsehoods.

> The commandment of the Lord is pure, enlightening the eyes.
>
> Psalm 19:8b

Here, David reminds us that God's Word is the "commandment of the Lord." A commandment is to be obeyed. Because God loves us with a perfect love, obedience to His commandments blesses us by "enlightening the eyes," by leading us in the path of righteousness, the path of truth and light. Anything pure is free from impurities, not polluted with mistakes or fallacies.

> The fear of the Lord is clean, enduring forever.
>
> Psalm 19:9a

When David called Scripture "the fear of the Lord," he didn't mean we should feel terror when we read it. We may feel terror when we violate its precepts, but fear in this verse means reverence coupled with awe-filled respect. God's Word gives us a fear that encourages us to know His plan and essence so we can be clean, cleansed of sin, able to serve Him. Cleanliness ensures that the Bible, our source of knowledge about being clean, is free itself from the contamination of uncleanness. No untrue, unclean thought can be found in its words. Because His Word is pure, it will endure forever. What we learn about God's Word today will remain true forever and remain in our souls forever also. Learning God's Word impacts us both now and forever.

> The judgments of the Lord are true; they are righteous altogether.
>
> Psalm 19:9b

God's Word can legitimately be viewed as "the judgments of the Lord." Judgments separate right from wrong, truth from error, good from evil, the beneficial from the damaging, the righteous from sin, and that which God loves from that which destroys what He loves. While not every word in Scripture is a judgment, its essence leads us to know and love God so He can produce His righteousness in us. The Bible is righteous because it meets the ultimate standard of God Himself. Because His Word is as true as its divine Author, it can reveal the truths that lead to Christ's favorable judgment of us at His Judgment Seat. True judgments contain no lies or deceptions.

How clearly this passage presents the inerrancy of Scripture! How sad if we pretend God's Word is not "alive and powerful," if we question its accuracy because the world demands we do! Again, the decision is clear. Do we believe infallible God, the Creator, or fallible man, the creature who would not exist apart from Him?

Other passages reaffirm the perfection of Scripture. For example, Psalm 119:140 describes the Word of God as "pure." Psalm 119:172 calls it "righteous." James 1:25 describes it as "perfect." Jesus Himself declared His Word "true" (John 17:17). In an interesting statement recorded in Matthew 5:18, Jesus repeated the truth of the inerrancy of the Scriptures:

> For truly I say to you, until heaven and Earth pass away, not the smallest letter or stroke shall pass away from the Law, until all is accomplished.

Fulfillment of prophecy, "until all is accomplished," proves the eternal nature of Scripture. We need to examine Jesus' words more carefully to understand the eternal durability of God's Word. What did He mean by the "smallest letter or stroke"?

FIGURE 7 SMALLEST LETTER, YOD IN HEBREW

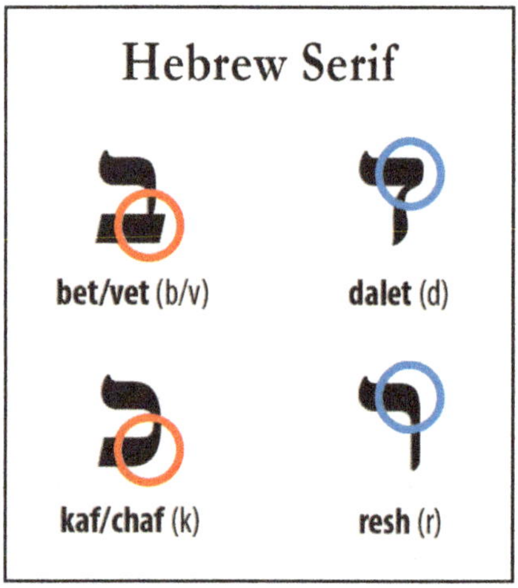

FIGURE 8 STROKE, HEBREW SERIF

The "smallest letter" likely refers to the Hebrew letter *yod*, see Figure 7 (Remember Hebrew reads from right to left), which is significantly the smallest letter in the Hebrew alphabet.[13]

The word "stroke" refers to the Hebrew serif, a small projection on single letters in the Hebrew alphabet that distinguish them from other letters. In Figure 8, notice the Hebrew letter *bet* with red circles, like our "b," as compared to the letter *kaf* like our "k." Also, the difference between the

13. The original language of most of the Old Testament was Hebrew.

letter *dalet*, like our "d," and *resh*, like our "r," is the little serif. What a difference a stroke makes! God's Word is reliable even in these smallest parts of its letters.

THE TAKE AWAY: God is omniscient, so the book He inspired is without error. Do you think He is unaware of history, that He doesn't carefully preserve crucial historical truths in His Word? Is scientific truth beyond His intellect? Does the One who named every star, all the hundreds of billions of them, not know everything that can be known about mathematics, botany, and chemistry since He created their principles? Shouldn't we expect Him to reveal accurate details about His creation in His Word? What about economics, language, politics, sociology, psychology, music, and even sex? Is anything beyond His understanding? Shouldn't we rejoice that our loving Father has generously presented truths about every area in life so that we can learn from them and live better, happier lives as a result?

The Bible is an inerrant book that gives us accurate and necessary insights into every area of life. I hope that, when we are done, you will agree and then use the scriptural ammunition you have acquired to convince others.

CAN THE RELIABILITY OF SCRIPTURE BE PROVEN IN OTHER WAYS?

Scripture declares itself inerrant, but can those claims be verified by outside sources? Can Scripture's internal evidence be corroborated by external evidence? The answer is a resounding yes. Archeology, history, and textual criticism offer proofs. Newly discovered data continually validates all biblical claims. Entire books on apologetics document this data in detail.[14]

The historical science of archeology by itself proves the Bible's inerrancy. Over the years, archaeological sites excavated at ancient Old and New Testament locations overwhelmingly verify historical details and persons referred to in the Bible. Jewish archaeologist Nelson Glueck wrote:

> It may be stated categorically that no archeological discovery has ever controverted a single biblical reference. Scores of archeological findings have been made which

14. Besides the apologetics books referenced in this section, you can consult the works of Josh McDowell, John Warwick Montgomery, Cornelius Van Til, and John M. Frame.

> confirm in clear outline or in exact detail historical statements in the Bible.[15]

Amazing! Not a single archeological find contradicts the Bible! Despite the howls of secular commentators to the contrary, archeology overwhelmingly and completely corroborates the accuracy of the biblical record.

But how many sites have been excavated? Is the archeological evidence truly exhaustive? Archaeologist and Assyriologist Donald Wiseman noted that "the geography of Bible lands and visible remains of antiquity were gradually recorded until today more than 25,000 sites within this region and dating of the Old Testament times, in their broadest sense, have been located."[16] Not only does every scrap of archeological evidence support the validity of the Bible, but those trace data are numerous. More than 25,000 sites for the Old Testament alone—and that was written in 1976! The figure in 2016 is at least 35,000 active sites, though hundreds are disappearing under newly constructed roads and buildings.[17]

Indeed, all of archeology's many finds support the Bible's claim of absolute reliability.[18] Never has any find that seemingly contradicts the Bible stood up under scrutiny.

Historians also affirm the Bible's inerrancy. Many skeptics attempt to discredit the Bible by alleging that historical errors abound. They claim its stories are myths, not accurate recordings of history. I am glad to report the historical integrity of the Bible has been confirmed as historians have uncovered more and more secular evidence supporting its accuracy. Many ancient secular documents mention major events and/or people referenced in the Bible. Thus, many unbelieving and non-sympathetic writers, who didn't realize they were giving historical confirmation, do, indeed, support the Bible's authenticity.

The main person referred to in these extra-biblical documents is Jesus Christ. His death and resurrection were recorded by several writers though they did not understand the spiritual significance of their accounts. They recorded them because they were historically important even in their pagan cultures. In Annuls 15.44 written to Emperor Nero,

---

15. https://bibleevidences.com/archaeological-evidence/. Glueck, Nelson, *Rivers in the Desert*, 1960, 31.
16. Geisler, Norman L., *Christian Apologetics*, Baker Book House, 1976, 322.
17. https:// www.hadassahmagazine.org/2016/06/08/dig-archaeology-israel/.
18. https://answersingenesis.org/archaeology/does-archaeology-support-the-bible/.

Publius Cornelius Tacitus explicitly mentioned Jesus' death. "Christus, the founder of the name, was put to death by Pontius Pilate, procurator of Judea in the reining of Tiberius." [19]

The full paragraph reads:

> Hence to suppress the rumor [of Nero burning Rome], he falsely charged with the guilt, and punished with the most exquisite tortures, the persons commonly called Christians, who were hated for their enormities. Christus, the founder of the name, was put to death by Pontius Pilate, procurator of Judea in the reign of Tiberius: but the pernicious superstition, repressed for a time broke out again, not only through Judea, where the mischief originated, but through the city of Rome also (AnnalsXV.44).[20]

Biblical scholar F. F. Bruce referred to the eminent first century AD Jewish historian Josephus, when he wrote:

> Here, in the pages of Josephus, we meet many figures who are well-known to us from the New Testament; the colorful family of the Herods; the Roman emperors Augustus, Tiberius, Claudius, and the procurators of Judea, the high priestly families—Annas, Caiaphas, Ananias, and the rest; the Pharisees and the Sadducees; and so on.[21]

God the Holy Spirit inspired Luke, the author of the Gospel of Luke and Book of Acts, to record many specific geographic locations, Roman leaders, and historical events. At nearly the same time, other writers with unbelieving worldviews recorded the same information, thus verifying the biblical account. In the first and second centuries, Greek satirist Lucian, Roman historians Tacitus and Seutonius, leader Pliny the Younger, historian Thallus, and others included information that supports the biblical account.[22]

Besides these unbelieving historians, early church fathers and apologists included similar information in their writings. Though their testimonies

19. Ibid, 323.
20. Ibid.
21. Bruce, F.F., *N. T. Documents: Are They Reliable*, Eerdmans, 1960, 104.
22. Check *Christian Apologetics*, 322-325, and *Championing the Faith*, 158-161, for quotations from these writers.

are not as weighty as those of secularists since these men were biased believers in the Bible, they are valuable since they wrote soon after the Bible's completion. Many believers wrote voluminously, especially Polycarp (one of the earliest, AD 69-155), Justin Martyr, Clement of Alexandria, Tertullian, Origen, and Eusebius. Their transformed lives and high regard for the Scriptures indicate its credibility and reliability.

The more we learn from archeology and history, the more the reliability of the Bible is confirmed.

> The best data we have are exactly what I would have predicted, had I nothing to go on but the five Books of Moses, the Psalms, the Bible as a whole.
>
> ARNO PENZIAS, 1978 NOBEL PRIZE IN PHYSICS

WHAT OUTSIDE EVIDENCE DOES THE SCIENCE OF TEXTUAL CRITICISM SUPPLY?

Many skeptics attempt to discredit the Bible by asserting that internal corruption occurred during the years following the original writings. How could it not be corrupted over so many thousands of years, they ask? Others allege innumerable changes were deliberately made to make the Bible more relevant to current generations. Still others laugh at us for pretending that any of the original manuscripts are still available. Are any of these charges true? Are skeptics justified in questioning the Bible's reliability?

To answer these questions, we will utilize another science, that of textual criticism. You have heard of the sciences of physics, chemistry, and biology, but have you heard of the science of textual criticism? Yes, it is a science, mainly an observational science, meaning, as you know, its findings are repeatable. Historical science plays a part, too, making textual criticism very much a science. Anyone who has confidence in science should have confidence in the conclusions of the science of textual criticism. Follow the science, as they say.[23]

---

23. If you are interested in textual criticism, you could begin with Bruce Metzger's *The Text of the New Testament* and J. Harold Greenlee's *Introduction to New Testament Textual Criticism.*

> Textual criticism is the study of copies of any written work of which the original autograph [the original copy] is unknown, with the purpose of ascertaining the original text.[24]

The works of ancient Greek philosophers like Plato and Aristotle as well as the Scriptures are among its subjects.

Critics of the Bible are correct in their assertion that the original writings, called original autographs, were lost soon after they were written. Only copies in a long line of copies exist today. That includes both the Old and New Testaments. Does that make the Bibles we read today less than accurate?

FIGURE 9 VATICANUS CODEX, ONE OF THE MOST IMPORTANT MANUSCRIPTS, IS A COPY IN A LONG LINE OF COPIES.

To answer that question, we will think like textual scientists. Comparing the Bible with ancient Greek pagan literature will help us gain a realistic perspective on this issue of accurate transmission. Like the Bible, no original ancient Greek written manuscripts have survived, only copies of copies far removed from the originals. Even so, these books, written by unbelievers, are so universally accepted that they are used in all classical literature and philosophy departments in all major and probably minor universities in the world. These departments are dependent on copies of copies of these books for their very existence. Through textual criticism, we can decide for ourselves if their acceptance of these ancient documents as accurate doesn't mandate we accept the Bible as accurate also.

Archaeologist and textual critic Steven Collins studied the writings of thirteen ancient authors.[25] I have selected three well-known ones for our study, historian Herodotus and philosophers Aristotle and Plato.

---

24. Greenlee, J. Harold, *Introduction to New Testament Textual Criticism*, William B. Eerdmans Publishing Co., 1964, 11.
25. Collins, Steven, *Championing the Faith: A Layman's Guide to Proving Christianity's Claims*, Virgil W. Hensley Inc Publishing, 1991, 68. These three examples are from this book.

In the Ancient Text Comparison chart below, the Date of Writing column shows that the three unbiblical works were written between BC 480 and BC 332 but, according to the Earliest Manuscript column, no copies were found until AD 900 at the earliest. The third column shows that over 1200 years passed before any copy was found. The last column shows that fewer than eight copies of these ancient works have ever been found. That's not many copies and even those were discovered over a millennium after the original works were written. Had too many years passed for the copies to accurately represent the originals?

Notice the extreme difference between the non-biblical writings and the divinely inspired New Testament.[26] The entire New Testament was written between AD 44 (Matthew and James) and AD 95 (Revelation). The earliest existing manuscript copies of the New Testament date to within decades of the original writing, rather than a millennium later as for the Greek classics. Multiple copies of entire New Testament manuscripts date to about AD 400, half the time of the classics. Logic dictates that the Bible must be at least as accurate as the Greek classics that so many in academia unquestioningly revere.

ANCIENT TEXT COMPARISON

	Date of Writing	Earliest Manuscript	Span	Copies
Herodotus	480-406 BC	900 AD	1300 years	8
Plato	427-347 BC	900 AD	1200 years	7
Aristotle	384-322 BC	1100 AD	1400 years	5
New Testament	44-95 AD	125 AD (fragment) 400 AD (NT)	Decades 300 years	1000's

The greatest contrast is in the number of existing copies available for textual criticism scientists to study. A huge number of copies of the Bible have been found, nearly six thousand, but that doesn't tell the whole story. More manuscript fragments continue to be discovered almost yearly. Compare that number to the twenty surviving copies for all three Greek classics combined. The difference is significant, isn't it?

26. Similar comparisons would be true of the Old Testament as well.

Many scientists working in textual criticism affirm that the Bible we have today is virtually the same as the original manuscripts of the original authors. Dr. Geisler wrote, "There are more manuscripts copied with greater accuracy and earlier dating than for any secular classic from antiquity."[27]

Kenneth Kitchen, Ancient Near Eastern scholar and Egyptologist from the University of Liverpool, concurred:

> The New Testament, how different and how vastly superior is the manuscripts evidence. Some 5,000 [now ~6,000] Greek mss. (whole or fragmentary) are known, not a mere eight or ten...Thus, the manuscript-attestation for the New Testament is of the highest quality in terms of date, and the sheer wealth of mss. Also enables textual scholars to determine very closely indeed the correct readings of the New Testament's text.[28]

Though the six thousand available manuscript copies contain usually insignificant variants, the entire original New Testament inspired text, all twenty-seven books, is contained within them.

One of the greatest Greek scholars of all time and the author of the most complete Greek grammar, A. T. Robertson, similarly pointed to the powerful manuscript evidence:

> There are some 8,000 manuscripts of the Latin Vulgate and at least 1,000 for the other early versions. Add to that over 4,000 [now ~6,000] Greek manuscript copies of portions of the NT. Besides all this, much of the NT can be reproduced from the quotations of the early Christian writers.[29]

To say the Bible does not have sufficient textual basis to be accurate is to bring into question all the literature and philosophy departments of all major universities in the world because the works on which they base their credentials have magnitudes less textual credibility. To say

27. *Christian Apologetics*, 306.
28. Kitchen, Kenneth A., *The Bible in Its World*, Inter-Varsity, 1978, 131. At the time of its writing, the numbers of manuscripts were as quoted.
29. Robertson, A.T., *Introduction to the Textual Criticism of the New Testament*, Broadman, 1925, 29.

the Bible is not a reliable document is to say every university should shut down those departments, preferably yesterday. Science proves a significantly greater probability that the Bible is more accurate than other, equally old, pieces of Greek literature. Follow the real science!

DOES THE SURVIVAL OF THE BIBLE ADD TO ITS RELIABILITY?

That leads us nicely into another area we should touch on briefly, the physical survival of the Bible. Few books endure past the lifetimes of their authors, much less for millennia as is true of the Bible. Most books are out of date within a few years of their writing, but the Bible remains as relevant today in every culture as it was when first written. Its survival testifies to its absolute uniqueness and enduring reliability.

Though many debate the precise dates various books of the Bible were written, no one disputes that the Bible is thousands of years old. The conservative date for the completion of the five books of Moses is about 1400 BC.[30] The Book of Job may be older, but its exact date of origin is unknown. The last book of the Bible, the Book of Revelation, was written in AD 95. That means the Bible was written over a span of about fifteen hundred years (1,400 + 95 = about 1,500). That makes the completed Bible over two thousand years old (from the last writing in the first century to today) with parts over thirty-four hundred years old (1,400 + 2,000 = 3,400). That is ultimate survival!

THE TAKE AWAY: Our Lord has sovereignly preserved His Word for us today. God is God after all. He wants us to know Him. Throughout history, many have attempted to destroy, ban, censor, and certainly suppress the Bible. Entire ideological and political systems have tried to eradicate it. God's Word has not only survived every attack but has been embraced by more and more people because of them. Historically, the persecution of Christians always leads to a revival in Christianity.

Even the Old Testament records one example of the wanton destruction of Scripture and that by a Jewish king who should have known

30. Moses may have written the book of Genesis before the Exodus from Egypt in 1445 B.C. If the date of the Exodus was 1445 B.C., he wrote Deuteronomy forty years later in 1405 BC, near his death and near the beginning of the conquest of the land of Canaan. The other books of Moses are Exodus, Leviticus, and Numbers.

better. After Jeremiah had his prophetic book, named after him, delivered to King Jehoiakim of Judah, the king cut it in pieces and burned it.[31] That was the only copy in existence. Rest assured Jeremiah rewrote it under God's inspiration, and God preserved it in Scripture. Let me repeat: God is God after all. He wants us to know Him.

A well-known example of the Bible's survival is from the Roman period. In AD 303, Emperor Diocletian attempted to destroy Christianity. He issued a series of edicts ordering the destruction of church property, the demoting of all Christians from positions of honor, and the burning of every Bible and theological book.[32] As would be expected, this persecution led to the growth of the church. Not only did the church grow, but the Bible survived intact to be treasured continuously down to our time.

History records many attempts to physically destroy the Bible, but other kinds of scriptural destruction have been even more damaging. A relentless campaign has been waged to undermine the reading and understanding of the Bible. Because such attempts to negate the power of Scripture are insidious and difficult to detect, they have been and continue to be even more damaging than all physical attacks.

Though the Bible is by far the bestselling book of all time, how many read, understand, and obey it? This hindering of understanding has come in a variety of forms. Many neglect to read it with a prayerful desire to understand it. Even more don't read it at all! Critics attempt to discredit it. False teachers distort it. Many are confused when reading it by faulty theological systems like deism and rationalism from people like Voltaire or Spinoza of the past or Antony Flew and even Neil Armstrong of moonwalk fame[33] in more recent times. Defective hermeneutic principles destroy the integrity of the text. Many interpretations are deliberately or ignorantly wrong. Do we need to point out that even the science community has rejected, distorted, and ignored the rich source of data contained in Scripture?

In spite of all attempts to suppress His Word, God has faithfully preserved it for our edification. The Bible is still available today to tell us how to have an intimate relationship with the Creator, of

31. Jeremiah 36:23.
32. Latourette, Kenneth Scott, *A History of Christianity*, Harper Row Publishers, 1953, 90.
33. https://en.wikipedia.org/wiki/List_of_deists.

course, but also to give us a foundation for gaining knowledge in all areas of study, including science.

Not only has the Bible survived all historical attacks, but it has had more influence on mankind than any other piece of literature or system of thought. Over history, billions of believers have not only been helped but radically transformed through the truth it contains. Its transforming impact has led many to pay a heavy price for defending it. Some of those have sacrificed their lives rather than renounce the God of Scripture. Some today continue to make that ultimate sacrifice for the God whom they adore. Through every attack, God has insured the survival of His Word. Shouldn't we revere it as one of the greatest gifts He has ever given us? Shouldn't we read and study it daily? Shouldn't we obey it?

> Nature...is inexorable and immutable; she never transgresses the laws imposed upon her, or cares a whit whether her abstruse reasons and methods of operation are understandable to men. For that reason it appears that nothing physical which sense-experience sets before our eyes, or which necessary demonstrations prove to us, ought to be called in question (much less condemned) upon the testimony of biblical passages which may have some different meaning beneath their words. For the Bible is not chained in every expression to conditions as strict as those which govern all physical effects; nor is God any less excellently revealed in Nature's actions than in the sacred statements of the Bible.
>
> **GALILEO GALILEI, ASTRONOMER, PHYSICIST, ENGINEER**

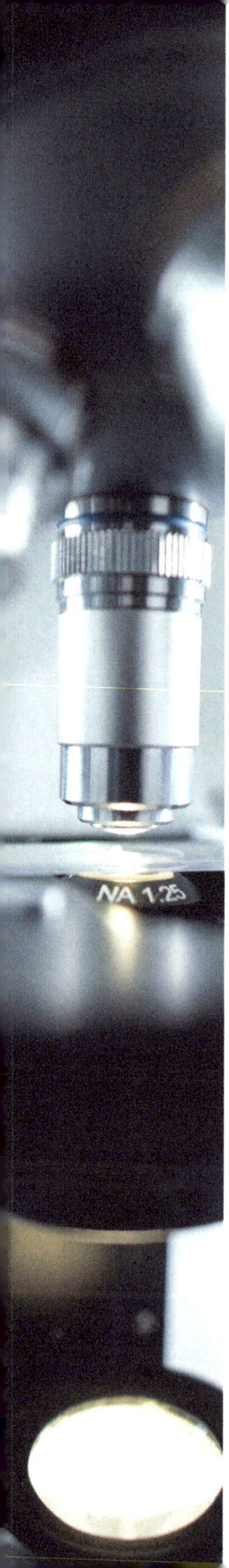

The Lord by wisdom founded the earth.
By understanding He established the heavens.

PROVERBS 3:19

"Who has ascended into heaven and descended?
Who has gathered the wind in His fists?
Who has wrapped the waters in His garment?
Who has established all the ends of the earth?
What is His name or His son's name?
Surely you know!"

PROVERBS 30:4

SOLOMON, WISEST MAN ON EARTH AND KING

CHAPTER 8

ABSOLUTE TRUTH AND SCIENCE

Our culture promotes an anti-biblical view of absolute truth. Post-modernists, who can be found teaching on every college campus, pontificating in every think-tank, and avoiding "God-talk" in many pulpits, declare absolute truth cannot and does not exist. They believe truth is relative. You've probably heard their slogan: "Your truth is not my truth, and my truth may not be your truth." They see truth as individual and personal, not testable or repeatable and certainly not absolute.

Note that post-modernists vehemently deny the existence of objective, absolute truth—except for their own absolute truth statement that absolute truth does not exist! All "truths" not based on God and His Word are equally irrational and would be laughable if not so dangerous.

A common view of absolute truth is that science strives for and often finds it because science self-corrects. In discussing mathematics, an indispensable aspect of science, mathematician Philip E.B. Jourdain (1879–1914) summarized this perverted idea of absolute truth. "The nature of Mathematics is independent of us personally and of the world outside, and we can feel that our own discoveries and views do not affect the Truth itself, but only the extent to which we or others see it. Some of us discover things in science, but we do not really create anything in science any more than Columbus created America…Some philosophers have reached the startling conclusion that Truth is made by men, and that Mathematics is created by mathematicians, and that Columbus created America; but common sense, it is refreshing to think, is at any rate above being flattered by philosophical persuasion that it really occupies a place sometimes reserved for an even more sacred Being."[1]

In a revealing quote, law professor and intelligent design advocate Philip E. Johnson pointed out that, in secular science, naturalism is absolute, but truth is not. This prevalent scientific worldview declares truth can change, but the naturalistic approach is inviolate!

> Truth as such is not a particularly important concept in naturalistic philosophy. The reason for this is that 'truth' suggests an unchanging absolute, whereas scientific knowledge is a dynamic concept...Only naturalism itself and the unique validity of science as the path to knowledge are absolutes...Scientific knowledge is by definition the closest approximation to absolute truth available to us.[2]

Did you get that? Truth is not important. Manmade ideas are sacrosanct, but God's Word is a myth. Beware because that blasphemous philosophy controls science today.

Materialistic and naturalistic worldviews claim the physical, material realm is all that exists. If they are right, then observations of that realm are adequate to come to scientific conclusions, but, if a realm exists outside the natural realm, outside the see-hear-taste-smell-touch realm, we can and should learn about it and apply that learning to the physical realm. Secular science rejects all data from such a realm because, for them, the immaterial, the supernatural, does not exist. The material realm is their only source of truth.

> The Holy Scripture cannot err and the decrees therein contained are absolutely true and inviolable. But its expounders and interpreters are liable to err in many ways.
>
> **GALILEO GALILEO, FOUNDER OF EXPERIMENTAL PHYSICS**

1. https://archive.org/stream/TheWorldOfMathematicsVolume1/Newman-TheWorldOfMathematicsVolume1_djvu.txt.
2. Johnson, Phillip E., *Objections Sustained: Subversive Essays on Evolution, Law, & Culture*, InterVarsity Press, 1998, 29-30.

While science is an acceptable, though limited, means of learning about the physical world, it cannot give a complete picture of it because, today at least, it ignores the equally important unseen realm that influences it. Naturalistic knowledge is not only inadequate but deficient to explain the creation and answer basic questions, such as, where do we come from? Where are we going? How can we best utilize the creation? Why does anything die? Should science consider the impact of the invisible when making scientific enquiries?

I was able to find a rare quote in a science book that at least allows for the possibility of truth outside the physical realm. The context indicates the author was not referring to the spiritual realm but simply to factors like beauty. It comes from my college chemistry text, so it is as ancient as I am. I wonder if science editors today would allow an author the latitude to make even this innocuous statement:

> Man's knowledge and his experiences fall into two distinct classes. First, he gains knowledge of the material things about him by the physical experiences he undergoes through contact with those things. Second, he accumulates knowledge of the non-material or intangible things by the mental experiences associated with them.[3]

Secular science, the supposed bastion of critical thinking, has a history of coming to wrong conclusions despite the clarity of the evidence. Even though the work of science is to observe the data in the creation and develop theories and laws of nature that are true to reality, and even though science attempts to self-correct by rejecting faulty theories, science continues to make mistakes and to delay correcting them. Why? Scientists are fallible. Moreover, those who reject God and His Word must base their conclusions on assumptions divorced from reality. They have rejected absolute truth, and mistakes, sometimes tragic ones, are a consequence.

THE TAKE AWAY: Though secular science says otherwise, knowledge from the natural realm is not the totality of knowledge. It is not even close to the most accurate source of knowledge. If we do not understand the nature of truth, we cannot properly understand science

3. Quagliano, James V., *Chemistry*, Prentice-Hall, 1964, 2.

or efficiently perform it. In fact, we should consider absolute truth a prerequisite for accurate science.

> I am very astonished that the scientific picture of the real world around me is very deficient. It gives a lot of factual information, puts all our experience in a magnificently consistent order, but it is ghastly silent about all and sundry that is really near to our heart, that really matters to us. It cannot tell us a word about red and blue, bitter and sweet, physical pain and physical delight; it knows nothing of beautiful and ugly, good or bad, God and eternity. Science sometimes pretends to answer questions in these domains, but the answers are very often so silly that we are not inclined to take them seriously.
>
> **ERWIN SCHRÖDINGER, 1933 NOBEL PRIZE IN PHYSICS**

WHERE CAN WE FIND ABSOLUTE TRUTH?

As we have seen, the source of absolute truth is God's Word.[4] The Bible presents truth as an objective reality, existing outside of and even separate from humankind. It also makes clear that God has given us the ability to acquire, understand, analyze, and process truth and knowledge and explain it to others.

The idea of truth and knowledge in relation to the natural realm appears early in Scripture. Can you guess where? If you said Genesis, you are right, but can you be more specific? Did you say Genesis 3 where Satan distorted truth? Not quite! We can go back even farther to Genesis 1:3 where God spoke ("And God said…"). Speaking is the communication of truth and knowledge, at least when God does the speaking.

God spoke the entire creation into existence, implying that all creation is intimately related to His speaking, to truth. Psalm 33:6-9 emphasizes this powerful evidence of truth residing in creation:

> By the word of the Lord the heavens were made, and by the breath of His mouth all their host...For He spoke, and it was done; He commanded, and it stood fast.

Even farther back in Genesis 1:1, the first sentence in God's Word, we see an indication of truth. "In the beginning God created the heavens and the

4. Refer to Appendix A for a word study of truth and knowledge.

earth." What an awesome statement of truth, a one-sentence refutation of evolution! God created, and so the heaven and the earth were perfect.

ELEMENTS OF AN UNBELIEVING WORLDVIEW

- GOD → None/Distorted
- ULTIMATE REALITY → Natural/Spirit
- KNOWLEDGE → Reason
- ETHICS → Relativism
- HUMANKIND → Basic Goodness
- DESTINY → Varies

The phrase "the heavens and the earth" is a merism, a common Hebrew figure of speech that puts two extremes together to refer to the totality of something. The merism in Genesis 1:1 puts the extreme of that which is farthest from us, the heavens, with that which is closest to us, the earth, thus indicating the completeness of the universe. Because the Hebrew language has no single word for the universe, the phrase "the heavens and the earth" designates the entire universe, both the material and the immaterial, part of which is truth and knowledge.

Another early and important example of truth and knowledge is God's clear instruction to Adam in Genesis 2:16-17:

> And the LORD God commanded the man, "You are free to eat from any tree in the garden; but you must not eat from the tree of the knowledge of good and evil, for when you eat from it you will certainly die."

This truth statement tested the first man with a choice between two options, obedience and disobedience. God made His command clear—as He does all His commands. Adam made his choice knowing the consequences. If you know the story, you know that Adam, preceded by Eve, failed the test by disobeying God's clear command.[5]

In the description of the fall of mankind, we see the serpent, an agent of Satan,[6] questioning God's truthfulness and introducing a deceptive alternative. Adam's decision to accept Satan's falsehood over God's truth was at the center of mankind's original fall into sin. This ongoing battle between truth and falsehood can be traced throughout the rest of the Bible and world history. We are faced with Adam's same decision,

5. Genesis 3:1-6.
6. Revelation 12:9 and 20:2 equate the serpent with Satan himself, indicating Satan's control of the creature.

the choice between good and evil, truth and lie, obedience and disobedience every day. I could legitimately say every moment of every day.

WHAT IS THE RELATIONSHIP OF GOD TO TRUTH?

As we've seen, absolute truth has the qualities of perfection, eternality, changelessness, limitlessness, and ultimate reality. Logically, such truth must come from someone who possesses those same attributes. Clearly, no human does!

1. GOD THE FATHER IS TRUTH.

Truth not only comes from God but is one of His perfections. In chapter 3 verse 33 of the Gospel of John, John the Baptizer said God the Father "is true." God not only has the perfection of truth but is, in fact, Himself absolute truth.

Hebrews 6:18 gives lying as the antithesis of truth. It is impossible for God to lie. Impossible. That's an absolute truth if I ever heard one!

2. GOD THE SON IS TRUTH.

The second person of the Trinity, Jesus Christ, the self-existent Son of God, declared, "I am the way and the truth and the life; no one comes to the Father but through Me" (John 14:6).

In the beginning of his gospel, John emphasized that Jesus Christ is deity and, as such, has the attribute of absolute truth.

SCIENTIFIC TRUTH	ABSOLUTE TRUTH
• Objective	• Eternal
• Present	• Unchanging
• Not Absolute	• Unlimited
• Tentative	• Ultimate Reality
• Consensus	• Perfect
• From Man	• Omniscient Source

> And the Word [Jesus Christ] became flesh, and dwelt among us, and we beheld His glory, glory as of the only begotten from the Father, full of grace and *truth*...For the Law was given through Moses; grace and *truth* were realized through Jesus Christ.
>
> John 1:14, 17

Later, Paul declared truth an essential component of Christ's character. He wrote that in Christ "are hidden *all* the treasures of *wisdom* and *knowledge*."[7] In Ephesians 4:21, he repeated that truth about truth. "*Truth* is in Jesus."

3. GOD THE HOLY SPIRIT IS TRUTH.

Since God the Father is absolute truth and Jesus Christ, the Son of God, is absolute truth, we would expect the third member of the Trinity, God the Holy Spirit, to be described in similar terms. And He is. "And it is the Spirit who bears witness because the Spirit is the *truth*."[8] The Holy Spirit is also described as the Spirit of truth in John 14:17, 15:26, 16:13 and 1 John 4:6.

Each member of the Trinity is absolute truth; therefore, truth is personal, not simply an abstract concept.

4. GOD'S WORD IS TRUTH.

Because God is truth, His Word, which He inspired, is also truth. In Jesus' prayer for believers, He asked God the Father to "sanctify them [the disciples] in the truth; Your word is truth."[9] God assures us the entire content of His inerrant Word is absolute truth.[10]

5. THE GOSPEL IS TRUTH.

The gospel message is *truth*, absolute truth (Galatians 2:5, 14); therefore, it "is the power of God for salvation" (Romans 1:16).

---

7. Colossians 2:3.
8. 1 John 5:7.
9. John 17:17.
10. John 5:33; 2 Corinthians 6:7; Galatians 5:7; 1 Timothy 3:15; 4:3; 6:5; Titus 1:1; James 1:18; 1 Peter 1:22; 1 John 2:21; 3:19; 3 John 8, 12.

> In Him [Jesus Christ], you also, after listening to the message of *truth*, the gospel of your salvation—having also believed, you were sealed in Him with the Holy Spirit of promise.
>
> Ephesians 1:13

Believing the gospel message is our starting point for understanding absolute truth. Only after we have entered into a personal relationship with absolute truth, with God Himself, can we begin to understand reality. The Bible uses many words to describe our salvific relationship with God: justification, salvation, regeneration, redemption, conversion, being born again. Only when our hearts are regenerated (justified, saved, etc.) by faith in the work of Christ on the cross can we begin to understand the absolute truth God has revealed in His Word and in His creation.

This may be a place in your reading, I hope not the first one, when you need to put this book down because you are compelled to simply worship the Lord. You should not be surprised that many of the founders of modern science also worshipped God as they worked.

Johannes Kepler interrupted his writing of scientific descriptions to pen this worshipful acknowledgement of God's authorship of all scientific reality:

> With a pure mind I pray we may be able to speak about the secrets of His plans according to the gracious will of the omniscient Creator, with the consent and according to the bidding of His intellect. I consider it a right, yes a duty, to search in cautious manner for the numbers, sizes, and weights, the norms for everything He has created. For He Himself has let man take part in the knowledge of these things and thus not in a small measure has set up His image in man...By examining them we observe to some extent the goodness and wisdom of the Creator.[11]

11. Nickel, 115-116. Nickel quoted from Kepler's *Epitome of Copernican Astronomy & Harmony of the World.*

WHAT IS THE RELATIONSHIP OF MAN TO TRUTH?

Paul's letter to the Romans tells us that absolute truth does not come from any human. Far from it! Since science is based on the discoveries of men and women, the Bible's view of mankind in general is important. In simple terms, God, who knows everything, including our every thought, does not think much of our inherent worth.

> There is none righteous, not even one; There is none who *understands*. There is none who seeks for God…There is not even one.
>
> Romans 3:10-12

> For they exchanged the *truth* of God for a lie and worshiped and served the creature rather than the Creator, who is blessed forever. Amen.
>
> Romans 1:23

Man's tendency is to suppress the truth, not seek it. "For the wrath of God is revealed from heaven against all ungodliness and unrighteousness of men, who *suppress the truth* in unrighteousness."[12] That sobering statement indicates we will surely suffer when we ignore the truth. Because methodological naturalism in science suppresses the truth of Scripture, its conclusions too often lead to inaccurate and inadequate understanding. May this be the reason the foolishness of evolution continues to be accepted even though scientific evidence overwhelmingly proves it a distortion of truth?

Not only are we not the source of absolute truth, but we tend to depart from it. Paul reported that many had "gone astray from the truth" (2 Timothy 2:18). Why do so many believers depart from the truth? They are "always learning and never able to come to the knowledge of the *truth*. And just as Jannes and Jambres [pagan Egyptian priests] opposed Moses, so these men also oppose the *truth*, men of depraved mind, rejected as regards the faith" (2 Timothy 3:7-8).[13] Believers who

12. Romans 1:18.
13. 2 Timothy 4:3-4, "Also, For the time will come when they will not endure *sound doctrine*, but wanting to have their ears tickled, they will accumulate for themselves teachers in accordance to their own desires; and will turn away their ears from the *truth*, and will turn aside to myths."

do not choose to learn truth, to learn God's thoughts, as revealed in Scripture, are no more spiritual than pagan unbelievers.

We don't have to remain divorced from truth, for God "desires all men to be saved and to come to the knowledge of the *truth*" (1 Timothy 2:4), knowledge of Himself. Such truth must be available and knowable if we are to fulfill His desire.[14]

> Thus says the Lord, "Let not a wise man boast of his wisdom, and let not the mighty man boast of his might, let not a rich man boast of his riches; but let him who boasts boast of this, that he *understands* and *knows* Me, that I am the Lord who exercises lovingkindness, justice, and righteousness on earth; for I delight in these things," declares the Lord.
>
> Jeremiah 9:23-24

> God is Truth. There is no incompatibility between science and religion. Both are seeking the same truth. Science shows that God exists.
>
> **SIR DEREK BARTON, 1969 NOBEL PRIZE IN CHEMISTRY**

WHAT IS THE RELATIONSHIP OF CHRISTIANS TO TRUTH?

God designed life so spiritual regeneration is a prerequisite for obtaining accurate knowledge. In Jesus' ministry, He focused on bringing people into a saving relationship with Himself because He is not only God but absolute truth. A regenerated person is a believer in Jesus Christ as Savior who, as a result, has a personal relationship with the Creator of the universe. In fact, God the Father is his Father. The Creator of the universe chooses to be every believer's Father! Try to wrap your mind around that awesome truth!

Every believer in Christ has the ability, a potential only, to continually absorb more absolute truth. Notice Jesus' words to new believers in John 8:31-32:

---

14. Also Hosea 6:6, "For I [God] delight in loyalty rather than sacrifice and in the *knowledge* of God rather than burnt offerings." 2 Timothy 2:25, "...with gentleness correcting those who are in opposition, if perhaps God may grant them repentance leading to the knowledge of the *truth*."

> Jesus therefore was saying to those Jews who had believed Him, "If you abide in My word, then you are truly disciples of Mine; and you *shall know* the *truth*, and the *truth* shall make you free."

Jesus meant that abiding or remaining connected to or joined to Him and His Word, to absolute truth, results in further knowledge of absolute truth, which results in true freedom, beginning with spiritual freedom.

A main command for the Christian life is to grow in our knowledge of the Lord and His principles for living.[15] Jesus told His first disciples that God the Holy Spirit would teach them absolute truth—if they desired it enough to listen and learn.

> But when He, the *Spirit of truth*, comes, He will guide you into all the *truth*; for He will not speak on His own initiative, but whatever He hears, He will speak; and He will disclose to you what is to come.[16]
>
> John 16:13

This passage primarily describes the inspiration of the New Testament scriptures but also implies the illuminating work of the Holy Spirit for all believers. The Holy Spirit continues to give each believer the power to gain knowledge and understand truth.

Absolute truth does not come to us automatically. In fact, our sinful nature resists all knowledge so radically contrary to our pre-salvation worldview. As new believers, we must immediately begin recognizing and overcoming the unbelieving worldview that was all we knew before salvation. Those old thought patterns do not vanish just because God regenerated us. We must choose to be renewed in our thinking through learning the absolute truth found only in Scripture.

---

15. 2 Peter 3:18.

16. Also John 3:21, "But he who practices the *truth* comes to the light, that his deeds may be manifested as having been wrought in God." Ephesians 6:14, "Stand firm therefore, having girded your loins with *truth*, and having put on the breastplate of righteousness..." 2 Peter 1:12, "Therefore, I shall always be ready to remind you of these things, even though you already *know* them and have been established in the *truth* which is present with you." 3 John 3-4, "For I was very glad when brethren came and bore witness to your *truth*, that is, how you are walking in *truth*. I have no greater joy than this, to hear of my children walking in the *truth*."

> But you did not *learn Christ* in this way, if indeed you have heard Him and have *been taught* in Him, just as *truth is in Jesus*, that, in reference to your former manner of life, you lay aside the old self, which is being corrupted in accordance with the lusts of deceit, and that you be *renewed* in the spirit of your *mind*, and put on the new self, which in the likeness of God has been created in righteousness and holiness of the truth.
>
> Ephesians 4:20-23

> And do not be conformed to this world, but be transformed by the *renewing* of your *mind*, so that you may prove what the will of God is, that which is good and acceptable and perfect.
>
> Romans 12:2

When we are connected to Christ, who is absolute truth, and are continually renewing our minds, we can see the world from a totally new and precisely accurate perspective. That sounds like a change in worldview, doesn't it? Seeing things from God's perspective means we have the "mind of Christ," rather than the thinking of the world that permeated our every thought before salvation.

> Now we have received, not the spirit of the world, but the Spirit who is from God, so that we may know the things freely given to us by God, which things we also speak, not in words taught by human wisdom, but in those taught by the Spirit [absolute truth], combining spiritual thoughts with spiritual words. But a natural man does not accept the things of the Spirit of God, for they are foolishness to him; and he cannot understand them, because they are spiritually appraised. But he who is spiritual appraises all things, yet he himself is appraised by no one. For "*who has known the mind of the LORD, that he will instruct Him?*" But we have the mind of Christ.
>
> 1 Corintians 2:12-16

We have absolute truth available to us because God revealed, inspired, transmitted, and preserved it; however, because of our fallibility and sin nature, we don't always grasp it. (Be reminded of our earlier discussion about the nature of man.) We cannot begin to plumb the depths of God's unending knowledge. Even if we lived millions of years and spent every minute studying His Word, His perfection and infinite knowledge would remain beyond our finite comprehension. But we need to learn as much as we can, don't we?

THE TAKE AWAY: In summary, the Trinity—God the Father, Jesus Christ, and the Holy Spirit—are absolute truth. The source of absolute truth is God Himself. Scripture is God's revelation of absolute truth. We dare not suppress His revelation.[17]

Because God makes truth and knowledge available, we can accurately evaluate all scientific data. Does it not make sense that science would be greatly enhanced if it started with the sound foundation of absolute truth, the very mind of Christ? In the words of Sir William Herschel (1738-1822), discoverer of the planet Uranus but more famous among astrophysicists as the father of stellar and galactic astronomy, "All human discoveries seem to be made only for the purpose of confirming more and more strongly the truths in the sacred scriptures."[18]

WHAT INFLUENCE DOES THE BIBLICAL WORLDVIEW HAVE ON SCIENCE TODAY?

THE TAKE AWAY: We absolutely cannot acquire an accurate understanding of the creation apart from knowing Scripture. How's that for an absolute statement! Any scientific pursuit, like methodological naturalism, that excludes God's absolute truth is deficient of truth if not completely in error. Science will always be in danger of collapsing without that sound foundation.

Even in today's anti-God environment, some scientists agree. Werner von Braun (1912-1977), the father of space science and acclaimed rocket scientist, said:

> Although I know of no reference to Christ ever commenting on scientific work, I do know that He said, "Ye shall know the truth, and the truth shall make you free." Thus

17. Romans 1:18.
18. Halley, H.H., *Halley's Bible Handbook,* 24th ed., Zondervan, 1965, p.19.

> I am certain that, were He among us today, Christ would encourage scientific research as modern man's most noble striving to comprehend and admire His Father's handiwork. The universe as revealed through scientific inquiry is the living witness that God has indeed been at work."[19]

Henry "Fritz" Schaefer, five-time Nobel Prize nominee for chemistry, wrote:

> "The significance and joy in my science comes in those occasional moments of discovering something new and saying to myself, "So that's how God did it." My goal is to understand a little corner of God's plan."[20]

Isn't that the goal of all true scientists, to understand a small part of creation that can be combined with the small parts uncovered by other scientists? Together, those small parts, acquired over lifetimes of study, can lead to stunning conclusions, ones that benefit everyone.

Owen Gingerich, research professor of astronomy at Harvard University, is another scientist who understands the necessary connection between God and science:

> Nevertheless, just as I believe that the Book of Scripture illumines the pathway to God, so I believe that the Book of Nature, with its astonishing details—the blade of grass, the *Conus cedonulli*, or the resonance levels of the carbon atom—also suggest a God of purpose and a God of design. And I think my belief makes me no less a scientist.[21]

One last example of many (though not nearly as many as on the secular side): Mathematical physicist Frank Tipler, one of the two founders of the famous anthropic principle regarding the physical constants necessary for the origin of life, wrote:

> When I began my career as a cosmologist some twenty years ago, I was a convinced atheist. I never in my wildest dreams imagined that one day I would be writing a

19. *Religious Implications of Space Exploration: A Personal View*, Belmont Abbey College, North Carolina, November 22, 1971.
20. Williams, Stephen, *What Your Atheist Professor Doesn't Know (But Should)*, RFH, 2013, as quoted in https://quotefancy.com/quote/1636307/.
21. https://godevidence.com/2010/08/quotes-about-god-atheism/.

> book purporting to show that the central claims of Judeo-Christian theology are in fact true, that these claims are straightforward deductions of the laws of physics as we now understand them. I have been forced into these conclusions by the inexorable logic of my own special branch of physics...From the perspective of the latest physical theories, Christianity is not a mere religion, but an experimentally testable science.[22]

If all you have read are secular science books and all you have seen is the Discovery Channel, you may not have heard of the biblical approach to science and wonder if it can be implemented. Only a few scientists, engineers, and physicians do, in fact, all their scientific work from the biblical worldview. Certainly, today, such scientists are a tiny minority in the scientific establishment, but they do exist and thrive.

I am fortunate to have two of them as friends. Both retired from the two prestigious national laboratories here in New Mexico, both are world class scientists, both have made huge contributions to the study of creation science, and both are continuing to do creation science research. For many years, Dr. Russell Humphreys worked for Sandia National Laboratory as a physicist. Dr. John Baumgardner worked for Los Alamos National Laboratory as a geophysicist. Both have long records of scientific accomplishments, which we will see in more detail later in this book. I was honored to have them write the forwards to this book.

Throughout the world, other scientists apply divine principles to their work. Here in Albuquerque, as in other areas, we have a local creation science group composed of biblically oriented scientists and engineers. Some work at nearby labs. All attempt to do their work from the biblical worldview.

Some creation scientists and engineers work with the four major creation science organizations, the Creation Research Society (creationresearch.org), the Institute for Creation Research (icr.org), Answers in Genesis (answersingenesis.org), and Creation Ministries International (creation.com). If you consult any of their resources or hear any of their speakers, you will be impressed by their knowledge, logic, and

---

22. https://quotes.yourdictionary.com/author/quote/551934.

achievements. Their enthusiasm for the application of the biblical worldview to science will encourage you to want to know more.

THE TAKE AWAY: Again, God is the Creator. Only He knows the intricacies of His creation. How can we possibly study science successfully if we don't make His Word the foundation for our scientific endeavors? Distortions and miscalculations haunt every scientific study that ignores absolute truth.

THE TAKE AWAY: Worldviews have consequences. Those of secular science are not only devastating to society but also thwart beneficial investigations. Consider the impact evolution has had on medicine alone. How many new treatments were ignored because evolution deemed vestigial organs unnecessary, man a mere extension of apes, and mutations beneficial? Scientists who close their eyes to absolute truth limit science's impact.

> As we conquer peak after peak we see in front of us regions full of interest and beauty, but we do not see our goal, we do not see the horizon; in the distance tower still higher peaks, which will yield to those who ascend them still wider prospects, and deepen the feeling, the truth of which is emphasized by every advance in science, that "Great are the Works of the Lord."
>
> **SIR JOSEPH J. THOMSON, 1937 NOBEL PRIZE IN PHYSICS**

WHAT HAVE WE LEARNED THUS FAR AND WHAT CAN WE EXPECT TO LEARN NOW?

Before we get into the details of the foundation stones for a biblical worldview on science, let's summarize what we have already mentioned, include some personal applications, and set the stage for the rest of our discussion.

- The Creator of the world is also its Savior.
- God designed the earth and universe to be discovered. By doing science, we learn about Him.
- Science points directly and powerfully to a Creator whose intricate designs reflect His perfection and His plan for a relationship with those He created in His image.

> My greatest desire is that I may perceive the God whom I find everywhere in the external world, in like manner also within and inside myself.
>
> **JOHANNES KEPLER, PIONEER OF ASTRONOMY, MATHEMATICS**

- Satan believed he could evolve into a new category, god, the first use of the failed "scientific" idea of evolution (Isa. 14:4-14).

- Genesis tells us God finished the complete creation of heaven and earth in six days, and it was perfect, lacking nothing. While sin causes natural decay, and environment influences changes in adaptation, nothing uniquely different has been added to creation since Day Six of that first week. Nothing has evolved into some new species.

- In the Garden of Eden, Adam and Eve had every bit of knowledge they needed to know God's desire for them. God's Word reveals that same knowledge to us. We, too, are without excuse when we sin, including using science for evil purposes.

> The fabric of the universe is most perfect and the work of a most wise Creator.
>
> **LEONHARD EULER**
> **PIONEER IN MATHEMATICS PHYSICS, ASTRONOMY, AND GEOGRAPHY**

- The Bible is revelation from the God who always has known everything. God gives us commands in His Word because He wants the best for us. He knows we can't find the best apart from His guidance. In His Word, He gives guidelines for science for our benefit, too.

> When I was young, I said to God, "God, tell me the mystery of the universe." But God answered, "That knowledge is for me alone." So I said, "God, tell me the mystery of the peanut." Then God said, "Well George, that's more nearly your size." And he told me.
>
> **GEORGE WASHINGTON CARVER**
> **PIONEER IN AGRICULTURAL SCIENCE, AND INVENTOR**

- We must know the truth to discern the lie. Thought renovation requires much time and work. Old, false preconceptions must be recognized and discarded to make room for the truth. Truth always prevails—though not always in our lifetime or in Satan's world. Lying is his specialty.

- Every atom and wave of creation holds a divine lesson. Light and water are obvious ones, each giving many lessons. Darkness is another. Even the insignificant ant, grasshopper, and ostrich are profound teachers. Science has much to uncover about God and life everywhere it looks!

> Nothing in the whole system of nature is isolated or unimportant. The fall of a leaf and the motion of a planet are governed by the same laws....It is in the study of objects considered trivial and unworthy of notice by the casual observer that genius finds the most important and interesting phenomena. It was in the investigation of the varying colors of the soap-bubble that Newton detected the remarkable fact of the fits of easy reflection and easy refraction presented by a ray of light in its passage through space, and upon which he established the fundamental principle of the present generalization of the undulatory theory of light....The microscopic organization of animals and plants is replete with the highest instruction; and, surely, in the language of one of the fathers of modern physical science, "nothing can be unworthy of being investigated by man which was thought worthy of being created by GOD."
>
> **DR. JOSEPH HENRY, PHYSICS**

- Adam and Eve decided they didn't want to follow God's rules but, instead, satisfy their own desires. Eve wanted to eat the fruit that looked good to her. Adam wanted to please the woman who looked good to him. Today, our society caters to our desires. We are encouraged to please self by giving into our sinful lusts, no matter how destructive or ridiculous. Just do it!
- As with Adam and the forbidden fruit, God expects us to obey His Word just because He, our Creator, says so, no explanation necessary. If we don't know and follow His Word, we heap grief on our own heads.

> No religion except ours has taught that man is born in sin; none of the philosophical sects has admitted it; none therefore has spoken the truth.
>
> **BLAISE PASCAL, PIONEER IN PROBABILITY THEORY**

- Nature in rebellion against us teaches that we remain in rebellion against God.
- Did the DNA of Abel's blood seeping into the earth condemn Cain (Genesis 4:10)? Was that first murder case also the first use of forensic science? Did God institute

forensic science in the first generation of mankind after introducing all science at creation?

- God instituted natural laws like gravity so we can survive in this world. Their predictability allows us to plan our days rationally. Actions must have foreseeable consequences if we are to make wise choices. When Jesus walked on water, the constant was not gravity but His eternal essence. Peter could defy gravity and walk on water because all creation bows before its Creator. When Peter remembered gravity and took his eyes off his Creator, he fell.

> The pursuit of science is a grand adventure, driven by curiosity, fueled by passion, and guided by reason.
> **JAMES CLERK MAXWELL, PIONEER IN PHYSICS**

- God knows everything. That He foreknew Jesus as the Messiah, the Deliverer, before the foundation of the world should not surprise us. His surpassing love has always motivated His plan to redeem us, a plan He set in motion the moment Adam and Eve sinned. The woman's seed, Jesus Christ, would be victorious over death, meaning the resurrection was foretold as early as Genesis 2:17. We all sinned in the one man Adam so we can all be saved in the one Man Jesus Christ.

> Yet even in earthly matters I believe that 'the invisible things of Him from the creation of the world are clearly seen, being understood by the things that are made, even His eternal power and Godhead,' and I have never seen anything incompatible between those things of man which can be known by the spirit of man which is within him, and those higher things concerning his future, which he cannot know by that spirit.
> **MICHAEL FARADAY, ELECTROMAGNETISM**

- History will remain unfinished until the Man Jesus Christ comes and rules perfectly on Earth, the divinely ordained job Adam failed to complete. Every miracle Jesus performed while living on earth pointed to Him as the promised King, who will cure every ailment, cast out every demon, and vanquish death.

- Every person is composed of billions of cells that must work together in precise rhythm. God is our master composer. Evolutionary chance had nothing to do with our complex composition.
- Scientists in Darwin's time had no idea of the complexity of the smallest cells of the human body. The language of DNA, the interconnection of billions of components to all others, makes the gradual change proposed by evolutionists as an antidote to God, not just improbable, but scientifically impossible.

> I find it quite improbable that such order came out of chaos. There has to be some organizing principle. God to me is a mystery but is the explanation for the miracle of existence, why there is something instead of nothing... It was my science that drove me to the conclusion that the world is much more complicated than can be explained by science. It is only through the supernatural that I can understand the mystery of existence.
>
> **DR. ALLEN REX SANDAGE, ASTRONOMY, FIRST TO IDENTIFY A QUASAR**

- Order cannot be a by-product of chaos. Order comes out of chaos only if an outside agent works on it. Evolution is based on an impossibility.
- Even though no tenet of evolution can withstand scientific scrutiny, evolutionists cling to the lie because they can't stand the thought of being under God's authority. We don't need to make up fairy tales like the theory of evolution because we have the sure Word of God, His direct and perfect inspiration. Only the Creator of time and eternity can reveal their workings to us.
- The Mount Saint Helens eruption was God's gift to a skeptical world. No one viewing that new landscape can legitimately deny the truth of a worldwide Genesis Flood or the certainty of divine judgment. The worldwide Genesis flood was as real as this morning's sunrise.

> The best way to show that a stick is crooked is not to argue about it or to spend time denouncing it, but to lay a straight stick alongside it.
>
> **DWIGHT L. MOODY, THEOLOGIAN**

- Everyone would agree that science has been and continues to be a valuable tool in our culture. Early modern scientists looked to God for wisdom and guidance. Because they adhered to the biblical foundation for science, scientific advances exploded both in number and benefit. To make beneficial scientific strides once again, we must return to basing all investigations on the solid biblical foundation extolled by the Isaac Newtons and George Washington Carvers of the past.

> If I have seen further it is by standing on the shoulders of Giants.
> **ISAAC NEWTON**

- No scientist can know tomorrow with certainty, but he can infuse it with the faith and confidence he gained through his study of creation and God's Word today.

> Philosophy and science have not always been friendly toward the idea of God, the reason being they are dedicated to the task of accounting for things and are impatient with anything that refuses to give an account of itself. The philosopher and the scientist will admit that there is much that they do not know; but that is quite another thing from admitting there is something which they can never know, which indeed they have no technique for discovering.
> **A.W. TOZER, THEOLOGIAN**

HOW SHOULD SCIENCE FUNCTION IN THIS SIN-FILLED WORLD?

Science is as flawed as the scientists who inhabit its laboratories. Too many scientists allow personal biases, greed, religious idols, and personality conflicts to blind them to reality. Equally disastrous, the philosophy of naturalism that controls science today stifles creativity and advancement. We are left with a conundrum. How can sinners, which we all are whether we admit it or not, put their sinfulness to the side when they confront a microscope or telescope, pick up an archeological spade, put on a forensic cap, or try to imagine enduring scientific formulas and laws?

The answer is simple though its execution is not. The best way for us to live in a secular world and not become part of it is to develop a Christian culture that advances the sciences and arts in service to God. Scientists must begin by changing their focus from the fallen theories prevalent today to the One who is absolute truth. If they do, they will discover a Creator perfect in design and execution, One who desires a relationship with those He created in His image. He wants us to know Him and admire His creation. He wants us to dedicate all scientific study to Him as the Creator of it.

> The Earth reminded us of a Christmas tree ornament hanging in the blackness of space. As we got farther and farther away it diminished in size. Finally it shrank to the size of a marble, the most beautiful marble you can imagine. That beautiful, warm, living object looked so fragile, so delicate, that if you touched it with a finger it would crumble and fall apart. Seeing this has to change a man, has to make a man appreciate the creation of God and the love of God.
>
> **JAMES IRWIN**
> **AERONAUTICAL ENGINEER, EIGHTH PERSON TO WALK ON THE MOON**

The Bible deals with reality in all realms, not merely the spiritual. Its principles provide the foundation stones needed to succeed in every area of life. If implemented, those biblical guidelines will infuse every area of life, including science, with momentum and integrity. We could legitimately say a fulfilling and eternally successful approach to life demands a biblical worldview.

The rest of this book is devoted to explaining that worldview from the perspective of science. If scientists can be persuaded to embrace the following ten biblical principles, they will once again have an accurate picture of reality. They will realize that the sovereign Lord of the Bible, the Creator, is intimately involved in His creation. He has every proton, neutron, and electron under His control. He wants us to examine every one of them, the forests as well as every cell in every tree, so we can know Him and appreciate His plan for us.

Science can and should be on the cutting edge of discovering the workings of God's creation and developing solutions and technologies that improve our lives in it. In other words, science should uncover every scrap of information we need to wisely rule creation as God commanded our progenitor Adam so many centuries ago.

> Let us...not set aside God's works, but, guided by them, revere the Master!
>
> **CARL LINNAEUS, PIONEER IN BIOLOGY**

PART IV

TEN FOUNDATION STONES FOR TRUE SCIENCE

He said to them, "I am a Hebrew, and I fear the Lord God of heaven who made the sea and the dry land."

JONAH 1:9

JONAH, PROPHET TO NINEVEH

"Arise, bless the Lord your God forever and ever! O may Your glorious name be blessed and exalted above all blessing and praise! You alone are the Lord. You have made the heavens, the heaven of heavens with all their host, the earth and all that is on it, the seas and all that is in them. You give life to all of them and the heavenly host bows down before You."

NEHEMIAH 9:5-6

LEVITES, LEADERS OF WORSHIP IN ISRAEL

CHAPTER 9

SCIENTIFIC FOUNDATION STONE #1

GOD AS THE STARTING POINT NOT NATURALISM

The first foundation stone for effective and enduring science emphasizes knowing the God of Scripture. I can't overemphasize the importance of having our thoughts, our desires focused on the Lord. If we don't know Him, His plan, and His character, we'll miss many opportunities to understand the order, stability, purpose, and intricacies of His creation. Our science will fall far short of its potential.

> If you don't know what's meant by God, watch a forsythia branch or lettuce leaf sprout.
>
> **DR. MARTIN H. FISCHER, MEDICINE**

We begin by remembering that if God is the Creator of heaven and earth, His fingerprints will be all over His creation, and they are. If He created man in His image, mankind will have a special authority over the rest of creation, and he does. True science proves the truth of both.

The designer of a building knows all aspects of the building he designed. Shouldn't he be the first we consult if a problem arises or a change is needed? If he is unavailable, shouldn't we inspect the blueprints he produced? Similarly, the best starting point for science is the Creator of science Himself in consultation with the Scripture He inspired. If we begin with anyone or anything else, we will quickly get lost in our own vain imaginations. Soon, we will envision scientific rooms that don't exist and follow unscientific hallways into blank walls. We will lose sight of the truth because we have failed to consult the Designer of truth and science.

Think back to our discussion of worldviews in Chapter 3. The starting point in understanding any worldview is its assumptions about God. Since every aspect of the biblical worldview begins with God, we shouldn't be surprised that the biblical view of the creation and, therefore, of science also begins with its Creator, the immutable, self-existent, infinite God. We must see the creation through His eyes to reach an accurate understanding of it, to begin to plumb the depths of the science imbedded in it.

> Both religion and science need for their activities the belief in God, and moreover God stands for the former in the beginning, and for the latter at the end of the whole thinking. For the former, God represents the basis, for the latter–the crown of any reasoning concerning the world-view.
>
> DR. MAX PLANCK, 1918 NOBEL PRIZE IN PHYSICS

The first words of the Bible practically hit us over the head with the importance of beginning with God. "In the beginning, God..." (Genesis 1:1). Not only is God the Creator, but He remains sovereign over His creation and over all things. Understanding His love for those He made in His image and the omniscience and omnipotence He used to express that love through His creation adds necessary depth to our understanding of the creation and, thus, of science.

The better we know the God of Scripture, the better we will understand and appreciate everything in His creation.[1] God, not His creation, is infinite, eternal, absolute truth, self-existent, immutable, and sovereign. Let's begin by glorying in these aspects of His unique and perfect essence while comparing His perfection to the limitations of the created realm.

---

1. I recommend studying the biblical revelation about God's nature and His perfections (attributes). I teach an entire seminary course on this called Theology Proper. It can be accessed at http://www.forhisglorynm.com/index.php/seminary-courses/theology-proper.

> We ought, whenever we speak of God, and of His attributes, to stand in great awe.
>
> ROBERT BOYLE, FIRST MODERN CHEMIST

GOD IS INFINITE

LIMITLESS, ENDLESS, IMPOSSIBLE TO MEASURE

The biblical viewpoint states that God is infinite, that He transcends all boundaries.[2] He is endless perfection. The creation is not infinite and is observably not perfect but has limits and boundaries. Some doubt these boundaries since none have been observed, even with the most powerful telescopes, but their view is marred by their lack of omniscience, a trait God has in abundance. His omniscience is as infinite as all His perfections.

THE TAKE AWAY: God's power, presence, knowledge, existence, in fact His every perfection (attribute) are unending but complete, encompassing all perfection eternally.

> Our knowledge can only be finite, while our ignorance must necessarily be infinite.
>
> SIR KARL POPPER, PSYCHOLOGIST

GOD IS TRANSCENDENT

SEPARATE AND INDEPENDENT

God as Creator is separate from, distinct from, and independent of the creation. The creation is separate from but not independent of its Creator. Eternal God has always existed but the creation had a beginning. God as omnipresent exists within but also outside of His creation.

> The statistical probability that organic structures and the most precisely harmonized reactions that typify living organisms would be generated by accident, is zero.
>
> DR. ILYA PRIGOGINE, 1977 NOBEL PRIZE IN CHEMISTRY

THE TAKE AWAY: God could exist without the creation, but the creation could not exist without God.

2. Psalm 145:3; Isaiah 40:15, 17; Romans 11:33-34.

GOD IS ETERNAL

OUTSIDE OF TIME, WITHOUT BEGINNING OR END

Only the one true God, the God of the Bible, is eternal, without a beginning or end.[3] Scientists today have conflicting views about the origin of the creation. Many believe that everything began with the Big Bang, a popular theory that has serious difficulties.[4] Many also believe that nature is eternal. (I don't know about you, but the thought of eternal poison ivy and mosquitoes is not appealing!) The creation is not eternal, nor did it begin with a self-initiated bang. God created it at a point in history and will end it as we know it at the right time in the future. If we want to understand eternal truths, we must begin by understanding eternal God and His teachings.

THE TAKE AWAY: Only God is eternal, having neither a beginning nor end. In fact, He is the beginning and the end![5]

> Astronomers now find they have painted themselves into a corner because they have proven, by their own methods, that the world began abruptly in an act of creation to which you can trace the seeds of every star, every planet, every living thing in this cosmos and on the earth. And they have found that all this happened as a product of forces they cannot hope to discover.... That there are what I or anyone would call supernatural forces at work is now, I think, a scientifically proven fact.
>
> **DR. ROBERT JASTROW, NASA PHYSICIST**

GOD IS ABSOLUTE TRUTH

IN ACCORDANCE WITH REALITY

Truth in science is based mainly on observations of the natural realm. Even scientists seldom claim that scientific truth is absolute. In contrast, Scripture states, as we have seen, that God[6] and His Word[7] are truth, absolute truth. Because God Himself is truth, truth is personal, not abstract. Scientific truth can be a reflection of divine truth but can

---

3. Psalm 90:2, 93:2, 102:12, 24-28; Isaiah 57:15; Micah 5:2; Romans 16:26.
4. An excellent critique of the big bang theory from a biblical worldview is the book *Dismantling the Big Bang: God's Universe Rediscovered*, Alex Williams and John Hartnett, Master Books, 2005.
5. Revelation 21:6.
6. John 3:33, 14:6; Romans 3:4, 7; Hebrews 6:18; Colossians 2:3.
7. John 17:17, 5:33; Galatians 5:7; 1 John 2:21.

never be absolute truth in itself. We can learn scientific truths that reflect God's absolute truth, but truth remains subject to God. The scientific truths that we come to understand are but a fragment of the vast knowledge God has imbedded in creation. Remember that 95% unknown estimate in astrophysics?

> Revelation from God is necessary for the understanding of absolute truth (otherwise absolute truth is not obtainable–as there will always be the likelihood of unknown factors that could affect it).
>
> **DR. JOHN G. LESLIE, BIOCHEMISTRY, MEDICINE, BIBLICAL ARCHEOLOGY**

THE TAKE AWAY: We can learn scientific truths that reflect God's absolute truth, but truth remains subject to God.

GOD IS SELF-EXISTENT

EXISTING TOTALLY INDEPENDENT

Only God is self-existent,[8] needing nothing outside of Himself. He created the universe, not because He needed anything, but simply because He was pleased to create it. Had He not created it, He would still be as perfect and complete as He always has been. He existed forever before He created us. While He doesn't need us, we definitely do need Him. Creation would not and could not exist without Him. The creation is not self-existent.

> Science is incompetent to reason upon the creation of matter itself out of nothing. We have reached the utmost limit of our thinking faculties when we have admitted that because matter cannot be eternal and self-existent it must have been created.
>
> **JAMES CLERK MAXWELL, PHYSICIST, MATHEMATICIAN**

THE TAKE AWAY: Everything in nature, including mankind, is dependent and inter-dependent and must be upheld and maintained by its Creator.

8. Isaiah 40:13-17; Exodus 3:14; Acts 17:25; Colossians 1:17.

GOD IS IMMUTABLE

UNCHANGING

Only God is immutable, unchanging.[9] Only He cannot and will not ever change. We can trust Him to remain forever perfect. The man-made laws and physical constants of nature are subject to His will. They are not as fixed as most scientists assume but fixed only as the Creator desires. He can intervene at any point and change them slightly or even radically. The miracles recorded in Scripture represent instances He chose to override the constancy and processes we normally observe in the natural realm. Scriptures such as 2 Peter 3:3-6 reveal times He intervened miraculously to override the constancy normally observed in the natural realm. He promises to intervene even more dramatically in the future.[10] During different time periods, which theologians call dispensations, not only did He make radical spiritual changes, but in some cases, radical physical changes as well.[11] These changes have far-reaching implications for science.

> The laws of physics that we regard as 'sacred,' as immutable, are anything but.
>
> DR. JOHN ARCHIBALD WHEELER, THEORETICAL PHYSICIST

THE TAKE AWAY: Creation changes when God changes it, but He never changes.

GOD IS SOVEREIGN

POSSESSING ULTIMATE POWER AND AUTHORITY

Scripture teaches that God is the only sovereign ruler over all things, including His creation and history.[12] Both are subservient to Him. Nothing in the creation operates without His orchestrating or at least permitting it. The implications of God's sovereignty have far-reaching consequences for science. Scripture records hundreds of times the Creator already has and will in the future exercise sovereign control over His creation, altering it radically to reflect His desires.[13] Secular

9. Psalm 102:26-27; Malachi 3:6; Hebrews 1:11-12, 6:17.
10. 2 Peter 3:7, 10
11. 2 Peter 3:3-6.
12. Psalms 103:19, 47:7, 115:3; Isaiah 46:10-11; Daniel 4:25, 35; Revelation 19:6, 16.
13. Some of these examples will be described in following chapters and are detailed in Appendix B.

science reaches wrong conclusions because it assumes natural processes operate independently of a god or forces outside the natural realm. Their conclusions are suspect because their foundational assumptions ignore the Creator. Only He has absolute control over all He created.

THE TAKE AWAY: We have a creation that has temporary laws and physical constants and is dependent on forces outside itself. God is the One who sovereignly orchestrates world history as well as all that takes place in His creation.

> Science is the process of thinking God's thoughts after Him.
> **JOHANNES KEPLER, PIONEER ASTRONOMER, MATHEMATICIAN**

You may have noticed that you do not need to read past Genesis 1:1 to see all of these perfections of God taught or implied in that one verse. "In the beginning God" informs us that God was there before the "the heavens and the earth" (universe) existed, thus teaching us that He is transcendent, eternal, self-existent, and sovereign. The verse at least implies that as Creator ("created") He is also infinite, immutable, and absolute truth. Also implied is that God is omnipotent, omniscient, omnipresent, wise, good, and even gracious. The rest of Genesis and Scripture simply gives us confirmation and the details of these truths.

Clearly, we cannot understand reality unless we first have a working understanding of the sovereign Creator. You will not find this truth in physics or chemistry textbooks. Can you see what you missed during your school years because those who worship naturalism hid it from you?

GOD

- Infinite
- Transcendent
- Eternal
- Absolute truth
- Self-existent
- Immutable
- Sovereign

CREATION

- Limited in all ways
- Separate and distinct
- Created and transformed
- Reflection of truth
- Dependent and upheld
- Temporary Constants
- Servant to Creator

> It seems to me that when confronted with the marvels of life and the universe, one must ask why and not just how. The only possible answers are religious...I find a need for God in the universe and in my own life.
>
> **DR. ARTHUR L. SCHAWLOW, 1981 NOBEL PRIZE IN PHYSICS**

HOW DOES SECULAR SCIENCE COMPARE WITH THE BIBLICAL WORLDVIEW?

A fundamental concept of secular science is that nature has certain laws and physical constants that are unchanging and unchangeable. A scientific law (a law of science) is a statement based on repeated experiments or observations that describe or predict natural phenomena.[14] An example of a well-known one is the law of conservation of energy, which states that energy can be changed into many forms but cannot be created or destroyed. This scientific observation confirms that God is the only Creator. Other examples of the many currently accepted scientific laws include Archimedes' Principle of Buoyancy, Dalton's Law of Partial Pressures, Hooke's Law of Elasticity, Bernoulli's Law of Fluid Dynamics, and Hubble's Law of Cosmic Expansion.

> Those laws [of nature] are within the grasp of the human mind. God wanted us to recognize them by creating us after his own image so that we could share in his own thoughts.
>
> **JOHANNES KEPLER, ASTRONOMER**

A physical constant, also known as a universal or fundamental constant, is that which is believed to be universal in nature and has an unchanging value at all times. The speed of light (186,000 miles per second) is one of the best-known physical constants. Another is the gravitational constant (G = 6.67408 x 10-11 N m2 kg-2) which, as we have seen, still has some mystery attached to it.

Scientists consider scientific laws and physical constants to be fixed, meaning they are always consistent and repeatable. They believe them to be independent and distinct from everything else we know about existence. They are just there. For example, gravity keeps our feet

14. https://en.wikipedia.org/wiki/Scientific_law.

planted on the Earth. We don't understand how it works. That is just the way it does work. Liquid water takes the form of its container, at least on earth. Again, that is just the way water always works.

Thus, in seeking truth, secular scientists believe they can look at nature and discover enduring principles that explain why things always have and always will work as they do today. They then pretend that these principles explain why nature is the way it is. They conclude truth is found through careful observation of the natural realm. They are wrong because their exclusion of the Creator from their observations taints their conclusions.

> A universe without law would be a universe without order, without the possibility of science and the manifestations of an intelligent governor and creator.
>
> **JOSEPH HENRY, PHYSICIST**
> **FIRST SECRETARY OF THE SMITHSONIAN INSTITUTION**

Because secular scientists assume the validity of evolution, they discount all divine accounts of ancient history as myths. They are blind to every piece of data that proclaims evolution's flaws. Secure in their own brilliance, they refuse to even consider that a creator might exist who would know more about his creation than they do. They have an a priori (a fancy term for a preconceived, unsubstantiated) belief in naturalism.

The secular viewpoint, at best, thinks of God as independent of the natural realm, that the laws and physical constants of science exist independent of Him, if He exists at all. They attempt to isolate the

SECULAR SCIENCE

- Nature has fixed laws and physical constants
- Truth is found through observing natural realm
- God is independent of natural realm

BIBLICAL VIEW

- Nature has temporarily fixed laws and physical constants
- Truth is found through study of both the spiritual and natural
- Natural realm is dependent on God

physical realm from all other possible realms, especially any spiritual realm. They claim ideas relating to God and the spiritual realm are merely personal opinions since such claims cannot be verified or tested. Sadly, most people, even most Christians, accept this anti-God perspective without question. It should be noted that many secular ideas cannot be tested either, especially in the area of historical science.

> Electricity is often called wonderful, beautiful; but it is so only in common with the other forces of nature. The beauty of electricity or of any other force is not that the power is mysterious, and unexpected, touching every sense at unawares in turn, but that it is under law, and that the taught intellect can even now govern it largely. The human mind is placed above, and not beneath it, and it is in such a point of view that the mental education afforded by science is rendered super-eminent in dignity, in practical application and utility; for by enabling the mind to apply the natural power through law, it conveys the gifts of God to man.
>
> **MICHAEL FARADAY, PHYSICS, CHEMISTRY SPECIALIZING IN ELECTRO-MAGNETISM**

THE TAKE AWAY: The scientist with a biblical worldview has an advantage over those who are oblivious to or unbelieving in God. If he chooses to see the world through the clear lens of Scripture, he knows that God recorded in Scripture an accurate account of the ancient past and that every mention of the creation in Scripture is grounded in reality. His viewpoint and approach begin with his dependence on God. By scouring the Bible for the divine view of the particular scientific area he is studying, he seeks to let God set the parameters for his scientific study. He places his scientific observations within that divine framework and comes to accurate conclusions.

> The day science begins to study non-physical phenomena, it will make more progress in one decade than in all the previous centuries of its existence.
>
> **DR. NIKOLA TESLA, INVENTOR**

I think you will find the chapters that follow fascinating because they explain the joy and success that come to those who follow the true scientific method.

WHAT ARE THE UNDERLYING ASSUMPTIONS OF SECULAR SCIENCE?

A Christian scientist needs to know the beliefs of secular science, so he won't inadvertently accept false premises as true and waste his time on erroneous postulates. As we have seen, a major foundation of secular science is naturalism, the belief that the natural world is the whole of reality and that no supernatural or spiritual realm exists or affects it. This religious viewpoint begins and ends with nature. Most scientists err by basing their assumptions solely on the natural realm, on the religion of naturalism. This theory is faulty because its main assumptions are faulty.

1. *Uniformitarianism*, the foundation stone for naturalism, asserts that present, observed processes must have been the same in the ancient past. Because this assumption that the present indicates the past cannot be established scientifically and is refuted by Scripture, we should view all scientific pronouncements based on naturalism with a healthy dose of skepticism.

> Things are as they are because they were as they were.
> DR. THOMAS GOLD, ASTROPHYSICIST

2. *Man's reasoning* can discover all truth. Though naturalism extols the perfection of the human intellect, Scripture shows man's reasoning is fallible and, therefore, not a dependable foundation for scientific study. Modern science is biased against the supernatural because it defies human reason. Not only is divine revelation rejected out-right, but many major biblical events are discounted or considered mythical, non-historical, and impossible. To reject such relevant data in scientific inquiries is unscientific, even deceptive.

3. *Evolutionism* builds its foundation for interpreting the past on the failed theory of mutations.[15] The more we learn about evolution, the more weaknesses and inconsistencies we uncover.

15. Refer to Chapter 6.

> On the table in front of me is a reproduction of a beautiful landscape painting of a country scene, dating from about 1770, by Paul Sandby...Evolutionary naturalism is asking us to believe in just such a scenario: a picture without a painter, art without an artist. Let us remember, also, that however good it is, the painting is only a dead, two-dimensional representation of a far more wonderful three-dimensional living reality: the landscape itself, trees, horses, children, sky, sun, and clouds! How wrong and how foolish to praise the work of a human hand and eye yet deny the work of the divine artist who put all things in place and gave existence and skill to that same human hand and eye! Not only does a design imply a designer, a design says something about its designer. As we consider the vastness of deep space, the intricacies of the human brain, the powerful forces holding the nucleus of each atom together, we may conclude that God is indeed immense, great in intelligence and in power. If God is thus, why shouldn't creation take six 24-hour days? He could have done it in six hours or six seconds if He had chosen to. Such a God can do whatever He chooses, whenever He likes, consistent with His own nature.
>
> **DR. STEPHEN TAYLOR, ELECTRICAL ENGINEERING**

WHAT ARE THE UNDERLYING ASSUMPTIONS OF SCIENCE FROM A BIBLICAL VIEW?

One of the main benefits of beginning all scientific studies with God and His revelation in Scripture is that all the fundamental assumptions of science make sense. We know we can depend on them because God's perfect nature underlies them.

THE TAKE AWAY: Through His Word, God establishes the parameters of reality, thereby revealing what to expect when we study the natural realm.

Because we know God is Creator, we would expect the creation to exhibit evidence of being created and to reflect something of its Creator. Because creation had a Creator, we would expect to find some of God's characteristics in it. Because we know mankind was created in His image, we would expect to find a relationship between man and the rest of the creation. Let's see if these expectations are met in the following foundational assumptions accepted by virtually every scientist.

THE PRINCIPLE OF OBSERVATION

GOD MADE THE PHYSICAL REALM OBSERVABLE.

Since God is the Creator, the universe exists as the product of His creation. He made the universe both real and observable. Take a walk outside if you aren't sure.

> My own studies in numerous biochemical control mechanisms, especially in the control of steroid hormone formation (for which I was awarded the higher doctorate, D.Sc.) convince me that all these processes are ordered precisely. This order and the extraordinary complexity are entirely consistent, in my own opinion, with the existence of a Creator, who himself must be capable of creating with such design.
>
> **DR. D.B. GOWER, BIOCHEMISTRY**

THE PRINCIPLE OF GENERALIZATION

GOD MADE IT POSSIBLE FOR US TO OBJECTIVELY OBSERVE THE PHYSICAL REALM.

He formed us to be part of but separate from and above the creation. Because He created us in His image. we have, in a finite measure, some attributes similar to His that allow us to learn from His creation. The Creator is personal, and He created us as personal beings who can have a relationship with Him and His creation.

Because He wants us to use the creation wisely, He gave us an intellect and, thus, the ability to objectively observe the creation, making experimentation possible. Our imaginations lead us to enjoy the process of experimentation. Even so, because of our human limitations and sin natures, including the human tendency to suppress truth, to want to be right even when we are not, our experiments can potentially fail.

Even though God endowed us with the ability to study and understand small portions of His creation, we can never know anything completely because His work is as comprehensive and incomprehensible as He is. He assures us, though, that we can understand and wisely utilize parts of it if we choose to take time to understand His plan and purpose.

> Experiment is the interpreter of nature. Experiments never deceive. It is our judgment which sometimes deceives itself because it expects results which experiment refuses. We must consult experiment, varying the circumstances, until we have deduced general rules, for experiment alone can furnish reliable rules.
>
> **LEONARDO DA VINCI**

THE TAKE AWAY: Our finite senses are always incomplete and limited. God's perfections are always complete and unlimited. The more we see His image revealed in ourselves, the more willing we are to turn to Him for truth about His creation.

THE PRINCIPLE OF PREDICTABILITY

GOD BUILT PREDICTABILITY, CONSISTENCY, STABILITY, MEASURABILITY, AND QUANTITATIVE CHARACTERISTICS INTO THE PHYSICAL REALM, MAKING SCIENCE POSSIBLE.

Even a child knows the sun comes up in the morning, flowers die, and snow is cold. Where does this predictability come from? From its Creator, who has the perfections of faithfulness, immutability, stability, and omniscience. He designed the creation to reflect those characteristics.

Because He is faithful, we can trust Him to maintain His creation. We need never fear flying off into space or having our tomato seeds produce pumpkins. Because He is omniscient, we can rest assured He will always do the ultimate, eternal best for us and all creation. Because He is a lawgiver, we can rejoice that His creation shows law-like characteristics that promote stability.

C. S. Lewis, the brilliant apologist of *Chronicles of Narnia* fame, showed the contradiction in not accepting the Creator as the foundation for scientific assumptions:

> If the solar system was brought about by an accidental collision, then the appearance of organic life on this planet was also an accident, and the whole evolution of Man was an accident too. If so, then all our thought processes are mere accidents—the accidental by-product of the movement of atoms. And this holds for materialists' and astronomers' as well as for anyone else's. But if their thoughts—i.e., of Materialism and Astronomy—are merely accidental by-products, why should we believe them to be true? I see no reason for believing that one accident should be able to give a correct account of all the other accidents.[16]

16. Lewis, C.S., *God in the Dock*, Wm Eerdmans Publishing, 1970, 52-53.

THE TAKE AWAY: Accidents are unpredictable and chaotic. God alone is the Author of the predictability and order so evident in the creation. Again, night always follows day, ocean tides follow the moon's schedule, and the reawakening of spring follows a dormant winter. God wants us to learn about Him through the creation, and so it reflects His consistency and stability.

> Religion, with its theological reflection, builds on faith. Science too builds on faith. How? For successful science of the type we know, we must have faith that the universe is governed by reliable laws and, further, that these laws can be discovered by human inquiry. The logic of human inquiry is trustworthy only if nature is itself logical. Science operates with the faith that human logic can in the long run understand nature's laws and that they are dependable. This is the faith of reason. We scientists work on the basis of a fundamental assumption regarding reason in nature and reason in the human mind, an assumption that is held as a cardinal principle of faith. Yet this faith is so automatically and generally accepted that we hardly recognize it as an essential basis for science.
>
> **DR. CHARLES TOWNES, 1964 NOBEL PRIZE IN PHYSICS**

THE PRINCIPLE OF LOGIC

GOD BUILT LOGIC, ORDER, AND MATHEMATICS INTO THE PHYSICAL REALM.

Since the Creator is absolute truth and knows all things (omniscience), we would expect to see His creation exhibit logic, order, and predictability.

THE TAKE AWAY: Why do we find order everywhere we look in creation from the rotation of the earth to the time of the tides and the length of gestation for every creature? First, God created man, the crowning glory of creation, with the intellectual capacity to reason and process information logically. Second, order and logic are observable in all creation. Is any area of study more logical and precise than mathematics? When has 2 + 2 ever accurately equaled anything but 4?

> ...what is physical is subject to the laws of mathematics, and what is spiritual to the laws of God, and the laws of mathematics are but the expression of the thoughts of God.
>
> **THOMAS HILL, MATHEMATICIAN, ASTRONOMER, INVENTOR**

Mathematics is a great tool discovered by man, not invented by man but discovered by him. As Galileo said, "Mathematics is the language in which God has written the universe."[17] Where did the logic and order of mathematics and the rest of creation originate? From the God of order, of course. If an orderly Creator did not create all things, how could chaos change to orderliness as evolution declares? Logic tells us that order is never the result of chaos. Has your cat litter box or the grease spot on the garage floor ever cleaned itself? I think not.

Until scientists acknowledge that only the God of order could have created an orderly, predictable world, science will remain chaotic, always needing revisions as secular theories continually fall apart upon examination.

> The ghostly presence of virtual particles defies rational common sense and is nonintuitive for those unacquainted with physics. Religious belief in God, and Christian belief that God became Man around two thousand years ago, may seem strange to common-sense thinking. But when the most elementary physical things behave in this way, we should be prepared to accept that the deepest aspects of our existence go beyond our common-sense intuitions.
>
> **TONY HEWISH, 1974 NOBEL PRIZE IN PHYSICS**

THE PRINCIPLE OF VERIFICATION

INFORMATION EXISTS IN THE PHYSICAL REALM BECAUSE GOD PUT IT THERE.

The entire enterprise of science involves processing, organizing, and communicating information. God gave those made in His image the ability to observe, process, organize, and communicate the information He built into the creation, information that reveals something about Himself. He desires that we recognize His information as absolute truth and see His glory in it. Simply look out your window and behold the wonders of His creation—unless you live in the city, of course. Then you see man's not always wise utilization of the creation. Even in the worst ghetto though, you can still see the expanse of the heavens spread out above you and marvel at its beauty and diversity.

17. Jones, Tudor Powell, *Creative Learning in Perspective*, 1972, quoted in https://www.quora.com/Where-did-Galileo-say-Mathematics-is-the-language-with-which-God-has-written-the-universe.

> If the omniscient author of nature knew that the study of his works tends to make men disbelieve his Being or Attributes, he would not have given them so many invitations to study and contemplate Nature.
>
> **ROBERT BOYLE, THE FIRST MODERN CHEMIST**

The scientific method is based on these five elementary assumptions. The principle of observation assumes a real creation exists to be observed. The principle of generalization assumes relationships in the creation can be organized into hypotheses. The principle of predictability recognizes the orderliness of the creation, which makes scientific laws possible. The principle of logic assumes information exists to be processed and that man has the ability to formulate hypotheses. The principle of verification assumes a repeatability, consistency, and logic in the creation. Without these assumptions, scientific study would be impossible. Because of them, both observational and historical science can be quantified. Even though historical science deals with one time occurring events, the traces left by the events can be evaluated using the scientific method.

We could discuss other unprovable scientific assumptions, but these serve to make our point. Secular scientists accept these five divinely established assumptions by faith because they work; however, they refuse to recognize God as their Author. Most scientists don't even think about these assumptions. They simply base their work on them, never considering they are standing on a biblical foundation.

> I build molecules for a living. I can't begin to tell you how difficult that job is. I stand in awe of God because of what he has done through his creation. My faith has been increased through my research. Only a rookie who knows nothing about science would say science takes away from faith. If you really study science, it will bring you closer to God.
>
> **DR. JAMES TOUR, NANOSCIENTIST**

THE TAKE AWAY: Every secular scientist is a beneficiary of the early roots of modern science that were based on the biblical worldview with God as the ultimate authority. We could say secularists have borrowed our biblical assumptions and scientific advancements without thanking the Creator or us. They are welcome!

WHAT CAN WE CONCLUDE?

The purpose of all things, including science, is to glorify God.[18] Ignoring the Creator is at the heart of rebellion against Him and always results in distorted thinking. As we would expect, the study of science, when divorced from the Creator, inevitably results in a distorted view of the creation. Our speculations even about science are suspect if we exclude God from them.

This divorce from divine reality has long-range spiritual ramifications. Notice the Apostle Paul's evaluation:

> For even though they knew God, they did not honor Him as God or give thanks, but they became *futile* in their speculations, and their foolish heart was *darkened*.
>
> Romans 1:21

THE TAKE AWAY: When we exclude God from our thoughts, we condemn ourselves to trying to find truth in the dark without the only Light that can reveal it. On the other hand, the study of science with the Creator at the center not only results in superior conclusions but receives His blessing. We understand the creation better because the Creator reveals aspects of Himself to us. We glorify Him when we make Him the focus of our study.

Maybe, right now, we should meditate on our Creator's words as revealed to the prophet Jeremiah.

> Thus says the Lord, "Let not a wise man boast of his wisdom, and let not the mighty man boast of his might, let not a rich man boast of his riches; but let him who boasts boast of this, that he understands and knows Me, that I am the Lord who exercises lovingkindness, justice, and righteousness on earth; for I delight in these things," declares the Lord.
>
> Jeremiah 9:23-24

18. 1 Corinthians 6:20; 1 Peter 4:16.

All other scientific foundation stones are built upon this one. If we do not place Almighty God at the center of our every endeavor, we cannot succeed in life and the spiritual life.

> The clear evidence of deliberate design in living organisms strongly confirms our faith in God's Word. Psalm 104:24 states: "O LORD, how manifold are thy works! in wisdom hast thou made them all: the earth is full of thy riches." God's creation clearly reflects the infinite wisdom which He used to design and create it. The orderliness of living things and their mind-boggling complexity are surely unmistakable indications that this creation did not come about by random and disorderly chance processes.
>
> **DR. JOHN P. MARCUS, BIOLOGICAL CHEMISTRY**

Lift up your eyes on high
And see who has created these stars,
The One who leads forth their host by number,
He calls them all by name;
Because of the greatness of His might and the strength of His power,
Not one of them is missing.

ISAIAH 40:26

Do you not know? Have you not heard?
The Everlasting God, the Lord, the Creator of the ends of the earth
Does not become weary or tired.
His understanding is inscrutable.

ISAIAH 40:28

ISAIAH, STATEMAN AND PROPHET

CHAPTER 10

SCIENTIFIC FOUNDATION STONE #2

THE NATURE GOD CREATED NOT EVOLUTION

Our second foundation stone focuses on the reality that God created the universe. Again, we go to Genesis 1:1 those truth-packed first words of the inspired Scripture. "In the beginning God created the heavens and the earth." These words have been called "the simplest, yet most profound, words ever written."[1] So true! God alone created all things, and He did so through the power of His Word. "Then God said…"

Our first stone admonished us to remember that God is the source of everything, including all correct information. To understand our second stone, we will adhere to Stone #1 and again focus on Almighty God, this time in His role as Creator. The creation account of the universe in Genesis 1 will be our source of firsthand information, firsthand from God that is. His insights about the creation in this short chapter alone are astounding.

I've been looking forward to adding this truth to your understanding of real science. I remember my excitement when the correlation between Scripture and science began making sense to me. I hope you will find the following revelation of their co-dependency equally fascinating.

1. Morris, *The Biblical Basis for Modern Science,* 77.

WHY DOES GENESIS 1 HAVE SO MANY CONTRADICTORY INTERPRETATIONS?

Genesis 1 has been under attack for years, probably since soon after Moses wrote it. Modern, frequently blasphemous re-interpretations of the first chapter of the greatest book ever written—and the only one ever inspired—have often been in response to new scientific theories. We don't have time to detail them all, but we need to consider a few recent ones that help explain naturalism's stranglehold on science today.

When modern science began to depart from a biblical worldview and gain high credibility in the culture, it came into increasing conflict with the Bible in general and Genesis in particular. Before the eighteenth century, most theologians and scientists assumed a biblical worldview.[2] The founding of America showed that many in every level of society accepted and followed (at least in public) the principles of Scripture.

> It is error only, and not truth, that shrinks from inquiry....The Almighty lecturer, by displaying the principles of science in the structure of the universe, has invited man to study and to imitation. It is as if he had said to the inhabitants of this globe that we call ours, "I have made an earth for man to dwell upon, and I have rendered the starry heavens visible, to teach him science and the arts. He can now provide for his own comfort, and learn from my munificence to all, to be kind to all, to be kind to each other....The mere man of pleasure is miserable in old age, and the mere drudge in business is but little better, whereas, natural philosophy, mathematical and mechanical science, are a continual source of tranquil pleasure, and in spite of the gloomy dogmas of priests and of superstition, the study of these things is the true theology; it teaches man to know and admire the Creator, for the principles of science are in the creation, and are unchangeable and of divine origin.
>
> **THOMAS PAINE, SUPPOSEDLY THE LEAST CHRISTIAN PATRIOT LEADER IN THE AMERICAN REVOLUTION**

2. For an excellent summary of the history of this controversy of a biblical worldview and the development of secular science, see Terry Mortensen's paper "The Early 19th Century British 'Scriptural Geologist': Opponents of the Emerging Old-Earth Theories of Geology" in the Proceeding of the International Conference on Creationism, Volume 5, Article 45, pages 539-550. https://digitalcommons.cedarville.edu/icc_proceedings/vol5/iss1/45/?utm_source=digitalcommons.cedarville.edu%2Ficc_proceedings%2Fvol5%2Fiss1%2F45&utm_medium=PDF&utm_campaign=PDFCoverPages

During the 1700s, however, rationalism began creeping into all areas of the culture, in Europe first and eventually in America. Rationalism declares that autonomous man can figure out all things, including science, by simply exercising his mental capability without any help from God or His revelation in Scripture.

Rationalism gained momentum in the 1830s when Scottish geologist Charles Lyell (1797-1875) proposed the principle of uniformitarianism to interpret the geological record, a direct attack on the young age of the earth as recorded in Genesis. Though, at first, many leading scientists resisted his ideas and defended Scripture, ironically, many theologians began rejecting the young earth of Genesis and adopting the old earth idea of uniformitarianism. This satanic switch opened the door for Darwin's evolutionary processes to gain credibility later in the century. By then, a loosely tied group of "British Scriptural Geologists," a few in America, became virtually the only ones resisting this movement away from the Bible.[3]

HISTORICAL ATTACKS ON SCRIPTURE

Before 1700 → Biblical worldview
Mid 1700 → Voltaire attacks Christianity
About 1700 → Hume attacks design
1795 → Hutton: Principle of Uniformity
1830 → Lyell: Uniformitariansim
1860 → Darwin: Natural Selection

By the end of the 1800's, most scientists and theologians had abandoned the Genesis record of the creation to embrace Darwin's opinions. Many, even pastors, declared science proved the Bible had errors and was not reliable for history or science.[4] Science became increasingly confident in long ages and naturalistic ideas partly because theologians did not believe or trust the biblical descriptions in Genesis 1. They also did not have the benefit of the modern information compiled in this book.

> What you learn from a life in science is the vastness of our ignorance.
> **DR. DAVID EAGLEMAN, NEUROSCIENTIST**

In the early 1900s, theologians began scrambling for new interpretations of Genesis while still trying to find ways to uphold the inerrancy

3. Ibid.
4. For more details on the history of the rise of humanistic and evolutionary influence see Morris, Henry, *The Long War Against God*, Master Books, 2000, 93-150.

of Scripture. Attempts to make the Bible agree with secular science, a satanic trick called accommodation, remain the centerpiece of most denominational creeds today. Even some conservative theologians like Charles Hodge tried to accommodate Genesis with the ideas of naturalistic science. They believed in inspiration and inerrancy but did not have the answers to scientific issues that have been solved since their time. They didn't wait for science to catch up to Scripture but imposed science over Scripture. We must be careful not to make the same mistake.

Theistic evolution is the best-known early compromise that theologians made with secular science. It combined the Bible on one hand with so-called evolutionary science on the other by declaring that God used evolution as His means of creation. They ditched Scripture to accommodate popular secular opinions. Specifically, they attacked the truth that God created the universe in six literal solar days.

The most prominent accommodating view, the day-age theory, said that the days of Genesis 1 were not literal days but long geological ages. This reinterpretation of Scripture allowed God to be Creator but over long, evolutionary ages of time. It denied that God meant a literal day when He said so in the clearest Hebrew possible. By doing so, it challenged God's communication ability. How can God be God if He can't communicate a simple fact like the meaning of a day? Satan seemed to be winning the battle for men's minds.

> Science tells me that *evolution is certainly not scientific*, while *creation is not in disagreement with what is truly scientific*. Hence, creation is the more acceptable account of origins. Since creation requires a supernatural, omnipotent Creator, and the Bible is the only convincing source of who this creator God actually is, then the biblical account of creation must be accurate in every detail, including six 24-hour days for completion from beginning to end.
>
> **DR. EDWARD A. BOUDREAUX, THEORETICAL CHEMISTRY**

Another attempt to solve these conflicts between secular science and Scripture was the gap theory, which put all geological and evolutionary ages into a period of undeterminable length between Genesis 1:1 and 1:2. This view allowed the Bible to be taken literally in its traditional interpretation by sweeping billions of evolutionary years under the rug. The Scofield Reference Bible popularized this theory in an explanatory note. For many decades, most Bible believing churches and theologians accepted this accommodation theory, never stopping to realize its insidious attack on the inerrancy of Scripture. Besides undermining God's Word, this theory violates a fundamental hermeneutical principle, that of avoiding reading ideas not there into the biblical text (called eisegesis).

While these views still persist in some quarters, they have lost much support. Today, the most accepted apostate view is progressive creationism, which is held by most old earth creationists.[5] This view claims to reject evolution and holds that God created the universe; however, it rejects the days in Genesis 1 as literal twenty-four-hour days and allegorizes most of the chapter. It holds that God periodically interjected acts of creation over billions of years. It not only accepts the current opinion of a 4.5-billion-year-old earth and a thirteen to twenty-billion-year-old universe but also accepts much of naturalistic historical science. That means it denies Adam's original sin and its curses, the universal Flood, and the tower of Babel.[6]

> Evolutionists argue that given sufficient time proteins could form by chance. But, for a protein of just the minimum 100 amino acids (many are tens of thousands long), for all to be left-handed, and assuming that it did not matter which order they appeared in, it's like flipping a coin 100 times and resulting in all heads. The number is astronomical.
>
> **DR. KEN FUNK, PEPTIDE CHEMISTRY**

THE TAKE AWAY: All accommodating views, including these, have two fundamental problems. First, they abandon sound exegesis (long-accepted interpretive procedures) and the plain meaning of the words of Scripture while twisting the biblical text to agree with current secular science. Second, even secular science rejects their accommodations as unscientific! For these two reasons alone, we should reject every attempt to accommodate Scripture with current scientific pronouncements.

5. A good description of this view along with the views called theistic evolution and young earth creationism is *Three Views on Creation and Evolution*, J.P. Moreland & John Mark Reynolds, general editors, Zondervan Publishing House, 1999.

6. One of the main proponents of this view today is Hugh Ross, an astronomer turned theologian. One of his books that I take issue with is *The Fingerprint of God*, Hugh Ross, Promise Publishing Co., 1989. See Chapter 19 of this book.

ACCOMMODATING VIEWS OF CHRISTIANS

THEISTIC EVOLUTION → God used evolution
DAY AGE THEORY → Geological ages
GAP THEORY → Gap in Genesis 1:1-2
PROGRESSIVE CREATION → Creation injections

Those who interpret Genesis 1 by forcing it to fit current scientific theories give scientists more authority than God and science more authority than Scripture. This approach creates its own problems because scientists are fallible, and science can and will change. Every one of these scientific changes causes every subjective view of the Bible based on them to change as well.

THE TAKE AWAY: Worldly theologians must change their Scriptural interpretations often now that science has uniformly rejected the biblical worldview. Since they don't allow the Bible to provide the foundation and parameters for science, they don't understand either the Bible or science.

Every scientific study must begin and conclude with God and His revelation, not with secular science (Scientific Foundation Stone #10). Scientists must allow Scripture to speak for itself and avoid imposing current scientific ideas on the biblical text. We, too, must maintain the inspired Scriptures, God's revelation, as our authority over all areas of life, including science. We must trust the absolute truth in Scripture rather than the speculations of any fallible human—even if his lab coat makes him look brilliant!

Decision time is here. Will we adhere to the foundation and parameters of Scripture and succeed in science as well as in every area of life, or will we pursue satanic lies and fail miserably? God gave us the volition and intelligence to make well informed decisions. Will we choose to believe Him or fallible scientists and theologians made in His image? Our decisions will affect us now on earth and for all eternity.

> I would rather have questions that can't be answered than answers that can't be questioned.
>
> **DR. RICHARD P. FEYNMAN, 1965 NOBEL PRIZE IN PHYSICS**

To be clear, no accommodation theory is supported by either good science or sound biblical exegesis. I am writing this book to not only show why we must reject all these accommodating views but also to give the best approach to understanding both science and Scripture. If we don't allow the Bible to provide the foundation and parameters for science, we can never understand either the Bible or science.

Many other attempts have been made to harmonize the Bible with the scientific theories of the day, but none are satisfying either theologically or scientifically. You will need to keep reading to get the solution to this problem.

WHAT IS THE BEST WAY TO INTERPRET THE BIBLE?

Over the centuries, Genesis 1 has been variously interpreted, often in response to new scientific theories, evolution being the most popular current one. With so many theologians and laymen distorting Genesis 1 in so many ways, you are right to ask, "How can I know I have a proper interpretation?" The answer is simple. If the interpreter uses the same sound interpretive principles to interpret Genesis 1 as he uses to interpret every other passage in Scripture, you have an accurate interpretation. So, what are these sound interpretive principles?

A time-tested method of interpreting the Bible, the Grammatical-Historical-Contextual (G-H-C) approach,[7] has produced accurate, reliable, and unchanging interpretations for many centuries. Rare differences in interpretations have arisen over a few passages and words, but overall, conservative Christians agree with the interpretations arrived at through G-H-C, especially those concerning the fundamentals of the faith.

G-H-C attempts to understand the original author's intended meaning by using all the conventions of language (grammatical), the facts of history (historical), and the framework of context (contextual). Let's not forget that while human authors wrote the Bible, a divine Author inspired every word. Inspiration means the Holy Spirit inspired the content while the author chose the words to convey it. Of course, we

7. This approach is abbreviated as "Grammatical-Historical" in much of the literature relating to hermeneutics. I teach an entire seminary course called "Hermeneutics and Bible Study Methods" that details this approach. It can be accessed from my website ForHisGloryNM.com.

can trust the Bible to tell us the truth! God inspired every word to be accurate and complete.

GRAMMATICAL-HISTORICAL-CONTEXTUAL METHOD

1. Laws of Grammar 2. Facts of History 3. Framework of Context

The G-H-C approach, which is objective and, therefore, scientific, is often referred to as the literal approach, literal in the sense of understanding the text according to the normal, everyday usage of the language, including the possibility of metaphors when clearly indicated. G-H-C uses well-established interpretive principles, known as biblical hermeneutics, plus a process for applying these principles to any biblical text, known as biblical exegesis.

Exegesis is an objective and, therefore, scientific process that attempts to extract from the text the meaning intended by the original author.[8] Fallible man's tendency is to avoid accurate exegesis and make the biblical text say what he wants it to say so he can add weight to his own opinions. This subjective process, called eisegesis,[9] inserts into the text a meaning unintended by the original author but desired by the interpreter. Eisegesis introduces the interpreter's opinions, thus violating the G-H-C method.

In both biblical hermeneutics and science, the first step after prayer is observation.

1. Biblical hermeneutics begins by making many observations about the words, syntax, history, and context of the biblical text. This step mirrors the first step in the scientific method, observation of the creation.

2. The second step of hermeneutics is formulating generalizations based on those observations. A good interpreter derives a preliminary interpretation from a biblical text just

8. The term exegesis comes from a Greek term (ἐξηγέομαι), meaning to explain or illuminate. The preposition ἐκ attached to the beginning of the word means out or out from, giving the sense of extracting the meaning out from a passage. The term occurs in John 1:18, which speaks of Jesus *explaining* the nature of the invisible God the Father.

9. The term eisegesis has the Greek preposition εἰς, meaning in or into, and gives the idea of reading a preferred meaning into a text.

as a good scientist generates generalizations about the data under investigation.

3. The third step is verification. In Bible study, any preliminary interpretation must be substantiated and refined through an examination of commentaries and, more importantly, similar information in other verses throughout the Bible. This substantiation helps ensure that the final interpretation of a passage accurately reflects the meaning of the original writers. In science, this third stage includes testing and retesting every hypothesis against all others to ensure accuracy.

THE TAKE AWAY: Anyone who carefully uses the G-H-C approach can arrive at the intended meaning of not only the original author but the divine one, the Holy Spirit. Because of divine inspiration, the meaning of both is the same. When this method is used properly, theologians reach a high level of agreement because exegesis scientifically and objectively tests every interpretation.

As a real-life example of correct exegesis, those early scientists mentioned in Chapter 3 took the three major steps of their study of Scripture (observing, hypothesizing, and verifying) and applied them to the study of God's revelations in the creation. By applying their biblical hermeneutic to their scientific endeavors, they developed remarkable, life-changing scientific conclusions. Both their biblical and scientific conclusions were objective and verifiable. Not surprisingly, both agreed.

> I am proud to be a Christian. I believe not only as a Christian, but as a scientist as well. A wireless device can deliver a message through the wilderness. In prayer the human spirit can send invisible waves to eternity, waves that achieve their goal in front of God.
>
> **GUGLIELMO MARCONI, 1909 NOBEL PRIZE IN PHYSICS**

WHAT DOES GENESIS 1 REALLY MEAN?

Many interpreters of Genesis 1 read pseudo-scientific theories into the text and, therefore, misinterpret it. Even many excellent evangelical scholars who interpret from the G-H-C method are inconsistent when it comes to the early chapters of Genesis. They seem so intimidated by secular science that they try to fit unproven and unprovable scientific theories into the text. Such secular explanations cannot be found in the biblical text and were not intended by the original authors. They are examples of the eisegesis that the exegetical process attempts to prevent.

We must apply the proper approach, G-H-C, to the interpretation of Genesis 1 just as we do to every other part of God's Word. When we do, we can accurately determine the intended meaning of the original author, Moses, under the inspiration of the Divine Author, the Holy Spirit. Once we are confident in our understanding of the text, we can adjust our understanding of science to correspond to it. This puts God's revelation as the authority over science rather than the other way around. It allows absolute truth to help us understand scientific truth, which is never absolute. Remember once again that only the Creator thoroughly understands His creation.

> Truth is the same thing to the understanding as music to the ear and beauty to the eye. The pursuit of it does really as much to gratify a natural faculty implanted in us by our wise Creator as the pleasing of our senses: only in the former case, as the object and faculty are more spiritual, the delight is more pure, free from regret, turpitude, lassitude, and intemperance that commonly attend sensual pleasures.
>
> **DR. JOHN ARBUTHNOT, EARLY MODERN MATHEMATICIAN, PHYSICIAN**

As I have explained in many classes over the years, Genesis 1 is not complicated. Moses wrote it in a relatively simple, clear, and straightforward style. I encourage you to consult a twelve-year-old when you study it because he/she can better explain it to you than most with a PhD in theology or science. Try it.

If you are a twelve-year-old, study Genesis 1 like you would read your video game instructions. You say that you do not read the instructions? Read Genesis 1, pretending you are reading the game instructions because your parents may need your help. You'll quickly realize that God wants you to understand Genesis as well as the whole canon of Scripture, and so He made it easy enough for a thoughtful child to understand and appreciate it.

I am not going to give an exegetical exposition on the entire text since many excellent commentaries do just that. (Some of you can heave a sigh of relief. The rest can study one or more of these books.[10]) We will limit our discussion to Genesis 1 as the foundation for all science.

Let's begin with a quick overview of the chapter so we have the context for understanding the science contained in it. After the introductory and summary statements of Genesis 1:1 and the pre-creation conditions of Genesis 1:2, Genesis 1:3 to 31 expands the six days of God's creative work. Genesis 2:1 to 3 record that God rested on the seventh day. No, He was not tired! We would have been, but He is God. Creation was perfect and complete, so He stopped creating.

What distinctive acts did God perform on those six days? On the first day of creation week, He created light (1:3-5). On the second day, He created an expanse to separate the water covering the earth (1:6-8). Your twelve-year-old will understand immediately that outer space separates the water on earth from the water somewhere out on the edge of the universe. He is right.

On the third day, God rearranged the waters on earth and created a landmass (1:9-10). In a second creative act, He fashioned plant life to fill the landmass (1:11-13). On the fourth day, He created the two major lights that we see in the sky as well as stars (1:14-19). Your twelve-year-old can tell you God created the sun and the moon and, as a spectacular addition, a septillion stars as well (that's at least 1,000,000,000,000,000,000,000,000 stars filling our skies!).[11] On the fifth day, He created all the creatures that live in the water and fly in the air (1:20-23). On Day Six, He gave life to every other living creature, including man (1:26-27). Notice how helpful and uncomplicated the G-H-C approach is!

God could have spoken all creation into existence with one word. He could have thought it into existence even. Why didn't He? Within each

10. I will recommend only four among many excellent commentaries in the order of priority. *Creation & Blessing: A Guide to the Study and Exposition of Genesis*, Allen P. Ross, Baker Books, 1988. *Genesis: A Commentary*, Bruce K. Waltke, Zondervan, 2001. I differ from Allen Ross on some conclusions on Genesis 1 but he does take the word for "day" literally in his commentary. Bruce Waltke accepts the old earth view of creation in Genesis 1 and I would differ with him on some other points as well. Apart from these disagreements these two are excellent commentaries. Two young earth commentaries have a strong science emphasis. *The Genesis Account* referenced earlier and *The Genesis Record: A scientific & devotional commentary on the book of beginnings*, Henry M. Morris, Baker Book House, 1976.

11. Star basics, science.nasa.gov/universe/stars; accessed 20 March 2024. Also: How many stars are there in the Universe? esa.int; accessed 20 March 2024 as found at creation.com/stars-purpose-people?utm_campaign=infobytes_us&utm_content=%5BDigest%5D%20Space%20telescope%20images%20contradict%20big%20bang%20predictions&utm_medium=email&utm_source=mailing.creation.com&utm_term=Fortnightly%20Digest%20-%202024.05.03

day, He taught us profound lessons about His character, plan, purpose, and love for us. Further, He used creation week to set a pattern for a six-day workweek with one day set aside for rest and worship. Feel free to take time for worship right now.[12]

> The only question I have is: why did He take so long–'six days'? Throughout Scripture God is shown as an instant Creator, not in six days or 20 billion years. To give but a few examples: a fish to swallow Jonah, the sun turning back, the parting of the Red Sea, water turning into wine, stilling of the storm, raising the dead, healing, etc. Therefore, why did He use six days? Scripture gives us the answer. He established a blueprint for our life, six days of work and one day of rest (Exod. 20:8). It must have been painstaking for God to slow down to our pace, to six days. It would have been more His nature to create everything instantly. I have often considered belief in Genesis chapters 1 to 11 as the "acid test" of believing in God, and in the salvation through Jesus Christ.
>
> **DR. JOHN K. G. KRAMER, BIOCHEMISTRY**

The chronicle of those six creative days (Genesis 1:3-31) follows a clear pattern. Five elements unify all six days, and two elements unify several of them.

1. The account of each day begins with the words "And God said...." Several of the accounts repeat that phrase, thus emphasizing God did the creating by speaking things into existence. He created from nothing, something only He can do. (1:6, 9, 11, 14, 20, 24, 26)

2. A divine decree described God's creative work on each of the six days. Each "let there be..." is followed by the creative act (1:3, 6, 9, 11, 14, 15, 20, 22, 24, 26).

3. Each of these divine fiats was precisely fulfilled. Usually, the description of God's creative command is followed by the words "and it was so" (1:7, 9, 11, 15, 24, 30). God always accomplishes His desires.

4. All six days contain a second creative act. On Day One, God separated the light portion of the day from the dark portion (1:4). On Day Two, He separated the waters above from the

12. This is expounded in the fourth commandment in Exodus 20:8-11.

waters below (1:7). On Day Three, He named the dry land and the waters (1:10). On Day Four, He placed the two lights in the expanse (1:17). On Day Five, He "made the stars also" (1:16). On Day Six, He created mankind (1:26).

5. On all except Day Two, God evaluated His creative work. On Day One, He declared "the light was good" (1:4). On Days Three, Four, and Five, God saw that "it was good" (1:10, 18, 21). On Day Six, He made His final evaluation that all was "very good" (1:31).

6. The sixth element, which occurred only on the last two of the six days, was a divine blessing. God blessed the creatures He created on Day Five and the creatures and man He created on Day Six (1:22, 28).

7. A terminating phrase identified each day. Each account ends with and "there was evening and there was morning" (1:5, 8, 13, 19, 23, 31).

THE TAKE AWAY: We should note that only mankind was created in God's image, an extremely significant and wonderful differentiation. We get the sense mankind was the most important part of God's creation because He also gave a purpose to those He created in His image (1:28-30). He finished by evaluating all six days of creation as very good (1:31).

Finally, on the seventh day, God rested from His creative work (2:1-3) because nothing could improve it. As we've seen, He rested to set a pattern for a six-day workweek with one day set aside for rest and worship. Feel free to take time for worship right now.

Now you know the essence of that seemingly daunting first chapter of the Bible. Why have so many made this simple description so complicated? Because they don't think as God desires them to think, with the simple, straightforward, humble faith of a twelve-year-old.

> A child in a Christian school class asked her teacher, "How could anybody create everything in six days from nothing?" Another very discerning young student blurted out, "But God is not just *anybody*!"
>
> KEN HAM, *THE LIE: EVOLUTION*

ARE SOME BIBLICAL STATEMENTS MORE THAN THEOLOGICAL STATEMENTS?

Before we look at the scientific aspects of Genesis 1, we need to discuss the nature of the statements in the Bible. Obviously, the Bible is about God, so every statement, in that sense, is theological. But what about those statements that touch on the natural realm? Since the Bible is inspired and inerrant, every statement concerning the natural realm must also be true. Since God is the Creator of all things, He knows every detail of every aspect of the His creation. Besides, He is omniscient, knowing all things about everything. By definition, then, any mention of the natural realm in Scripture must describe reality in an absolute sense. Each description is more accurate than anything science can discover.

Though the Bible gives no scientific formulations like $E=MC^2$ or formal statements as in a textbook, it does contain scientific descriptions as well as scientific statements. To prove its scientific accuracy, we'll examine the Bible's first words, Genesis 1:1, as we would any scientific statement.

"In the beginning God created the heavens and the earth." This comprehensive statement is not only theological in that it tells us about God as Creator but also scientific in that it tells us about the origin of the natural realm.

- Matter and materials are major areas of scientific study. Genesis 1:1 mentions the totality of all matter, the universe.
- Science often studies processes. Genesis 1:1 mentions the ultimate process, the creation of all things.
- Science studies the agency of a scientific process. Genesis 1:1 introduces the ultimate agent of creation, the Creator Himself.
- Often in science, energy is a factor. Genesis 1:1 reveals the totality of energy built into the universe by its omnipotent Creator.
- Often in science, time is a factor. Genesis 1:1 describes the beginning of all time relative to the natural realm.

Does the Bible talk about science? Clearly. Consult your twelve-year-old.

Genesis 1:1 tells us the universe had a beginning. Since no universe existed until its creation, the study of that universe did not exist until then either. Obviously then, science has not always existed. We could say God was not only creating a universe but also building objective information into it any scientist could have observed had he been present.

GENESIS 1:1 IS A SCIENTIFIC STATEMENT

Matter → Universe
Energy → Omnipotence
Processes → Creation
Agency → God
Time → "In the beginning..."

THE TAKE AWAY: Science has not always existed, but God has. "In the beginning..." He created science by embedding all scientific knowledge into the creation. Why? He wants us to understand and utilize all His gifts to us. Only biblically knowledgeable scientists can do that with enduring success.

WHY DO WE START WITH THE CREATOR?

A scientific inquiry that begins with God as the Creator of all that is good helps put the material realm in its proper perspective. Creation is the product of a Creator who spoke everything into being, a Creator who designed incredible details into His creation. Genesis 1 begins unfolding those details.

THE TAKE AWAY: Please note that God is distinct from His creation. Theologians use the fancy word transcendence to describe His existence outside the material realm. We cannot understand science if we don't understand this reality. God existed before creation and, therefore, is not part of it or dependent on it. Creation exists only because God chose to create it. Every detail of creation is as it is because God designed it that way. Nothing would exist had God not designed and created it.

As we study the natural realm, we need to remember the sharp distinction between the Creator and His creation. We are God's creation. We are not God and never can be, much as some might wish to be.

This concept of transcendence is contrary to most religions and philosophies. The pagan cultures described in the Bible, including the Sumerians, Egyptians, Canaanites, Assyrians, Babylonians, Greeks, and Romans, had gods that were co-existent with the creation, that existed within the creation. For example, the ancient Egyptian culture worshipped many aspects of the creation itself, creating gods they believed were behind each of them. They worshipped the Nile River, the sun, the moon, and various creatures like frogs, snakes, cows, and locusts. They even worshipped dung beetles! We laugh, but is that lunacy any more absurd than believing that man evolved from apes, which evolved from reptiles, and on and on?

GOD AND CREATION

NON-BIBLICAL VIEWS

- gods co-existent with creation
- god is creation

BIBLICAL VIEW

- Creator distinct from creation
- Transcendent God
- Two realms of existence

This false concept of the continuity between pagan gods and creation has continued into modern times. In America today, the New Age religion with its prisms and occultic practices equates god with the creation. Their mantra is "god is all, and all is god," meaning nature should be worshiped. To quote King Solomon, "There is nothing new under the sun" (Ecclesiastes 1:9). Pagan ideas worm their way into every culture, often under the guise of science or erudition. If we are not to be fooled as so many have been down through the ages, we must remember that Creator God is separate and distinct from His creation. He is as superior to it as a potter is to the pot he molds and creates.[13]

> Genuine religion has its root deep down in the heart of humanity and in the reality of things. It is not surprising that by our methods we fail to grasp it: the actions of the Deity make no appeal to any special sense, only a universal appeal; and our methods are, as we know, incompetent to detect complete uniformity. There is a principle of Relativity here, and unless we encounter flaw or jar or change, nothing in us responds; we are deaf and blind therefore to the Immanent Grandeur, unless we have insight enough to recognise in the woven fabric of existence, flowing steadily from the loom in an infinite progress towards perfection, the ever-growing garment of a transcendent God.
>
> SIR OLIVER JOSEPH LODGE, PHYSICIST

13. Isaiah 29:16; 45:9; Jeremiah 18:6; Romans 9:21; Revelation 2:27.

God was present before the beginning not only because He is transcendent but also because He is eternal. The creation had a beginning, one God initiated. Genesis 1 gives the absolute beginning of the universe but not of God, who had no beginning. Hard to imagine having no beginning, isn't it! We need to keep reminding ourselves that God is superior to us in all ways.

"In the beginning, God…." Elohim, the Hebrew word for God used here, is a plural noun. We know from other passages that God is One.[14] How interesting that God presents Himself as both singular and plural.

Hebrew grammar requires that a plural subject have a plural verb (grammatical concordance), but the verb "created" in Genesis 1:1 is in the singular. Why? The Hebrew text often uses a singular verb but only with the plural "Elohim" when the reference is to God Himself.[15] Again, why? We know God doesn't make mistakes. He can't because He is God. So why does He seem to do so here?

God is a brilliant Teacher. He used these subtle grammatical oddities to introduce the difficult doctrine of the Trinity developed more fully later in Scripture. Slowly, as we read the entirety of the Bible over and over, He adds piece after piece to our knowledge of His person and plan. Each new piece should spur us to dig more deeply into His words, not merely for the precise meaning of each one but for its place in the panorama of truth He unfolds. The more He reveals His brilliance, the more we should yearn to know Him better.

Note that the word "created" (Hebrew: *bara*) has special significance in this context. It is used only with God as its subject. Here, it suggests God created the universe out of nothing (creation *ex nihilo*).

Note that Genesis 1:2 introduces another member of the Trinity, the "Spirit of God," as a Creator:

> The earth was formless and void, and darkness was over the surface of the deep, and the Spirit of God was moving over the surface of the waters.
>
> Genesis 1:2

---

14. Deuteronomy 6:4; Ephesians 4:4-6.
15. This is true throughout Genesis 1 (1:3, 4, 5, …) and elsewhere.

The New Testament tells us Jesus Christ, God the Son, was the primary Creator of the material universe. The apostles John and Paul both attest to this truth.

> In the beginning was the Word, and the Word was with God, and the Word was God. He [Jesus] was in the beginning with God. All things came into being through Him, and apart from Him nothing came into being that has come into being.
>
> John 1:1-3

> For by Him [Jesus] all things were created, both in the heavens and on earth, visible and invisible, whether thrones or dominions or rulers or authorities—all things have been created through Him and for Him. He is before all things, and in Him all things hold together.
>
> Colossians 1:16-17

So which is it? Was the Son or the Spirit the Creator? We will soon learn that both were. How fascinating and incomprehensible God is!

Consider also that the one and the many, the basis for the Trinity, is a foundational principle for mathematics. Could God have based mathematics on His own nature? I will let you think about that for a moment. God has so much for us to learn!

> Mathematics is the codified body of all logical thought.
>
> DR. GEORGE DAVID BIRKHOFF, MATHEMATICIAN

Other divine perfections that apply to all three members of the Trinity are introduced in Genesis 1. Not only is God transcendent, eternal, self-existent, and Trinitarian but also personal, holy, immanent, omnipotent, omniscient, omnipresent, sovereign, wise, and good. The mere mention of these perfections is enough to force us to our knees.

THE TAKE AWAY: Our God is magnificent. We should expect His design for creation to reflect His perfection.

How much more fun our early general science courses would have been had they included these essential truths. The fun would have intensified when the ACLU found out! Odd, isn't it, that, today, any mention of the Bible and God is considered blasphemy? All anti-God ideologies truly are faith-based religions that work to muzzle the one true God. Though these false religions may confuse many, they cannot silence God! He is shouting!

WHAT WAS THERE AT THE BEGINNING OF THE CREATION?

Note first that Genesis 1 describes no big bang. Jonathan Sarfati, with a PhD in physical chemistry, mockingly described the big bang theory, one of the main foundations of evolution, as "nothing exploded and became everything."[16] Terry Pratchett said it this way, "In the beginning, there was nothing, which exploded."[17] Either way, any twelve-year-old knows nothing cannot explode. In fact, Genesis 1 contains no natural processes, big bang or otherwise, only supernatural creative acts.

The earth is the first and most important part of the creation, except for mankind, of course, since we were made in God's image.[18] Secular science views our planet as a small speck in the massive Milky Way, which, as has been recently discovered, is only one of hundreds of billions of galaxies. From God's perspective, however, the earth is anything but insignificant. The planet that would become the home of His image bearers was the focus of His creation, not another planet, not even one of those ginormous galaxies, but our seemingly unimportant planet Earth. Don't waste your time looking for aliens on Mars or in UFOs. God placed all physical life on Earth. We don't want to go off on a rabbit trail or have time to discuss this issue adequately here. Just relax and believe that God created the universe for mankind and no one else.

---

16. Sarfati, Jonathan D., *The Genesis Account: A Theological, Historical, and Scientific Commentary on Genesis 1-11*, Creation Book Publishers, 2015, 155.
17. http://www.goodreads.com/quotes/63117-in-the-beginning-there-was-nothing-which-exploded.
18. Genesis 1 does not mention the creation of angels. Though Scripture doesn't reveal the date of their creation, they must have been around in time for the angel Satan to rebel against God and then tempt the first woman. Genesis 3.

> It is absurd for the Evolutionist to complain that it is unthinkable for an admittedly unthinkable God to make everything out of nothing, and then pretend that it is more thinkable that nothing should turn itself into everything.
>
> **G.K. CHESTERTON, AUTHOR**

Genesis 1:2 mentions Earth in its unfinished state. Thus, began geophysics—if anyone had been there to study it! We have difficulty envisioning the earth before God's creative acts. We have only the simple description of formless, void, and dark (1:2).[19] We can't impose our present scientific laws and physical constants on that time because God quickly began fashioning Earth to make it not merely habitable but welcoming for those He would soon make in His image.

SCIENCE IN GENESIS 1:1-2

- Absolute beginning
- No Big Bang
- Earth is priority
- Origin of physics and chemistry

Before God's six days of creation, the Earth was simply water. The Hebrew word for "the deep" (*tehom*) used in Genesis 1:2 refers in later verses to deep ocean waters.[20] I take this description in a literal sense. Since other Hebrew words describe ice (*qerach* in Job 6:16; 37:10) and water vapor (*aed* in Genesis 2:6; Job 36:27), the water here appears to have been liquid. Chemistry informs us that water molecules are composed of two atoms of hydrogen and one of oxygen (H_2O). Thus, in the beginning of Genesis, God introduces us to chemistry and hydrology (the study of the nature and mechanics of water).

I would guess that if our twelve-year-old were asked if this water was wet, he would say, "Of course, it's wet, silly. Who ever heard of dry water?"

Maybe we should consider this God's first hint about physics since the water molecules appear to have been held together by some force like gravity. Forces within the water molecules held the protons, neutrons, and electrons together while the molecules themselves held the mass

19. Many conservative commentators take these descriptive terms in the negative sense used elsewhere in the Old Testament. They conclude the fall of Satan and one-third of the angels caused this chaos (Allen Ross p. 106-07, Waltke p. 59-60). The original creation context does not require such a conclusion but could simply be describing the uncreated condition before God acted to make Earth the perfect habitat for mankind.
20. Exodus 15:5, 8; Job 38:16, 30; Psalm 107:25-26; Jonah 2:5.

of water together. Because of gravity, the earth as a water mass probably was spherical in this early stage. Notice that the details of the text are driving our conclusions about science, not the other way around.

At this point in Genesis, the Earth was the focus of the entire creation. Even spatially, the Earth was the center of God's creation. On this Earth, God would create man and eventually send His Son as the perfect Man to die for the sins of all mankind.

Humbling thought: Jesus Christ loved us so much He chose to die in His humanity rather than live without us eternally.

Even though most astrophysicists think of the earth as only an insignificant speck in an unspectacular location in an average galaxy among a trillion trillion others, the Earth remains the focus of God's attention. A humbling thought, isn't it?

DOES DAY ONE OF CREATION TELL US ANYTHING MORE ABOUT SCIENCE?

> Then God said, "Let there be light"; and there was light. God saw that the light was good; and God separated the light from the darkness. God called the light day, and the darkness He called night. And there was evening and there was morning, one day.
>
> Genesis 1:3-5

You may be surprised to learn that Day One of creation is filled with many scientific revelations. First and mainly, God created light. What is light? Your twelve-year-old might roll his eyes and take you to a light switch, flicking it up and down. You can help him by explaining that visible light is only a small portion of the entire electromagnetic spectrum, which includes x-rays, ultraviolet light, infrared light, microwaves, even radio waves.

The Bible teaches that God is light and clothed in light.[21] Does that mean this new light was His light? That cannot be since no aspect of His nature had an origin. God has always existed exactly as He is, perfect in all ways. Day One describes, not His light, but the physical light that

21. 1 John 1:5; Psalm 44:3; 104:2; 1 Timothy 6:16.

exists even today everywhere in the universe. This light contrasts with the darkness in Genesis 1:2. Thus, optics and electromagnetic waves, both aspects of physics, became a reality on Day One of creation.

> The tendency of modern physics is to resolve the whole material universe into waves, and nothing but waves. These waves are of two kinds: bottled-up waves, which we call matter, and unbottled waves, which we call radiation or light. If annihilation of matter occurs, the process is merely that of unbottling imprisoned wave-energy and setting it free to travel through space. These concepts reduce the whole universe to a world of light, potential or existent, so that the whole story of its creation can be told with perfect accuracy and completeness in the six words: "God said, Let there be light."
>
> **SIR JAMES JEANS, PHYSICS, MATHEMATICS, ASTRONOMY**

Earth in the form of water and light existed on Day One, but God didn't create the sun, moon, and stars until Day Four. You can further help your twelve-year-old by explaining that the sun is not the only source of visible light. Remind him about the light switch and ask the source of that light. Electricity isn't the only source of light either. You can explain about chemical reactions, including fire, nuclear reactions, and lightning (a form of electrical light).

Day One also informs us about time. The phrase "evening and morning" may tell us that a time frame exists independent of the sun. The sun simply gives us the means of measuring time. Can we also conclude that the earth was rotating, and the light source was localized? Or is time totally independent of any light sources? God, in His omniscience, always knows the time, but we need a way to measure it.

> We all know that there are regions of the human spirit untrammeled by the world of physics. In the mystic sense of the creation around us, in the expression of art, in a yearning towards God, the soul grows upward and finds fulfillment of something implanted in its nature. The sanction for this development is within us, a striving born with our consciousness or an Inner Light proceeding from a greater power than ours. Science can scarcely question this sanction, for the pursuit of science springs from a striving which the mind is impelled to follow, a questioning that will not be suppressed. Whether in the intellectual pursuits of science or in the mystical pursuits of the spirit, the light beckons ahead and the purpose surging in our nature responds.
>
> **SIR ARTHUR EDDINGTON, ASTRONOMER, PHYSICIST, MATHEMATICIAN**

How much fun we can have trying to imagine that which, like time, is beyond imagination! God gave us our imaginations, in part, to challenge and expand our thinking, to help us think like scientists.

THE TAKE AWAY: Note that all these creative acts were direct works by God. No long natural processes took place. God created everything supernaturally and instantly. He spoke, and light turned on. Each individual creative act was preceded by the words "and God said."[22] God's word created every glorious part of the creation, and His Word continues to cause it to function correctly today. If you don't believe me, ask your twelve-year-old if the sun came up this morning.

The phrase "and God said" shows that language originated with God, not with man. Language and communication are aspects of His nature and thus eternal. They presuppose His thoughts. Genesis 1:26 indicates that intra-Trinitarian communication existed before creation. In that verse, God said, "Let Us make man..." The "Us" shows that the three members of the Trinity communicate. While the Trinity is not explicitly taught here, this verse adds to the development of that doctrine begun in Genesis 1:1.

> God has strewn our paths with wonders and we certainly should not go through life with our eyes shut.
>
> **ALEXANDER GRAHAM BELL, INVENTOR OF THE TELEPHONE**

King David understood the importance of communication. He worshipped God verbally by connecting God's actions to His thoughts:

> Many, O LORD my God, are the *wonders* which You have done, and Your *thoughts* toward us. There is none to compare with You. If I would *declare and speak* of them, they would be too numerous to count.
>
> Psalm 40:5

We will see that God built language into humans as a crucial aspect of His image. He gave us the ability to form thoughts and then communicate them back to God or others. Would anyone deny thought and language are prerequisites for all scientific study? Notice God created not just the "visible" but the "invisible,"[23] including language and time.

22. Genesis 1:3, 6, 9, 11, 14, 20, 24, 26, 28-29. Also see Psalm 33:6-9.
23. Colossians 1:16 above.

God created an entire spiritual realm undetectable to us living in the material realm. He seems to have given angels the ability to form thoughts, communicate, and even express worship in music. God told Job that the angels observed and worshipped Him at the creation, singing hymns of praise to Him.

> "Where were you when I laid the foundation of the earth? Tell Me, if you have understanding, Who set its measurements? Since you know. Or who stretched the line on it? On what were its bases sunk? Or who laid its cornerstone, When the morning stars [angels] sang together and all the sons of God shouted for joy?"
>
> Job 38:4-7

Thus, with that little phrase "and God said," God began giving us hints about the beginnings of science, technology, engineering, and architecture. Isn't God's Word fascinating, and aren't His abilities incomparable?

> In the words of Psalm 19, "The heavens declare the glory of God and the firmament showeth his handiwork." Thus scientific research is a worshipful act, in that it reveals more of the wonders of God's creation.
>
> **ARTHUR SCHAWLOW, THE 1981 NOBEL PRIZE IN PHYSICS**

WHAT DOES DAY TWO ADD TO OUR UNDERSTANDING OF SCIENCE?

> Then God said, "Let there be an expanse in the midst of the waters, and let it separate the waters from the waters." God made (*asah*) the expanse and separated the waters which were below the expanse from the waters which were above the expanse; and it was so. God called the expanse heaven. And there was evening and there was morning, a second day.
>
> Genesis 1:6-8

The second day of creation gives us additional insights science can never discover. It cannot be stressed too often that we are wholly dependent on God's revelation to know anything about the creation week. We destroy scientific integrity when we try to impose our present understanding of science on it. And consider that four more days passed before God finished creating the vast and astonishing universe scientists study today!

Day Two describes the spreading out of the expanse of heaven. Most modern Bibles translate the Hebrew word *rakia* correctly as "expanse;" however, its incorrect translation as "firmament" in the King James Version (AD 1604) promoted the false idea of a flat earth. Incorrect translations do not produce correct science!

If we don't correctly understand the meaning of the Hebrew word for expanse, we cannot understand God's work on the second day. The Hebrew verb form of this word was used in a non-theological sense of something stamped[24] or beat out, spread out or beat thin into a sheet like gold or another soft metal.[25] In the context of Genesis 1, the noun *rakia* for expanse refers to something thinned or spread thin.[26]

The context of Genesis 1:6 helps us understand the meaning by naming the expanse "heaven," which refers to the interstellar space above the present earth (1:6-8, 14-17; Psalm 19:1; 150:1), including the atmosphere in which birds fly (1:20). Russ Humphreys, the world-class scientist we met earlier, described this expanse as not a vacuum but having some substance to it, something like the fabric of space.[27]

This spreading out of the expanse, using the same Hebrew word, is emphasized throughout the Old Testament.

> "Can you, with Him, *spread out* the skies, strong as a molten mirror?"
>
> Job 37:18

> Thus says God the Lord, Who created the heavens and stretched them out, Who *spread out* the earth and its offspring, Who gives breath to the people on it, and spirit to those who walk in it...
>
> Isaiah 42:5

---

24. Ezekiel 6:11.
25. Exodus 39:3.
26. Brown, Francis, Driver, S.R. and Briggs, Charles, *A Hebrew and English Lexicon of the Old Testament*, Clarendon Press, 1974, 955-96.
27. Humphreys lectures often on creation science. I have heard this description in several of them.

A surprising number of other Old Testament passages use three different Hebrew terms to describe this same spreading out of the heavens.[28]

> Covering Yourself with light as with a cloak, *stretching out* (*natah*) heaven like a tent curtain.
>
> Psalm 104:2

> Thus says the Lord, your Redeemer, and the one who formed you from the womb, "I, the Lord, am the maker of all things, stretching out (*natah*) the heavens by Myself and spreading out (*rakia*) the earth all alone."
>
> Isaiah 44:24

The text uses another creation word in Genesis 1:7, *asah*, to describe the creation of the expanse. This term is probably a synonym for *bara* in Genesis 1:1. It could indicate something made from previously created material rather than a creation out of nothing. After God created the universe out of nothing, He seemed to use this created material to make other aspects of the universe. The expanse was a creation of God as are all other components of the universe.

During Day Two, God not only created the expanse but also separated the waters of the initial earth containing them. The expanse separated and continues to separate the waters of the earth into two parts, the water that remained on the earth and the waters separated from the earth by the expanse. God spread the expanse away from the earth, forming interstellar space with water as its outside edge.

Astrophysicists have not yet observed the edge of the universe, but we would expect, based on the Genesis account, to find that its edge consists of a layer of water.[29] In this area, science has not yet caught up with God's revelation. God created the universe with a finite boundary, a water layer being that boundary.

In the past, creationists theorized the "waters above" referred to a water canopy in the atmosphere that God used to supply the forty days of rain during the Flood. Most creationists today have rejected that theory

---

28. Also note Job 9:8; 26:7; Isa. 40:22; 48:13 (טפח, *tapach*); 51:15; Jer. 10:12, 51:15; Zech. 12:1.
29. Humphreys holds to this theory.

because the text indicates the "waters above" from Day Two remain at the edge of the universe even though no man has yet observed them.

What hints about science does Day Two give us? For one, astrophysics began on this day. We could also include aeromechanics (the flow of air), atmospheric science, meteorology (short-term weather predictions), climatology (long-term weather patterns), and aeronomy (the physics and chemistry of the upper atmosphere).

These lists of the sciences involved in each day of creation are not meant to be exhaustive but to whet your interest to learn more about God's direct work in them. If you are preparing for a career, one of these areas may appeal to your imagination. Bringing the Christian viewpoint to any scientific specialty would not only be exciting for you but beneficial to many.

WHAT DOES DAY THREE OF CREATION SAY ABOUT SCIENCE?

> Then God said, "Let the waters below the heavens be gathered into one place, and let the dry land appear;" and it was so. God called the dry land earth, and the gathering of the waters He called seas; and God saw that it was good. Then God said, "Let the earth sprout vegetation, plants yielding seed, and fruit trees on the earth bearing fruit after their kind with seed in them;" and it was so. The earth brought forth vegetation, plants yielding seed after their kind, and trees bearing fruit with seed in them, after their kind; and God saw that it was good. There was evening and there was morning, a third day.
>
> Genesis 1:9-13

On the third day, God performed two major creative acts related to Earth. The first was another separation of water, this time from dry land. The second was the creation of plant life both on that land and in the water. Already, the earth that had been formless and void was beginning to take on a recognizable form.

This water separation created at least one continent. Some have called it Pangea. Such a continent may, in fact, have existed. The Genesis

Flood may well have broken up and separated that one landmass into the seven continents we know today. We will discuss that probability in more detail later.

> Then God said, "Let the waters below the heavens be gathered into one place, and let the dry land appear;" and it was so.
>
> Genesis 1:9

We could understand this creative act in two different ways. First, when God gathered the waters "into one place," submerged land may have come to the surface. That may be the reason the text says that "the dry land appeared" rather than was made or created. If so, God had already created all the elements making up the landmass before Day Three. Was that land under the water that covered the earth in verse 2? The second option is that God created the dry land from nothing or from the existing water molecules.

Where is your twelve-year-old? Is he still there? I'm not sure which of these two options he might pick. I'm inclined to choose the latter of the two. Why wouldn't omnipotent God speak the land into existence as He did light on Day One and the expanse on Day Two? The text does not give us enough details to be conclusive. Don't ask a scientist because none was there.

Either way, this day introduced the science of oceanography, complete with ocean floors and pre-Flood geology. We could even say the science of oceanography began earlier, on Day One, when the entire earth was one big ocean. I say pre-Flood because the worldwide Flood described in Genesis 6-9 radically transformed the entire crust of Earth. Geologists today would come to much different conclusions than those stated on Day Three since they discount the Genesis Flood. This day provided the material for the science of plate tectonics, even if only one surface continental plate existed in the beginning.

> Then God said, "Let the earth sprout vegetation, plants yielding seed, and fruit trees on the earth bearing fruit after their kind with seed in them;" and it was so.
>
> Genesis 1:11

God's second creative work on Day Three involved all plant life, including that needed in oceans and fresh water sources for the water creatures of Day Five. The text in 1:11 uses a vivid Hebrew construction that could be translated literally as "let the earth vegetate vegetation." This phrase includes both the noun and verb form of the word and gives definiteness and power to the action.[30] These were huge creative events.

Consider the varied colors, shapes, sizes, and materials of the vegetation that exists on Earth today. And all have fruits, nuts, seeds, or grains to propagate them easily. Many even have different scents! Further, God's creativity among plants unfurls throughout every season we experience today.[31] The bright green leaves of spring turn to the varied deep greens of summer, and then the vivid reds, yellows, and oranges of autumn, further examples of God's desire for us to see His lovingkindness in His creation. Our Lord seems to love beauty and variety as well as the joy they bring His image bearers.

This verse describes the creation of plants in the broadest of plant categories, using three different Hebrew terms. We can understand these three terms in two ways. First, they may refer to three different categories, grassy plants, smaller green plants or bushes, and larger seed-bearing plants or trees. The second, more probable interpretation is that the first term describes plant life in general followed by two categories of smaller green plants or bushes and larger seed-bearing plants or trees.

God took the Earth that had been formless and void on Day One, gave it form on Day Two, and then filled it with land and a dazzling array of plant life on Day Three, making possible the sciences of botany and genetics, to name only two.

The text tells us that God created every variety of plant life to reproduce "after their kind." The Hebrew term for kind, *min*, occurs ten times in Genesis 1 in reference to both plant and animal life, showing the importance of this principle.[32] The term indicates that interbreeding in plant and animal life are limited to within "kinds." The biblical kinds are a broader biological category than species and closer to the biological category of a family or genus. Species would have come about after

30. Technically, this Hebrew construction is called an infinitive absolute.
31. The biblical text suggests the seasons came about after the Genesis Flood (Genesis 8:22).
32. The word "kind" occurs thirty-three times in all the Old Testament, especially in the Creation and Flood narratives.

several generations of variation within the kinds. After time, some species can no longer interbreed outside their species. Different kinds cannot cross genetic barriers, meaning one kind cannot contaminate another kind's unique DNA. This divinely instituted limitation ensures the fixity and distinctiveness of the created kinds.

> All fish, living and fossil, belong to distinct kinds; "links" are decidedly missing. Incidentally, creationists have no reason to be committed to any particular classification scheme, nor to any particular taxa above the kind level. "Orders," "classes," and "phyla" must not be allowed to become hallowed by tradition. They may be correctly identified (higher taxa *are* real), but there again they may not. Some "missing links" have been artifacts of bad classification systems. Morphology (and now biochemistry) have dominated classification, but ecology may yet prove to be a better guide.
>
> **DR. ARTHUR JONES, BIOLOGY**

In the words of Carl Linnaeus (1707-1778), Swedish botanist, zoologist, physician, and father of the scientific system for defining and naming all varieties of life, "All the species[33] recognized by Botanists came forth from the Almighty Creator's hand, and the number of these is now and always will be exactly the same."[34] Fixity of kinds guarantees no evolution into new kinds. Those of the same kind can change slightly but never so dramatically that a new kind is created.

Plants within a kind may vary, such as green, lima, and pinto beans, but God does not allow the formation of new kinds different from those He created. We can be sure no peach tree will ever produce lima beans! We should praise God for the uniformity He incorporated in each type of seed. Imagine planting a garden and not knowing if flowers, weeds, or tomatoes would sprout. Would be hard to survive without nutritious vegetables, wouldn't it?

Modern genetics and geology confirm this rigidity within the created order. As we have seen, no evidence of transitionary kinds has been found in any geological layer—or in any other place, cells included. On Day Three, from the very beginning, God made transition from one plant kind to another impossible. Doesn't this truth alone eliminate

---

33. When Linnaeus used the word "species," he was referring to the "kinds" as described in Genesis 1, not to the way we use the word species today. Today, biblical kinds are closer to biological families that may include many species.

34. https://todayinsci.com/L/Linnaeus_Carolus/LinnaeusCarolus-Species-Quotations.htm.

the possibility of evolution? Our twelve-year-old probably rolled his eyes again, amazed that so many scientists have allowed themselves to be fooled into believing Satan's lie.

> Progressing in my studies, I slowly realized that evolution survives as a paradigm only as long as the evidence is picked and chosen and the great pool of data that is accumulating on life is ignored. As the depth and breadth of human knowledge increases, it washes over us a flood of evidence deep and wide, all pointing to the conclusion that life is the result of design. Only a small subset of evidence, chosen carefully, may be used to construct a story of life evolving from nonliving precursors. Science does not work on the basis of picking and choosing data to suit a treasured theory. I chose the path of science which also happens to be the path of faith in the Creator.
>
> **DR. TIMOTHY G. STANDISH, BIOLOGY**

WHAT SCIENCES DID DAY FOUR INTRODUCE?

> Then God said, "Let there be lights in the expanse of the heavens to separate the day from the night, and let them be for signs and for seasons and for days and years; and let them be for lights in the expanse of the heavens to give light on the earth;" and it was so. God made the two great lights, the greater light to govern the day, and the lesser light to govern the night; He made the stars also. God placed them in the expanse of the heavens to give light on the earth, and to govern the day and the night, and to separate the light from the darkness; and God saw that it was good. There was evening and there was morning, a fourth day.
>
> Genesis 1:14-19

God created all heavenly bodies on the fourth day, but focused on the two lights most important for the Earth and mankind. Scripture does not identify them as the sun and moon but as light-bearers, a Hebrew word that the major Hebrew Lexicon describes as "luminaries, light, light-bearer, luminary, lamp."[35] The word is used nineteen times in the Old Testament, five of those on Day Four. Here, the term refers to sources

35. BDB, 22.

of light, not to the light itself. Notice that the sun and moon were not the originators of light but simply the bearers of the light God created on Day One. The emphasis in the text is on the sovereign Creator of all things and the purpose of His creation in relation to mankind.

By using the Hebrew word for light-bearers, Moses may have been offering a polemic against ancient cultures like the Sumerians, Egyptians, Canaanites, and Babylonians that worshiped the sun and moon. The inspired text also serves as a polemic against our current religion of naturalism. R. Hooykaas, Dutch historian of science, wrote:

> In the 1st chapter of Genesis it is made evident that absolutely nothing, except God, has any claim to divinity: even the sun and moon, supreme gods of the neighboring peoples, are set in their places between the herbs and the animals and are brought into the service of mankind.[36]

Isn't it interesting that Moses only briefly mentioned the vast expanse of stars, galaxies, planets, and other objects in the expanse of interstellar space? In the Hebrew text, only two words[37] describe them all, literally saying "and the stars." Was this phrase added almost as an afterthought? Was Moses saying "Oh, by the way, the stars also" or "Only since you were wondering, the stars also" or "Because it is easy for God, the stars also." Again, the emphasis of Day Four was the Earth and the benefit it receives from these light-bearers, not the vastness of space.

> Those who study the stars have God for a teacher.
> **TYCHO BRAHE, EARLY MODERN ASTRONOMER**

On the second day, we saw the beginning of astrophysics. Here on the fourth day began astronomy and its related sciences. You may wonder if the science of the sun is called sunology. Sorry, but no. It is called heliology. Science tends to avoid the simple, doesn't it?

The Apostle Paul added insights into astronomy long before modern science made its discoveries.

36. Frans van Lunteren, "Geology and Christianity," ISIS—Volume 109, Number 1, March 2018.
37. In the Hebrew text at the end of 1:16, a waw is attached to a direct object marker (first word) followed by the plural of the word *star* with the article attached (second word). Or is that four words if you count the direct object marker and the article?

> There is one glory of the sun, and another glory of the moon, and another glory of the stars; for star differs from star in glory.
>
> 1 Corinthians 15:41

The Hebrew and Greek words for star can refer to other heavenly bodies like planets, asteroids, and comets.[38] Today, we see an amazing diversity of heavenly bodies, including planets that move relative to the fixed background of the stars. Stars alone come in many forms, red and blue super giants, hypergiants, white, brown, and yellow dwarfs, variable stars, pulsars, binaries, planetary nebulae, neutron stars, and on and on. God created all this beauty and order on the fourth day of creation. What a spectacular Creator! He is the real "star" of the story, isn't He?

Naturalistic theory believes the sun, moon, and stars had to have been formed before light, but Genesis 1 declares that God created light on Day One and the heavenly bodies to contain it on the fourth day. In another inconsistency with the naturalistic view, God created the earth before the stars and the galaxies. If only more scientists consulted God's revelation, their understanding would be greatly enhanced and their discoveries better benefit mankind.

> I love to think of nature as an unlimited broadcasting station, through which God speaks to us every hour, if we will only tune in...God is going to reveal to us things He never revealed before if we put our hands in His. No books ever go into my laboratory. The thing I am to do and the way of doing it are revealed to me. I never have to grope for methods. The method is revealed to me the moment I am inspired to create something new. Without God to draw aside the curtain I would be helpless...My prayers seem to be more of an attitude than anything else. I indulge in very little lip service, but ask the Great Creator silently, daily, and often many times a day, to permit me to speak to Him through the three great Kingdoms of the world which He has created–the animal, mineral, and vegetable Kingdoms–to understand their relations to each other, and our relations to them and to the Great God who made all of us. I ask Him daily and often momently to give me wisdom, understanding, and bodily strength to do His will; hence I am asking and receiving all the time.
>
> **GEORGE WASHINGTON CARVER, INVENTOR**

38. Daniel 8:10; Revelation 6:13; 8:10-11; and possibly Mathew 2:3, 7, 9-10; 24:29.

WHERE DID THE ZOOLOGICAL SCIENCES ORIGINATE?

> Then God said, "Let the waters teem with swarms of living creatures, and let birds fly above the earth in the open expanse of the heavens." God created the great sea monsters and every living creature that moves, with which the waters swarmed after their kind, and every winged bird after its kind; and God saw that it was good. God blessed them, saying, "Be fruitful and multiply, and fill the waters in the seas, and let birds multiply on the earth." There was evening and there was morning, a fifth day. Then God said, "Let the earth bring forth living creatures after their kind: cattle and creeping things and beasts of the earth after their kind"; and it was so. God made the beasts of the earth after their kind, and the cattle after their kind, and everything that creeps on the ground after its kind; and God saw that it was good.
>
> Genesis 1:20-25

During the final two days of creation, God focused on creatures that breathe, described as *living*. The creatures created on Day Five included two diverse categories, aquatic and flying. God created land animals and humans the next day. Scripture describes in detail only one category apart from mankind, the great sea monsters. Was this to show that even the largest creatures were His creation and under His control? Were those monstrous creatures dinosaurs? Again, ask your twelve-year-old.

On the sixth day, God performed two distinct creative acts. First, He created all land animals (1:24-25). Then, He created mankind, both the man and the woman. He gave both animals and mankind the ability to procreate only after their own kind (1:26-27). No ape has ever morphed into a man! Thank God.

> An essential feature of the creation model is the placement of considerable genetic variety in each created kind at the beginning. Only thus can we explain the possible origin of horses, donkeys, and zebras from the same kind; of lions, tigers, and leopards from the same kind; of some 118 varieties of the domestic dog, as well as jackals, wolves and coyotes from the same kind. As each kind obeyed the Creator's command to be fruitful and multiply, the chance processes of recombination and the more purposeful process of natural selection caused each kind to subdivide into the vast array we now see.
>
> **DR. LANE P. LESTER, GENETICIST**

We should note again that God created mankind in His own image, a significant distinction. We sense we were the most important part of His creation because He also gave us a purpose.

> "Be fruitful and multiply, and fill the earth, and subdue it; and rule over the fish of the sea and over the birds of the sky and over every living thing that moves on the earth."
>
> Genesis 1:28-30

God created no other creature in His image or gave it a high purpose.

Days Five and Six instituted zoology and all related sciences, including ornithology (birds), ichthyology (fish), mammalogy, herpetology (reptiles and amphibians), entomology (insects, which is divided into at least ten sub-specialties), acarology (mites and ticks), arachnology (spiders and scorpions), and on and on. The results of God's creativity go on and on, too, as science discovers hidden scientific gems that God made available during those first six days. Your twelve-year-old can set his career goal on pursuing one of these sciences and digging for more biblical insights in that area. So much is yet to be discovered! An exciting prospect, isn't it?

On Day Six, God introduced the most significant of all the sciences, anthropology. We have to be careful, though, because in promoting their idea of evolution, many have inserted primates into anthropology, which should be exclusively devoted to mankind. As we have seen, such accommodations are dangerous distortions of truth. Besides, they are unscientific.

> Andrew explained that in all of his practice and research as a doctor, he could not think of one single example where a belief in evolution had advanced medicine or the care of individuals. "It's just the opposite," he says, citing abortion and euthanasia as examples of the tragic influence of evolutionary beliefs. His own research is relevant to creation/evolution. It involves developing suitable replacement valves for the human heart. He says, "Most people think that we are closely related to the apes, but we use pig valves. Because of a fundamentally good created design (and for commercial reasons) they are an adequate match to human valves." He explained that scientists have tried to manipulate pig DNA to more closely match human DNA. But there has been a high rejection and failure rate using the organs from these engineered animals.
>
> **DR. ANDREW HODGE, CARDIOTHORACIC SURGEON**

DOES GENESIS 1 GIVE SCIENTIFIC INSIGHTS ABOUT MANKIND?

Humanity occupies a special place in God's creation. Theologians often describe humanity as the crowning achievement of creation. The great commentator, H.C. Leupold, wrote, "From every point of view man is seen to be the crown and climax of God's creation."[39] The earlier days of creation not only led up to man's creation but were prepared expressly to benefit him.

> And I believe there are many species in Nature, which were never yet taken notice of by Man, and consequently of no use to him, which yet we are not to think were created in vain; but it's likely...to partake of the overflowing Goodness of the Creator, and enjoy their own Beings...all the Creatures in the World may be some way or other useful to us, at least to exercise our Wits and Understandings, in considering and contemplating of them, and so afford us Subject of Admiring and Glorifying their and our Maker. Seeing them, we do believe and assert that all things were in some sense made for us, we are thereby obliged to make use of them for these purposes for which they serve us, else we frustrate this End of their Creation.
>
> **JOHN RAY, BOTANIST, FATHER OF NATURAL HISTORY IN BRITAIN**

The obvious and striking change in the description of this creative act clearly shows man's special place in God's eyes. On Day Six, following

39. Leupold, H.C., *Exposition of Genesis*, Vol I, Baker Book House, 1942, 86.

the usual "And God said," we find the words, "Let *Us* make man in *Our* image, according to *Our* likeness" (1:26). The words "Us" and "Our" have provoked considerable debate about who is in view. The best answer is that these carefully chosen words refer to the three persons of the Trinity, God the Father, God the Son, and God the Holy Spirit. Although Genesis has no explicit statement on the Trinity, as we have seen, the very first verse, 1:1, gives hints about it. The Trinity is in view because creation involved all three persons of the Godhead. This unity of purpose in the Godhead confirms the importance God placed on mankind, His final, culminating, and unique creative act. He created only mankind in His image.

God's image is evident in all mankind, in both males and females.

- God is spirit,[40] and He created man with a finite spiritual nature.
- God is omniscient. He built into man a finite measure of intellect, including a capacity to reason.
- God can communicate within His being and with all creation. He gave man the ability to communicate with God Himself as well as with image bearers.
- God has ultimate free will and has granted mankind the ability to make choices, including moral, intellectual, and spiritual ones.
- Only God is eternal, but He grants mankind eternal existence after a beginning in time. Believers in Jesus Christ as Savior will live with the eternal God into eternity. Those who reject Him will exist into eternity as well but separated from God forever, the greatest punishment of all.
- God is the ultimate creator of all things. He has given mankind the ability to be creative in several ways. First, a man and woman can create children after their own image, bearing their DNA. Second, man can take existing materials and create works of art and utility.

40. John 4:24, "God is spirit, and the people who worship him must worship in spirit and truth."

Man was not simply the last creature in a long evolutionary process but rather a special and separate creation. The Apostle Paul made this distinction clear.

> All flesh is not the same flesh, but there is one flesh of men, and another flesh of beasts, and another flesh of birds, and another of fish.
>
> 1 Corinthians 15:39

Separate flesh and not equal—we are foolish if we forget the distinction God created between humans and the rest of creation.

THE IMAGE OF GOD IN MAN

SPIRIT → Spiritual nature
OMNISCIENT → Intellect, reasoning
OMNIPOTENT → Abilities, volition
HOLY → Moral capacity
ETERNAL → immortality
LOVE → Relationships
SOVEREIGN → Rule earth, choice

Note also that God's creation of mankind included both male and female as detailed in Genesis 2. God created them with equality of personhood but distinctions in roles. That should remind us of the Trinity. Just as men and women are different physically, so they are different in their responsibilities. The man has authority and leadership in marriage while the woman has an equally important role as his helpmeet, giving him the support that allows them both to excel.[41]

> In chemistry, you need to first grasp the basic foundational principles. As a chemist, if I have that basic foundation of chemistry, then when I start actually doing the research, it's that foundation that makes things so much easier, so I can be much more creative. So, too, with Christianity; to be effective one must have a sound understanding of the foundations–and that means understanding Creation and Fall and their relevance to the Gospel. Sadly, teaching on Genesis is generally very poor.
>
> **DR. PATRICK H. YOUNG, CHEMISTRY, MATERIALS SCIENCE**

The sixth day concluded with the divine evaluation, "it was very good." This commendation applied to the entire creative work (1:31) because the text says that "God saw all that He had made." "All" means all and

41. Ephesians 5:22-3; 1 Peter 3:1-7.

refers to every act of creation. This statement has so many significant scientific insights that I will expand it in later chapters.

WHAT CAN WE CONCLUDE?

Genesis 1 simply but profoundly presents the biblical worldview on origins. The text not only presents theological statements but also scientific descriptions, including the origin of all areas of modern science.

Genesis 1 nullifies the present-day naturalistic view of origins, the heart of which is the idea of evolution. A summary of the major contrasts between evolution and Genesis 1 proves the fallacy of this modern-day pagan ideology.

1. God created everything directly and personally without a big bang.
2. God created the earth before He created outer space, the sun, moon, other planets, other stars, and the galaxies.
3. God created everything during six days (144 hours to be clear), not billions of years.
4. Every creative act was immediately performed and did not include long—or short—evolutionary processes.
5. God created vegetation, which needs photosynthesis to grow, before He created the sun as a light bearer.
6. God created light before He created light sources like the sun and stars.
7. God created fruit trees before He created the insects needed to pollinate them.
8. God created plant life with a fixity of kinds rather than with a continuous relationship between all plants.
9. God made a distinction between plant life and animal life, not a continuity.
10. God created aquatic animals and flying creatures at the same time, not sequentially over long evolutionary ages.

11. God created birds before reptiles rather than after.
12. God created mammals and reptiles on the same day, not after long evolutionary ages.
13. God created animal life with a fixity of kinds rather than with a continuous relationship between all animals.
14. God created mankind with a sharp discontinuity from the rest of the animal kingdom.
15. God created mankind in His image, occupying a place far above every other part of creation. No apes need apply!

As a bonus, I hope the following questions will make your discussions with evolutionists a lot more fun.

- Evolution teaches life originated from one single cell and then, through mutations, evolved into all other life forms. Ask an evolutionist whether that first cell was a plant cell or an animal cell.
- Either response will require a follow-up question: If that first evolutionary cell was a plant cell, did it evolve into an animal cell? If not a plant cell, did an animal cell produce all plants?
- Even further, did evolution require two original cells, a plant cell and an animal cell? If so, evolution requires not just one miracle origination but two. Two Big Bangs? Doesn't it take a lot less faith to believe the sovereign Creator supernaturally created each creature and all plant kinds separate and distinct from each other over a six-day period just as He said?

> If amino acids can only be made where there is no free oxygen in the atmosphere, and porphyrins can only be made when there is free oxygen, then these things needed by every cell could not have existed together to form the first cell! What's more, many of these compounds are antagonistic. They will combine and destroy each other—anywhere except within a living cell.
>
> DR. LAWRENCE O. RICHARDS, THEOLOGIAN

THE TAKE AWAY: The biblical worldview of creation is far superior to every other viewpoint. Imposing current scientific processes and present conditions on the six days of creation, which were "in the beginning," is not only a serious breach of scientific techniques (no observation or replication) but a disingenuous attempt at omniscience by fallible man. Why pretend to have been there billions of years ago when you can read the account of the One who truly was there a few thousand years ago, the Creator Himself?

The inconsistencies between evolutionary ideas and the descriptions of Genesis 1 are too drastic and numerous to attempt to harmonize. History has shown that all attempts end with the Bible being compromised. When pseudo-science is imposed on the biblical text, biblical fact is always twisted to fit popular opinions, resulting in a distorted interpretation of both the text and the natural realm.

You need to choose one or the other. Does the inerrant Bible with its absolute truth have greater authority, or should you believe human-derived scientific theories that continually change with the uncovering of new information?

Equally, you cannot hold to both naturalism and creation. You must choose either fallible man's evaluation or infinite God's revelation in Genesis 1. Even your twelve-year-old knows you should accept the inspired and inerrant divine revelation and interpret it using the Grammatical-Historical-Contextual approach. Not only will you gain a greater understanding of truth, but you will have an accurate foundation for all science.

> Why do I believe in God? As a physicist, I look at nature from a particular perspective. I see an orderly, beautiful universe in which nearly all physical phenomena can be understood from a few simple mathematical equations. I see a universe that, had it been constructed slightly differently, would never have given birth to stars and planets, let alone bacteria and people. And there is no good scientific reason for why the universe should not have been different. Many good scientists have concluded from these observations that an intelligent God must have chosen to create the universe with such beautiful, simple, and life-giving properties. Many other equally good scientists are nevertheless atheists. Both conclusions are positions of faith...I find these arguments suggestive and supportive of belief in God, but not conclusive. I believe in God because I can feel God's presence in my life, because I can see the evidence of God's goodness in the world, because I believe in Love and because I believe that God is Love.
>
> **DR. WILLIAM D. PHILLIPS, 1997 NOBEL PRIZE IN PHYSICS**

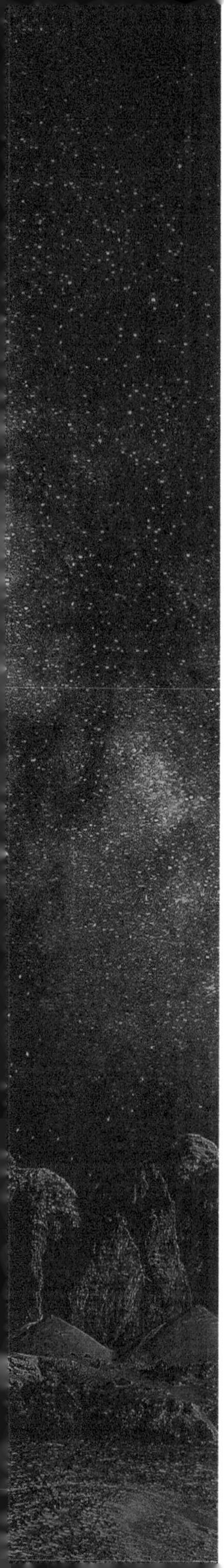

It is He who made the earth by His power,
Who established the world by His wisdom;
And by His understanding He has stretched out the heavens.
When He utters His voice, there is a tumult of waters in the heavens,
And He causes the clouds to ascend from the end of the earth;
He makes lightning for the rain,
And brings out the wind from His storehouses.

JEREMIAH 10:12-13

Ah Lord God!
Behold, You have made the heavens and the earth
by Your great power and by Your outstretched arm!
Nothing is too difficult for You.

JEREMIAH 32:17

JEREMIAH, WEEPING PROPHET

CHAPTER 11

SCIENTIFIC FOUNDATION STONE #3

THE FINITE CREATION

NOT ETERNAL

The third foundation stone for accurate and enduring scientific conclusions is the truth that God designed a finite creation. It has limitations. Only God is eternal and unbounded.

Most secular scientists assume the universe existed even before the Big Bang. They cannot conceive of an existence without matter or energy and, therefore, without time. Because they cannot observe limits to the universe, they also assume it never ends. They base their theories on their own limited understanding and experience. They are wrong. According to God as revealed in His Word, the universe not only has temporal limits but also spatial boundaries, and someday, it will end as we know it.

> Even if you think the Big Bang created the stars, don't you wonder who sent the flowers?
>
> DR. ROBERT BREAULT, ASTRONOMER

DID THE UNIVERSE HAVE A TEMPORAL BEGINNING?

Though the original creation was perfect, it was never infinite. It had a beginning (Gen. 1:1) and will have an end (Rev. 21:1; 2 Pet. 3:10-13).

The first Hebrew word in the Bible (*berashit*) requires three English words to translate it. "In the beginning..." This Hebrew word tells us that both time and the universe had a beginning. Before that beginning, nothing existed except God Himself. The universe is not eternal. As we cannot repeat too often, only God is eternal, having no beginning or end.

Note that this first Hebrew term appears in the emphatic position as the first word grammatically, the first word of all creation, the first word of all history, and the first word of all Scripture. This emphasizes not only the importance of time itself but shows that God is the creator of a time frame. The beginning of time marked the absolute beginning of the material universe.

The spiritual realm has always existed because God has always existed. He is spirit and truth (John 4:24); therefore, all spiritual truths, in fact all truth, has existed forever, too. We can understand unseen spiritual truths only when the Creator reveals them to us through His Word. The same is true for all events in history of which we have no direct knowledge. Only God has known and continues to know the entire truth about history.

If we are honest, we will admit we don't know all the details even about the events in our own lives. Just listen to sports commentators describe the same play. Each description is a little different because everyone's perspective is a little different, shaped by his background and prejudices, his strengths and weaknesses. Only God knows the absolute truth about history because only He is absolute truth—and only He is omnipresent and omniscient.

> I got myself into a lot of trouble when teaching Sunday school. At the beginning of the year, I said to the class, "Your opinion is irrelevant. My opinion is irrelevant." Holding up a Bible, I said, "It's only God's opinion that's going to count for anything." I was booed and jeered because they all felt entitled to an opinion. My response to that was, "Which verse is that?"
>
> **DR. SAAMI SHAIBANI, FORENSIC PHYSICIST**

Colossians 1:16 tells us that Jesus Christ created all things, including both the "visible and invisible." Time is one of those invisibles. A common and simple definition of time would be the progression of events from the past to the present and into the future. The concept of time arguably began in the far past with the creation of the heavens and earth. The beginning of time marked the absolute beginning of the material universe we inhabit. Certainly, time is an integral part of life on Earth. We cannot conceive of living outside the time frame of the universe because time has always circumscribed the boundaries of our days, weeks, and years.

Will time continue forever? Scripture provides strong assurances that time is an integral aspect of heaven, one that will continue without end. Without time, there would be no motion, no action, no communication, no worship; yet Scripture provides glimpses of heaven that include continuous worship of and service to God.[1]

The picture of the New Jerusalem in the new heaven and new earth in Revelation 22:2 includes a measure of time:

> On either side of the river was the tree of life, bearing twelve kinds of fruit, yielding its fruit every *month*; and the leaves of the tree were for the healing of the nations.

A month is a familiar measure of time. How will eternal time be different from the circumscribed time we know today? Only spending time in eternity will tell us.

Since the beginning of creation, time has progressed sequentially. At the end of Day One, Moses mentioned this sequencing of time with the sentence, "There was evening and there was morning, one day." A similar sequencing concluded each of the six creative days.

This time frame existed independent of the sun and moon, which God didn't create until the fourth day. Interesting. On the fourth day (a period of time indicating that God had already instituted time), God created the sun and the moon, in part for calculating time, "for seasons and for days and years" (Genesis 1:14-18). By doing so, He set the parameters for time in this universe. He also showed us how to categorize and measure time.

Man discovered time, uses it in calculations, and utilizes it, but he has no control over it. Today always comes after yesterday and before tomorrow. Every second that passes is gone forever. Time is under the control of its Creator alone. We can observe it, but we can't control any aspect of it no matter how hard we try. Turning back the clock is impossible. The Fountain of Youth will never be found for those of you who've

1. Revelation 4:8 ("day and night"), 10; 5:15; 21:25 ("in the daytime"); 22:2 ("every month"), 3 ("will serve Him"), 5 ("will reign forever").

thought about looking. Finalizing the details of our future is impossible, too. The passage of time is inexorable. Our control of it is nonexistent. As Creator, God can end it or change it according to His desire.

Another indicator of the sequence of time is found in the Hebrew grammar of the early chapters of Genesis. The Hebrew language in which the Old Testament was written often uses the waw (pronounced "vav") consecutive. Because English translates the waw consecutive with the simple word "and," we tend to overlook it in our translations. In Hebrew, however, the waw emphatically indicates a sequence of events, one event following another. Thus, this word is a major characteristic of Hebrew historical narratives. The presence of the waw consecutive in Genesis 1 is striking. Not only does it tell us these days of creation were historical events but that they happened sequentially. Obviously, time has been an essential element of history since its beginning.

Speaking of history, the Bible not only makes theological and scientific statements but also historical ones. Genesis 1:1 is one of those historical accounts.

- History records events in time. Genesis 1:1 gives a historical account of the beginning of time.
- History records major characters. Genesis 1:1 introduces the ultimate in personages, God Himself.
- History focuses on events. Genesis 1:1 describes the most important event in history, one without which history could not have happened, the creation of all things.
- History gives geographical locations. Genesis 1:1 gives a panorama of the entire universe.

THE TAKE AWAY: History is dependent on time, and history defines our lives. Yes, time is important, and, yes, time had a beginning.

DOES THE UNIVERSE HAVE A PHYSICAL BOUNDARY AND END?

Creation is also finite in terms of its physical extent. Its size is not endless. Though no one has yet observed the edge of the universe, its boundary is far out there in space somewhere. As we've seen, the divine narrative of the second day of creation describes that boundary as a layer of water.

Additionally, the creation, which had a supernatural beginning, will have a future supernatural end. The present creation is not eternal. It will cease to exist. God promised through the Apostle Peter that…

> …the day of the Lord will come like a thief, in which the *heavens will pass away* with a roar and the elements will be destroyed with intense heat, and the *earth* and its works *will be burned up*.
>
> 2 Peter 3:10

There really is a "big bang" in the biblical worldview, but it is at the end, not the beginning. God guarantees a worldwide conflagration will dramatically end the universe as it exists today. Will it also expose the spiritual realm? Like Elisha's servant (2 Kings 6:17), will we be able to see all that is hidden from us now? Because Scripture doesn't say, we can't know, but someday we will.

Peter described the heavens as passing away. He meant our present interstellar space will end. He also noted the earth will be burned up and "all these things are to be destroyed." In the biblical worldview, the big bang is at the end when God destroys creation, not at the beginning. (Once again, secular science with its fearmongering about global warming has it all wrong. The earth will have a fiery end caused by God, not a watery one caused by melting glaciers. Big difference! Don't you just love God's sense of humor?) The end of our present universe is as certain as its six-day beginning, making this present creation both temporal and finite.

> I love nuclear energy. I just want to be sure it stays where God put it—93 million miles away, in the sun.
>
> **WILLIAM MCDONOUGH, ARCHITECT, ENVIRONMENTAL ENGINEER**

Just three verses later, the apostle Peter pointed to an exciting further dimension to God's plan, a New Heavens and a New Earth.

> But according to His promise we are looking for a new heavens and a new earth, in which righteousness dwells.
>
> 2 Peter 3:13

In Scripture, righteousness means rightness or right standing before God. We should look forward to a time when all will be made right according to God's perfect standard of rightness. Imagine a new world in which only righteousness dwells, no evil, disease, decay, or death. We can praise Him for that future time now because we know He always keeps His promises.

THE TAKE AWAY: Secular science doesn't know the future and cannot give any insights into it. Only the revelation of the omniscient One Who exists separate from the universe can reveal its beginning and tell us His plans for its future. He is "the alpha and omega, the beginning and the end."[2] We would be wise to learn as much of His revelation as we can. The accuracy of our scientific discoveries depends on it. So do our eternal rewards.

> A scientific discovery is also a religious discovery. There is no conflict between science and religion. Our knowledge of God is made larger with every discovery we make about the world.
>
> JOSEPH H. TAYLOR, JR., 1993 NOBEL PRIZE IN PHYSICS

2. Revelation 1:8; 21:6; 22:13; Isaiah 41:4; 44:6; 48:12.

For behold, He who forms mountains and creates the wind
And declares to man what are His thoughts,
He who makes dawn into darkness
And treads on the high places of the earth,
The Lord God of hosts is His name.

AMOS 4:13

He who made the Pleiades and Orion
And changes deep darkness into morning,
Who also darkens day into night,
Who calls for the waters of the sea
And pours them out on the surface of the earth,
The Lord is His name.

AMOS 5:8

AMOS, SHEPHERD AND PROPHET

CHAPTER 12

SCIENTIFIC FOUNDATION STONE #4

GOD'S CREATION MANDATE

NOT ENVIRONMENTAL TECHNOLOGY

The fourth foundation stone for accurate and enduring scientific conclusions is the creation mandate of Genesis 1:28-30. Much theology as well as scientific information is embedded in it. This divine decree is often referred to as the dominion mandate. I prefer to call it the creation mandate since God gave it in the context of all creation. In it, God revealed His purpose for humanity.

> God blessed them [Adam and Eve]; and God said to them, "Be fruitful and multiply, and fill the earth, and subdue it; and rule over the fish of the sea and over the birds of the sky and over every living thing that moves on the earth." Then God said, "Behold, I have given you every plant yielding seed that is on the surface of all the earth, and every tree which has fruit yielding seed; it shall be food for you; and to every beast of the earth and to every bird of the sky and to everything that moves on the earth which has life, I have given every green plant for food;" and it was so.
>
> Genesis 1:28-30

God's creation mandate has two elements. First, a man is to marry a woman and have a family. Second, he is to rule creation with his wife at his side as his valued helpmate.

Why does God hold marriage and family in such high esteem? They were the first clearly spelled out divine institutions[1] and continue to be essential if mankind is to succeed both individually and collectively. History records that even seemingly invincible nations like Babylon, Greece, and Rome self-destructed following the degradation of marriage and family. The continuance of civilized societies depends on maintaining healthy marriages and stable family units. Americans would benefit from remembering that historical constant.

The second element of the creation mandate, to subdue creation and exercise a measure of sovereignty over it, is contingent on our choosing to obey God's commands about marriage and family. The limited sovereignty God gave the first man Adam in the Garden of Eden allowed him to take the resources of the earth and use them to help, sustain, and prosper his family.

THE CREATION MANDATE

FRUITFULNESS → Creating life, marriage, and family

RULING → Exercising sovereignty (through science, technology, manufacturing, vocations, etc.)

Imagine Adam's joy as he discharged his part of the mandate and uncovered the truths of botany, biology, zoology, agriculture, even physics, chemistry, mathematics, all the other areas of the creation that God had commanded him to rule. As he gained greater understanding of the creation, he was able to use its resources more wisely. His God-given sovereignty over all creation led him to develop technologies (the application of scientific knowledge for practical purposes). Even today, discoveries made during scientific studies logically lead to new products. This development of technology continues to be the essence of God's mandate that we subdue and rule over the creation.

Again, we see that science had its beginning in the beginning of Genesis. In fact, the creation mandate of Genesis 1 laid the founda-

1. Divine Institutions are the basic practices and customs (institutions) God (divine) set up for the entire human race (believer and unbeliever) in order to protect them, preserve them, and allow them to enjoy blessings on and from the earth. God instituted personal responsibility (Gen. 1:26-30; 2:15-17; Ps. 8:3-8; Heb. 2:5-8), marriage (Gen. 2:18-24; Matt. 19:6; 1 Cor. 7:2-4; Eph. 5:22-33; Col. 3:18-20; Titus 2:3-6; Heb. 13:4; 1 Pet. 3:1-7), and family (Gen. 1:28; 4:1-2; Lev. 19:3; Deut. 6:4-9; Ps. 127:3-5; Prov. 1:8; 6:20; 13:24; 23:13-14; 29:15, 17; Eph. 6:1-4) before man's fall into sin. He began government after the Flood (Gen. 9:5-6; Rom. 13:1-7; 1 Tim. 2:1-4; Titus 3:1; 1 Pet. 2:13-17) and nations after Babel (Gen. 9:25-27; 10:1-32; 11:1-9; Deut. 32:8; Acts 17:26).

tion and parameters for good science. The study of science and the development of technology have no philosophical basis in the secular worldview. Again, secular scientists unknowingly stole or assumed them from the biblical worldview.

> In a darkened world no longer illuminated by the light of this center [God], technical advances are scarcely more than despairing attempts to make Hell a more agreeable place to live in. This must be particularly emphasized against those who think that by spreading the civilization of science and technology even to the uttermost ends of the earth, they can furnish all the essential preconditions for a golden age. One cannot escape the Devil so easily as that.
>
> **DR. WERNER HEISENBERG, 1932 NOBEL PRIZE IN PHYSICS**

Let's take a closer look at the main words that describe the second part of the creation mandate, subduing and ruling, so we can better understand our God-given purpose in life. The Hebrew word for subdue, *kabash*, is a strong, even military word, often used for bringing something or someone into bondage.[2] In the Genesis 1 context, before the effects of sin, its sense would have been of positive strength with no negative aspects. The word in the sinless Genesis context best carries the idea of harnessing the resources God built into the creation. God commanded Adam to subdue the creation, specifically to harness and utilize all its resources. As long as Adam obeyed this command, he flourished because following God's commands always leads to blessings.

SUBDUE
kabash
to bring into bondage; harness

RULE
radah
to have dominion; sovereign rule

To obey God's command to control, harness, and utilize the resources of the creation, Adam had to have the ability to accomplish those tasks. God insured that he did. When God created him in His image, He gave him the mental capacity to make observations, analyze data, make generalizations, and test his conclusions about the creation. By using these God-given abilities, Adam became the first scientist and the first to utilize the scientific method. Adam without sin would have been a genius of a scientist far greater than an Einstein. God has given us those same abilities,

2. Joshua 18:1; 1 Chronicles 22:18; 2 Samuel 8:11.

though dumbed-down now through sin's cumulative degradation over time. In comparison to Adam, the brightest of scientists today would be considerably diminished.

Adam was not only the first scientist but also the first sovereign. The second command in Genesis 2:28, to "rule" (*radah*), is another strong Hebrew word. It connotes ruling or having dominion over something or someone. God gave Adam authority over a portion of His overall creation. Sinless Adam was to rule it like a wise and benevolent king. An effective king exercises his sovereignty carefully and thoughtfully by managing his kingdom well. He utilizes all he rules for the good of all in his kingdom.

> To invent, you need a good imagination and a pile of junk.
> **THOMAS A. EDISON, INVENTOR EXTRAORDINAIRE**

WHAT WAS ADAM'S FIRST JOB AS SUBDUER AND RULER OF CREATION?

In Genesis 2, we see Adam obediently carrying out the second element of the creation mandate, to subdue and rule. In doing so, he was both a scientist and a king.

> Out of the ground the LORD God formed every beast of the field and every bird of the sky and brought them to the man to see what he would call them; and whatever the man called a living creature, that was its name. The man gave names to all the cattle, and to the birds of the sky, and to every beast of the field, but for Adam there was not found a helper suitable for him.
>
> Genesis 2:19-20

Notice that God built into Adam the ability to name the animals. First, Adam was able to make observations on each kind of animal. From those observations, he made distinctions between the different kinds as well as generalizations concerning the unique distinguishing characteristics of each kind. Finally, after analyzing all the data, he was able to formulate appropriate, descriptive names for all the creatures that came before him. Does that remind you of the work of a biologist? It should.

In recording this event, God gave us a glimpse of Adam's work as a zoologist within the science of biology. Through his study of the first creatures, the first man established the foundation for the zoological classification system still used in biology today. In fact, Carolus Linnaeus (1707-1778), the biologist who structured and codified today's classification system, has been called a "second Adam."[3]

Like so many truly great scientists, Linnaeus was a creationist who was inspired by the Genesis account. A foundational belief underpinning his zoological and botanical breakthroughs was God as Creator. In his own words:

> Theologically, man is to be understood as the final purpose of the creation; placed on the globe as the masterpiece of the works of Omnipotence, contemplating the world by virtue of sapient reason, forming conclusions by means of his senses, it is in His works that man recognizes the almighty Creator, the all-knowing, immeasurable and eternal God, learning to live morally under His rule, convinced of the complete justice of His Nemesis.[4]

Adam reflected his Creator when he exercised creativity in the naming process. Notice that God did not coach or help Adam. The Bible pictures God as a passive observer who allowed Adam to fulfill the creation mandate independent of divine suggestions. This tells us God built into the first man not only the intellectual capacity but the creativity to interact positively and productively with the creation. Both attributes continue to be essential for accurate scientific study.

We also see from this passage that Adam had a ministry as the ruler of creation. In Scripture, the giving of a name not only indicates the ability to discern characteristics about the one named but also signifies sovereignty over it. An invading military leader often renamed the city and king he conquered as a sign of his sovereignty over them.[5] During the days of creation, God Himself gave names to different

3. *The Genesis Account*, 326.
4. https://libquotes.com/carl-linnaeus.
5. 2 Samuel 12:26-29; 2 King 23:33-34; 2 Chronicles 36:4.

aspects of the creation (light, heavens, seas), indicating His sovereignty over them as Creator.[6] Adam gave his wife the name Eve to indicate his God-given headship in their marriage.[7]

Archeologist Roland de Vaux summarized the significance of this naming motif:

> Among primitive peoples, and throughout the ancient East, the name denotes the essence of a thing: to name it is to know it, and consequently, to have power over it. In the earthly paradise, when God allowed man to name the animals, it was a sign that he was putting them under man's power.[8]

IS THE CREATION MANDATE STILL IN EFFECT?

God issued the creation mandate in the perfect environment that existed before the Fall of mankind into sin. While the mandate is still in effect, thanks to sin, we have to work much harder now to fulfill it even partially.

Today, environmentalists insist that mankind must leave the earth pristine, whatever that means. Their agenda, another example of the politicizing of science, is in direct conflict with the biblical worldview. The creation mandate gives no hint that we should abuse the environment in even the smallest way. Instead, it commands us to manage and use it wisely to God's glory. Contrary to environmentalists' claims that overpopulation (again, whatever that is) will destroy the earth and us with it, God promises that we will never run out of the resources necessary to sustain abundant life.[9] While it is true that sin has corrupted those resources and the scientists who determine their use, we can be sure we will continue to have all we need because God keeps His promises. He loves us.

> Education without religion is in danger of substituting wild theories for the simple commonsense rules of Christianity.
>
> **SAMUEL MORSE, DEVELOPER OF MORSE CODE**

6. Genesis 1:5, 8, 10.
7. Genesis 2:23.
8. de Vaux, Roland, *Ancient Israel, Volume 1: Social Institutions*, McGraw-Hill Book Company, 1965, 43.
9. 2 Peter 1:3.

Environmentalism insists on limiting the use of resources rather than wisely utilizing them. In contrast, God commands us to study, control, and develop every resource because all creation, including us, benefit from its careful management. In fact, our wise use of the creation, which God gave us for our benefit, allows us to not just survive but also thrive.

In summary, man's exercise of limited sovereignty over creation leads to science, which leads to technology, which leads to manufacturing, which leads to vocations. These vocations may include involvement in science, technology, manufacturing, farming, construction, education, communication, medicine, engineering, electronics, computers, every occupation that uses the creation wisely. Though our vocations have great diversity, they should share one common purpose, glorifying God. In the words of Johannes Kepler, best known for his laws of planetary motion, "I wanted to become a theologian; for a long time I was unhappy. Now, behold, God is praised by my work even in astronomy."[10]

The Bible gives us content and guidance for theology, of course, but also for doing good science and developing technology. To reach maximum success, scientists and technologists need to understand and heed the directions God gave in the creation mandate.

Are you beginning to hear God shout?

10. https://juicyecumenism.com/2016/11/15/10-quotes-johannes-kepler-god/.

The burden of the word of the Lord concerning Israel.
Thus declares the Lord
who stretches out the heavens,
lays the foundation of the earth,
and forms the spirit of man within him.

ZECHARIAH 12:1

ZECHARIAH, PROPHET

Do we not all have one father?
Has not one God created us?
Why do we deal treacherously
each against his brother
so as to profane the covenant of our fathers?

MALACHI 2:10

MALACHI, PROPHET

CHAPTER 13

SCIENTIFIC FOUNDATION STONE #5

THE "VERY GOOD" CREATION

NOT SURVIVAL OF THE FITTEST

The fifth foundation stone focuses on the original creation. If we understand the perfection of the original creation and the cause of its fall, we are less likely to enter and live in scientific la la land. With our feet firmly planted in God's creation reality, we will see creation as He designed it, not as naturalism and its proponents distort it.

We know God is perfect, so we know He had to have created a perfect "heaven and earth." God Himself told of its perfection when He judged it "very good," without the slightest imperfection (Genesis 1:31). Looking at the world around us, we have to ask ourselves, what happened to that "very good" creation? What changes took place in the natural realm to cause the chaos and decay so evident everywhere today? And, pertinent to our study, how have those changes affected science?

WHAT WAS THE NATURE OF THE ORIGINAL CREATION?

The fifth foundation stone begins with Genesis 1:31 when the Creator described the original six-day creation as "very good," not just good but "very" good. That "very good" declaration had far-reaching implications. It meant that, during the six days of creation, God not only brought the universe into being out of nothing but also built properties and physical constants into it to maintain its goodness.

The creation in all its intricacies was not complete until, at the end of the sixth day, He declared it "very good." He "rested" because His work was not only complete but also perfect. Nothing could make the heavens and earth better. After I finish any project, I always end up tweaking it after finding needed improvements. There were no tweaks by God on day seven.

We cannot project ourselves back in time to observe God's creative processes. We have no way of studying them either. Even if we could, they are far beyond our finite understanding, but we can be certain the creation we live in today is substantially different from the original creation. "Very good" is not the evaluation we would make of a home destroyed by a fire, a swath of land obliterated by a tornado, a population ravaged by disease or war, or a nation attacked by rampant immorality and violence.

Scientifically, when God declared the creation "very good," He put the first law of thermodynamics into place. Stated simply, this law, still observed by scientists today, informs us that nothing is created or destroyed but only transformed. One of my engineering textbooks states that

> ...one form of the first law is the conservation of energy principle (or the law of conservation of energy): Energy can be neither created nor destroyed although it can be stored in various forms and can be transferred from one system to another.[1]

It states that the first law is one of the best-established laws of science.

With regard to its truth and range of application, all we can say about the first law of thermodynamics is that many, many experimental measurements are in accord with it and all attempts to find an exception to it have so far failed.[2]

We have already seen that the universe will have a fiery end, so it appears this law will not be operating after God completes His plan

1. Jones, James B., and Hawkins, George A., *Engineering Thermodynamics: An Introductory Textbook*, John Wiley & Sons, Inc, 1960, 62.
2. Ibid, 56.

for the universe. Until then, this law cannot in any way limit the work of the sovereign Creator. We have already seen that He was sovereign in His six-day creative work. We will see in a later chapter that He will continue to be sovereign over the entire creation until its divinely predetermined end and beyond.

> I realized that it made sense to believe that in the beginning God created, as this did not violate the laws of thermodynamics. I noted that modern-day observations, as well as the evidence of the fossil record, indicate that both plants and animals reproduced after their own kind as stated in Genesis chapter 1. I also realized that there was no simple explanation for the evolution of the information content which is found in living systems. Contemplating the amount of information in living systems has caused two professors at my own university (Professor Chandra Wickramasinghe and Professor Sir Fred Hoyle) to make the famous analogy that if you believe the information content in living systems to be the result of chance, then you believe that a tornado can go through a junkyard and assemble a jumbo jet!
>
> **DR. A.J. MONTY WHITE, PHYSICAL CHEMISTRY, GAS KINETICS**

WHAT HAPPENED TO THAT "VERY GOOD" CREATION?

The creation before the Fall had no decay, no movement toward disorder, nor any disintegration of systems. Nothing deteriorated or died. Everything remained perfect.

But, you might ask, didn't God tell Adam he could eat from any tree but one? Didn't He make vegetation delicious for both humans and animals? Didn't those plants die when they were eaten? Good questions. Again, reflect on God's perfect character. Since God gave Adam and Eve plants and fruits as food to be freely eaten, we can safely conclude, though we can't point to a specific verse, that He views death as pertaining only to human and animal life, not to plant life. See how easily the complex becomes simple when viewed through God's eyes?

The better we know the Creator, the better we understand His creation. Maybe those words should be inscribed above the entrance to every science lab!

God declared to Adam that he alone could damage the perfect creation. In fact, God warned Adam that if he disobeyed his Creator, he would die, the greatest of all deteriorations and destructions (Genesis 2:16-17). In essence, God promised Adam He would keep the creation very good and free from death if Adam chose to continue obeying Him. Adam didn't listen. In the most disastrous event in human history, he chose to rebel against His Creator. Nothing on earth, except Jesus Christ in His incarnation, has been perfect since then.

Paul explained the effects of Adam's first sin in Romans 5:12-14:

> Therefore, just as through one man [Adam] sin entered into the world, and death through sin, and so death spread to all men, because all sinned—for until the Law sin was in the world, but sin is not imputed when there is no law. Nevertheless, death reigned from Adam until Moses, even over those who had not sinned in the likeness of the offense of Adam, who is a type of Him [Jesus] who was to come.

Adam's sin brought death to the human race, but Jesus Christ, the second Adam, brought life, just as God promised.[3]

Today, viruses and death, accidents and infestations, murders and deformities are obvious and increasing, making it difficult to envision a once-perfect Earth and universe. In this generation alone, newly reported illnesses such as SARS-CoV-2 (Covid-19), swine flu, bird flu, and monkeypox (whether they are natural occurrences or man-made distortions of the "very good" creation) should have convinced us that the creation is becoming worse.

Ironically, many people, including secular scientists, believe that science will eventually bring us to a utopian world. They assure us that we are not actually moving into greater chaos and disaster, as Scripture teaches, and our senses perceive. Again, secular scientists are not only blind but wrong, terms aptly applied to anyone who excludes Creator God from his thinking.

The obvious question is, why is the creation no longer "very good"? Something had to have happened to that original creation in all its

3. 1 Corinthians 15:45.

perfection to make it the mess it is today. Genesis 3 gives the answer. It explains that the fall of man into sin not only affected mankind and his relationship with the Creator but also radically damaged the entire universe. Again, secular science has no time machine to make observations on that perfect world or to see the original sin that ruined it. They choose to ignore everything related to God because they can't see and evaluate it personally. Their arrogance and ignorance ensure that their scientific conclusions about every area of creation are incomplete and often flawed.

The sin introduced by Adam in Genesis 3 affected not only mankind but all the creation. Commentaries, theologians, and Bible teachers stress appropriately the spiritual aspects of the Fall because those aspects are the most important both temporally and eternally. The equally dramatic physical changes mentioned in Genesis 3 are often neglected. We will try to rectify that deficiency because those physical changes directly impact the study of science. The biblical foundation for science will help us recognize and account for those changes.

> The yearning to know what cannot be known, to comprehend the incomprehensible, to touch and taste the unapproachable, arises from the image of God in the nature of man. Deep calleth unto deep, and though polluted and landlocked by the mighty disaster theologians call the Fall, the soul senses its origin and longs to return to its source.
>
> **A.W. TOZER, THEOLOGIAN**

WHAT DOES THE BIBLE SAY ABOUT THOSE CHANGES?

God's Word describes the effects of sin on the creation in simple but profound language. By observing the biblical text and interpreting it using the Grammatical-Historical-Contextual approach, we can understand the radical changes from the pre-Fall creation to the post-Fall era.

THE TAKE AWAY: All creation changed because of the first sin. God instituted changes in both the spiritual and the natural realms as judgment for that sin. Even some fundamental laws of science and physical constants changed, not through natural processes but through God's sovereign action.

Some readers get lost in the uncomplicated, straight-forward language of the biblical text. They are tempted to add the world's thoughts to the inspired text because they find God's thoughts too direct and uncompromising. We don't want to become equally confused because the physical aspects of those changes relate to both science and history.

First, the consistent application of the G-H-C approach to biblical interpretation shows that the Fall of man was an actual event, as real as any other event in history, as real as your birth (which only a few people observed I might add). The whole Bible, not only the Old Testament but also the New, often refer to the events in Genesis 3 as real and historical.[4] Jesus Himself referred to Adam and Eve as actual people who lived on Earth.[5] We can be sure that the creation and the Fall happened just as God recorded even if the angels were His only audience (Job 38:7).

Moses recorded the events of Genesis 3 as a simple but factual narrative conversation between the Lord God and the man and woman He created. Our twelve-year-old might suggest we think of the Genesis 3 account like a diary entry, only divinely inspired and, therefore, completely accurate.

> Now it is important to realize that the Lord Jesus Christ believed in the early chapters of Genesis as historical fact. Indeed, such was His high view of Old Testament Scripture that He called it the 'Word of God,' and that God's Word "was truth," affirming that "scripture cannot be broken" (John 10:35; 17:17). That the Lord Jesus Christ believed in Adam and Eve, Cain and Abel, Noah and a worldwide cataclysmic Flood is evident from such passages as Matthew 19:4, 23:35, and 24:37-39. In Mark 10:6 Jesus said, "But from the beginning of the creation God made them male and female." In these words of Jesus, we find that He teaches that Adam and Eve were created at the "beginning of creation," not millions of years after the beginning! This also implies that God had prepared a world for them shortly beforehand. Everywhere we find that the Bible is consistent with the view of a literal six-day, recent creation. This has been the almost universal teaching of the Christian Church until the last hundred years or so.
>
> **DR. STEPHEN TAYLOR, ELECTRICAL ENGINEERING**

Second, the biblical text describes in some detail the radical effects of the Fall on the physical realm. If we could time-travel back to the world before the Fall, its beauty and perfection would astound us. Similarly, the science understood by Adam before the Fall was

4. Luke 3:38; Romans 5:12-14; 1 Corinthians 11:7-12; 15:21-22, 45-47; 1 Timothy 2:13-15.
5. Matthew 19:3-8, Mark 10:6.

significantly different from the science we study today. We will examine those differences by studying more of Genesis. Just in case, have your twelve-year-old on stand-by.

Genesis is a historical document, not mythology or poetry or allegory as some more liberal theologians suggest. Its literary form, which is consistently historical narrative throughout, can be divided into two major sections:

1. Primeval history (Genesis 1:1-11:26)
2. Patriarchal history (11:27-50:26)

The first division, primeval history, can be divided into four subdivisions:

1. The history of creation (1:1-2:3)
2. The early history of mankind (2:4-3:24)
3. The early history of civilization (4-9)
4. The early history of nations (10-11)

Genesis 2-3, a part of the second subdivision and the one we will study, can be divided into two sections:

1. The creation of mankind in innocence (2:4-25)
2. The fall of mankind into sin (3:1-24)

Genesis 3 can be divided further into five parts:

3. Satan's temptation of man (3:1-5)
4. Man's transgression (3:6-7)
5. God's confrontation of man (3:8-13)
6. God's condemnation of man (3:14-19)
7. God's provision (3:20-24)

After expounding on the temptation of Adam and Eve in Gen. 3:1-5, Moses described the Fall of mankind through disobedience to and rebellion against God's clear command in 2:16-17:

> The Lord God commanded the man, saying, "From any tree of the garden you may eat freely; but from the tree of the knowledge of good and evil you shall not eat, for in the day that you eat from it you will surely die."

Adam chose to obey his wife, not God. That one sin adversely affected the entire universe both spiritually and physically and continues to gain momentum as sin has been added to sin. The cumulative effect of many sinners sinning many times over many generations has been devastating. Some have compared sin to a cancer that metastasizes and grows, eventually destroying its host. Take a look around at the signs of death, decay, and destruction that demonstrate sin's omnivorous power, and you will agree.

The portion of the text we call God's condemnation of man (3:14-19) gives the judgments the Lord God imposed on the first couple because of their disobedience. Those judgments affected not only Adam and Eve but also all succeeding generations because we are related to them genetically and we, too, live on a sin-ravaged Earth filled with billions of sinners just as prone to sin as we are.

After Adam's original sin, God didn't first judge Adam but the agent of his temptation, the serpent. The biblical text tells us that God significantly changed that messenger of deceit.

> "Cursed are you more than all cattle, and more than every beast of the field; on your belly you will go, and dust you will eat all the days of your life."
>
> Genesis 3:14

God caused physical changes in the anatomy and even the DNA of that serpent and all subsequent serpents of its kind. The text also hints at possible dietary changes. How does eating dust sound to you? I'd say that was definitely a curse! So would our twelve-year-old but maybe not a two-year-old!

What was going on with the serpent? How could any non-human creature communicate and perform actions related to moral choices? Without going into too much detail, this member of the animal kingdom had a close relationship with another personal being not mentioned in Genesis. Based on revelations from the book of Revelation and elsewhere,[6] Satan himself was directly involved in the serpent's deception. Revelation 12:9 and 20:2 indicate a close relationship between these two, "the serpent of old who is called the devil and Satan." God seems to have permitted Satan to use this creature to bring about the first temptation. Even so,

6. Isa. 14:12-21; Ez. 28:12-19. There also seem to be close relationships between Satan and human kings.

a distinction remains between the physical creature and the invisible satanic angel (Gen. 3:1-5, 14-15).

The phrase "more than" in Genesis 3:14, used twice, emphasizes that God also cursed the rest of the animal kingdom, not just the serpent. Because the text describes only representative creatures, it is reasonable to conclude that the scientific study of zoology became radically different—and for the worse—after the Fall.

Next, God judged the woman Eve, saying, "I will greatly multiply your pain in childbirth. In pain you will bring forth children" (3:16), implying she would have felt no pain had she had children before the Fall. The penalty for disobeying God made "very good" childbirth very painful. As any woman knows, other changes associated with childbearing also adversely affected Eve's body and every woman's body since. The science of anthropology, including the sciences of gynecology and reproduction, were fundamentally changed by this curse. The text suggests, though does not implicitly state, that the man underwent physical changes as well. Either way, the science of anthropology, including the science of reproduction, was fundamentally affected by that first sin.

Next, the Lord God judged the ground. "Cursed is the ground because of you" (3:17). God did not have Moses delineate every physical change or their extent, but they must have been far-reaching if the other judgments are any indication. Other passages (Romans 8) and other events (the Genesis Flood), both of which we will study, give more information about this judging of the ground.

Did the changes include earthquakes and tectonic shifts? What about hurricanes, tornados, and droughts? This curse informs us the entire Earth was affected by the Fall. So was the science of geophysics.

Genesis 3:18 reveals all botany also changed. "Both thorns and thistles it shall grow for you." This suggests no plants were weaponized against man before the Fall. Also, because thistles are prickly weeds that are especially hard to eradicate, all plants pre-Fall must have been beneficial in every way. Other botanical changes may have occurred though they are not detailed in the text. The plants in view were representative of all plants and the changes representative of a variety of curses. Our twelve-year-old might mention poison sumac or deadly nightshade about now.

Finally, God judged the man, the one with no excuse. "By the sweat of your face You will eat bread" (3:19). God's curse judgment on the male implied additional, unstated changes. Work before the Fall appears to have been delightful and fulfilling. God's judgment initiated adverse physical forces. Death, not just of man but also of all animals, insects, and everything with breath was a direct consequence of the Fall. Man must now deal with enemies visible (bugs, drought, too much water, evil men) and invisible (viruses, systemic weaknesses, wickedness). These changes made work that had been easy, enjoyable, and fulfilling before the Fall, laborious, unpleasant, and seemingly unending afterward. No doubt you can identify with that curse.

PHYSICAL EFFECTS OF THE FALL

ZOOLOGY → Serpent cursed

ANTHROPOLOGY → Woman's pain

GEOPHYSICS → Ground cursed

BOTANY → Thorns & thistles

PHYSICS → Toil, sweat, death, inefficiency

All these curses imply an imposition of some fundamental law upon the entire natural realm, a movement from organization to disorganization, the tendency of systems to unwind, the law of decay.[7] Even those with little scientific understanding can see that God's judgment of that first sin radically changed the laws and physical constants of physics and biology. Those changes continue to affect nearly all other physical constants and processes in the natural realm.

> I actually don't believe God created viruses as separate entities, I believe they were a part of the DNA in cells. Some evolutionists put viruses down as a predecessor of cells, but that doesn't work, because they need to have the machinery of cells to reproduce. I actually see viruses as genetic garbage, having escaped from cells way back, as a result of mutation, environmental damage—part of the Curse on creation [Genesis 3]. I would predict from that theory that we should find pieces of "virus" DNA in the human genome (DNA). And that's starting to be found.
>
> **DR. IAN MACREADIE, MICROBIOLOGY**

The operation of this imposed law of decay is obvious in the astrophysical realm as well as on Earth, implying the Fall affected the entire universe.

7. This law of science will be described in more detail below.

That makes sense, doesn't it, since God created the universe as a backdrop for both Earth and mankind and placed it under man's dominion?

Much more could be said about the changes brought about by that first sin, but these are enough to make the point the pre-Fall world was radically different physically from the post-Fall world. If we could study that pre-Fall world scientifically, we would be astounded by its perfection and saddened by the decay and disintegration sin brought. Is that ever an understatement! Sin is dangerous and damaging!

DOES THE NEW TESTAMENT DESCRIBE THESE EVENTS?

Genesis 3 is not the only passage in Scripture that details the radical changes that resulted from the Fall of mankind into sin. We have additional inspired documentation that confirms the historical account and makes clear both the theological and physical aspects of those changes. Romans 8 is one such passage.

Before we dive into the exciting details of Romans 8, let me set the context. First, concerning the broad historical context, Genesis and Romans are separated by over a millennium. Genesis 3 was written before 1446 BC by Moses, a Jewish leader who believed in God as Creator. He probably wrote Genesis to the Jews to prepare them to leave Egypt as His distinct people, a divinely established nation ready to reenter and rule the land He had promised them.

Moses reminded the Jews that their Creator had always been faithful in the past, starting with Creation, and, therefore, they could rely on His faithfulness as they faced their militarily superior Egyptian and Canaanite enemies. He assured them that since God gave the first man the power to rule all creation, and He did, He could certainly give the Jews, if they obeyed Him, the power to overcome His physical enemies. You can see that the creation mandate to subdue and rule has been relevant throughout history.

Some fifteen hundred years after Moses, in AD 57-58, the apostle Paul wrote a letter to the small groups of believers scattered throughout the city of Rome.[8] In it, he emphasized, as had Moses, the truth

8. Several, mainly small, church groups existed in Rome in the first century, some of them home churches. Romans 16:5, 10-11, 14-15.

of God's faithfulness, this time in the context of those who believe in Christ as Messiah and Savior. Paul had been an unbelieving Jew, a devout killer of Christians in fact. After a personal encounter with the risen Christ,[9] this renowned Christ-hater believed in Christ as his Savior and became His main apostle to the Gentiles. Don't you love hearing about God's mercy to even His greatest enemies? He truly loves each of us and wants to use and bless us in astonishing ways.

In his letter to the Romans, Paul explained the effects of the ancient creation and Fall events on Christians in the Church Age, the age that began at the Day of Pentecost and will end in the future with the Rapture of the Church.[10] We can be sure those ancient Genesis events affect us as much two thousand years later as they did his Roman readers.

Since God included both Genesis and Romans in His inspired Scriptures, both are equally true and important regardless of their different historical settings. Moses and Paul both explained absolute truths about the creation but from different perspectives. Paul confirmed that the ancient first-sin tragedy continued to affect his time just as it had Moses'. I think you will find that their different perspectives make an interesting study.

Second, the broad context concerns the nature of the book of Romans, arguably the most theologically intense book, not just in the New Testament, but in the whole Bible. The literary genre of the book is epistolary (a formal letter). In contrast, Genesis is a historical narrative. Romans is primarily didactic and explanatory, more interpretive than observatory. Genesis gives historical information about the creation and Fall of mankind. Romans 8 gives the interpretive, explanatory truths concerning those historical events along with their theological implications.

Third, the theological context concerns Paul's focus on God's righteousness as one of His perfections.[11] Because God is righteous, meaning perfect in all moral aspects, He is entirely holy and faultlessly just. He is the only standard for holiness, righteousness, and justice because He alone embodies those perfections perfectly. Paul explained to the Romans (and through them to us) that every one of us can

9. Acts 9:1-9; 22:4-11; 26:13-18.

10. The Church Age began at the Day of Pentecost when the Holy Spirit descended on the early disciples and will end with the Rapture of the Church before the seven-year Tribulation.

11. Isaiah 41:10; 45:19-24; John 17:25.

possess this righteousness that belongs only to Jesus Christ by placing our faith in Him alone as Savior.[12]

THE TAKE AWAY: God created mankind with a measure of righteousness. After all, mankind in the persons of Adam and Eve were part of that "very good" creation. In the beginning, man had a perfect standing and relationship with his Creator. At the Fall, he lost not only his standing before God but also his perfect physical nature. He fell from that perfect standing and condition into an unrighteous standing and an unrighteous nature. To put it bluntly, every human being is born without righteousness.[13]

In Romans, Paul explained we can gain a perfect relationship with the Creator, one like Adam's before he sinned, which includes receiving Christ's perfect righteousness at the moment we trust Him. At that moment, we are declared righteous in our standing before God but are not made righteous. Paul then explained the way we can live out and grow experientially in that righteousness in everyday life.[14]

Fourth, the context concerns the place of Romans 8 in the book of Romans. Romans can be divided into two parts:

1. The provision of God's righteousness to mankind (1-11)
2. The application of God's righteousness to everyday life (12-16)

The first division is predominantly theological while the second is more practical in its application to our lives.

The first division has five subdivisions:

1. The introduction to the book (1:1-17)
2. The condemnation of all mankind because of sin (1:18-3:20)
3. The justification of mankind through faith in Christ alone (3:21-5:21)

12. Ephesians 2:8-9; Romans 3:21-26; 1 Peter 4:17.
13. Romans 3:10-18.
14. The Bible and Paul explain this day-by-day process as sanctification in Romans 6-8.

4. The sanctification (growing in righteousness) of believing mankind by faith, including the promise of glorification in a future resurrection body (6-8)
5. The vindication of God's righteousness in relation to the nation Israel (9-11)

Fifth, the more immediate context of our verses is toward the end of the fourth subdivision. Paul had completed his explanation of the principles of experiential sanctification, which we can define as theological principles for living the Christian life. Beginning with Romans 8:17, as a transition to the next paragraph, Paul began discussing suffering. He explained that suffering is that part of God's sanctifying process that enables us to better appreciate and enjoy His blessings. Suffering is also part of our training for growth in righteousness.

In Romans 8:18-25, Paul gave the biblical perspective on suffering, that the believer should persevere during suffering by staying focused on eternity. He went all the way back to the origin of suffering, the Fall of mankind, and looked forward to a future when God will remove all the suffering sin brought. He taught that, for the believer in Christ, suffering is temporary. For those belonging to God, all suffering has an end, an encouraging truth to help us not lose hope, the major theme of the passage.

THE TAKE AWAY: Paul encouraged us to endure suffering by staying focused on God's assurance of our glorification, our perfect and eternal future. That eternal future perspective is the big picture perspective that helps us in the midst of the pain of suffering.

> Vanity of science. Knowledge of physical science will not console me for ignorance of morality in time of affliction, but knowledge of morality will always console me for ignorance of physical science.
> **BLAISE PASCAL, EARLY MODERN MATHEMATICIAN, PHYSICIST, INVENTOR**

Paul's teaching in this passage included not only the pain of suffering but also the negative characteristics of the post-Fall universe. He cited Genesis 3 as the origin of pain and suffering and explained the physical changes that occurred as a consequence of God's judgment on that first sin. He included scientific concepts and gave scientific descriptions. In fact, Romans 8 gives one of the clearest descriptions of the impact of the Fall on the physical realm, not just in all the Bible, but in all non-technical literature.

Romans 8:18-25 can be summarized as the hope of believers in the midst of suffering. We will break it down into four alliterative parts:

1. The antithesis to suffering (8:18)
2. The anticipation of glory (8:19-21)
3. The agonizing wait (8:22-23)
4. The application of faith (8:24-25)

The antithesis of the suffering of Romans 8:18 is the future glory described in 8:19-25, anticipation of which glory should give us hope as we apply God's promises by faith.

Within these helpful and encouraging insights into suffering and blessing, Paul included statements relating to the natural realm and science. He referred to the natural realm four times just in Romans 8:19-22, using the same Greek word for "creation" each time. Those four scientific statements will be our focus.

> The discovery of the laws of definite proportions is one of the most important and wonderful among the great and brilliant achievements of modern chemistry. It is sufficient of itself to convince any reasoning mind, that order and system pervade the universe, and that the minutest atoms of matter, and the vast orbs that move round the havens and equally under the control of the invariable laws of the creator.
>
> DR. JOHN LEE COMSTOCK, CIVIL WAR SURGEON

This section is only one of many examples of God's revelation of scientific data that scientific investigation could never discover. At the same time, any honest scientific observer can uncover some of its insights, thereby proving the validity of those that cannot be discovered scientifically. God is good to us, isn't He?

WHAT DOES SCIENCE HAVE TO DO WITH THE FALL?

With that overview, let's focus on Romans 8:18-23 and the scientific elements it introduces. We have seen that the Bible not only describes the physical realm but also makes scientific statements about it. This passage sets some parameters for some easily observable and unchanging laws of nature that have existed since the Fall. Science textbooks seldom include these parameters and, thus, present an always degenerating, depressing future without hope or remedy. At the same time, they predict a world gradually getting better through evolution. Secular thinking cannot be logical because it excludes God, the Author of logic and order.

> The exquisite order displayed by our scientific understanding of the physical world calls for the divine.
>
> **DR. VERA KISTIAKOWSKY, PHYSICS**

In His Word, God, the Creator and Sustainer of the universe, says He will free the world from its suffering. Even our twelve-year-old knows we should believe the One who has known the end from the beginning,[15] not listen to those who can't do more than make wild guesses about that which they cannot know.

> For I consider that the sufferings of this present time are not worthy to be compared with the glory that is to be revealed to us. For the anxious longing of the creation waits eagerly for the revealing of the sons of God. For the creation was subjected to futility, not willingly, but because of Him who subjected it, in hope that the creation itself also will be set free from its slavery to corruption into the freedom of the glory of the children of God.
>
> Romans 8:18-21

The sentence that introduces the paragraph (8:18) hints at scientific ideas. The opening phrase, "I consider," translates the Greek verb, *logizomai*, which is used several times in the New Testament, mainly in a theological context. You may have noticed that this word is related to another, better-known Greek word, *logos*, a noun used as a title for

15. Isaiah 46:10.

Jesus, THE Word.[16] It is also related to the idea of logic and reasoning, both essential to scientific investigation.

Sometimes Scripture uses *logizomai* in the theological sense of imputation or crediting something to one's account.[17] At other times, it means considering, regarding, and taking something into account.[18] Though often used in this accounting sense, it is also used in a more scientific sense of reasoning something out and coming to a logical conclusion after evaluating all the data.[19] Could Paul have been saying that he had looked at all the scientific data and come to the conclusions ("I consider") he would discuss in this passage? I think that assumption is reasonable.

In the portion that I call the anticipation of glory (8:19-21), Paul described a drastic change in the make-up of the entire physical realm that led to the suffering that has plagued mankind ever since. In it is one of the clearest descriptions of the fundamental nature of the creation. The last part of the passage (8:21) anticipates release from that drastic change, the glory anticipated. Both have huge and profound scientific implications. The long-ago change continues to affect virtually every science. When a scientific model is formulated, it often contains a predictive element as we have here, in the anticipation of a future outcome. The future change will be equally dramatic.

Not only is the description of the judgment in Genesis 3 about the entire creation, but Paul reiterated this comprehensive sense by referring to the entire creation four times in his scientific description. In only three sentences (8:19-22), he used the word creation (*ktisis*) four times, so we know the whole of creation was his focus. The entire natural realm has reeled in judgment since that first sin.

CONSIDER
logizomai
to reason and come to a logical conclusion

CREATION
ktisis
the natural realm

16. As we have seen, John 1:1 revealed Christ as the Word, the ultimate Logos.
17. In Romans 4:3-6, 8-11 and Galatians 3:6, righteousness is credited (imputed) to those who believe in Christ as Savior.
18. 2 Corinthians 10:2, 7, 11; 11:5.
19. 1 Corinthians 13:5, 11.

Throughout these verses, Paul used an interesting figure of speech, personification, to convey literal truth. Personification, which is not allegory, is common in Scripture as well as in our culture. It involves describing something inanimate as if it were a person. Today, we might say the White House issued an executive order. We understand the building itself cannot think, write down ideas, or issue orders. Instead, this personification more dramatically and effectively explains that the President who lives in the White House issued an important piece of information. His living in the White House, the ultimate symbol of American ruling authority, emphasizes his power to make such pronouncements.

In a comparable way, Paul personified the creation throughout this passage so that we can see his emphasis more clearly and authoritatively. In 8:19, he personified it as having strong desires, "the anxious longing of the creation." In 8:20, he gave the creation volition, "not of its own will." In 8:21, he personified it as a slave to be set free, "will be set free from its slavery." In 8:22, he personified it as experiencing pains like a pregnant woman giving birth, "groans and suffers the pain of childbirth." These personifications make the physical and scientific descriptions in the verses more vivid, visual, and relatable in the context of suffering. We can better understand creational suffering when we relate it to our own human suffering.

The use of personification is a common literary device that conveys literal truths. It is not allegory. Paul used personification to emphasize the horrendous outcome of that first sin. He convinced me sin is dangerous. What about you?

DOES ROMANS 8 DESCRIBE A LITERAL DRASTIC CHANGE TO THE NATURAL REALM?

The main verb in 8:20, "was subjected," reveals other important insights. This strong Greek word was often used in a military and conquering sense.[20] Scripture uses it mainly to refer to God's order within government and families.[21] In our verse, the word carries the idea of subjugation or, as translated, to be subjected. Paul used it two times in

20. 1 Chronicles 22:18.

21. Romans 13:1, 5; Ephesians 5:22; 1 Peter 2:13; 3:1.

this verse to emphasize a definite, powerful divine act. He wanted us to understand that the curses in Genesis 3 were definite, powerful acts of judgment by the Creator, which he described as "subjected to futility."

The Greek language has a variety of past tenses. The verb "was subjected" is in the aorist tense, which refers to a definite past event that took place at one point in time. It is not in the imperfect tense which indicates an action taking place indefinitely or over an extended time. Paul deliberately chose the aorist tense to make sure we understand that God's subjection of creation was an actual, one-time, historical event. Moses gives us the details of that historical event in Genesis 3.

SUBJECT
hupotasso
authoritative subjecting by God

FUTILITY
mataiotes
emptiness; purposelessness

The verb "subjected" is in the passive voice, which means the subject of the sentence was acted on by an outside entity. In this case, the creation did not choose to be subjected. Something outside of it performed some subjecting action on it. Paul's use of the passive voice emphasizes that God Himself did the subjecting, all because mankind, in the person of Adam and Eve, chose to defy His one prohibition.

The next phrase, "not willingly," confirms this truth. Paul used this personification to show that the creation did not want to be subjugated, that something outside the creation acted on it. The text adds a qualifying phrase "because of Him who subjected it." The New American Standard Bible (NASB) correctly capitalizes the pronoun "Him," indicating that God was the One doing the subjecting. The context indicates that God did every supernatural act described in the passage. Scientifically, this verse confirms that the entire universe experienced something related to subjection as a result of the Fall of mankind into sin. This change was radical and supernatural.

What was this subjecting action? The context refers to God's judgments in Genesis 3 that affected not only Adam and Eve, the ones who sinned, but also every aspect of the creation. The explanatory phrase, "because of Him who subjected it," not only repeats the verb for subjecting but also relates it to God, the One doing the subjecting, the cursing.

This subjecting was a supernatural action, a direct judgment from God. Only He has the power and authority to affect the entire creation at one moment in time.

God introduced this curse into the creation at this time of judgment, not during the creation process. Because of sin, God instantly destroyed the "very good" condition of the original creation.

God promised Adam that if he ate fruit from the Tree of the Knowledge of Good and Evil, he would most assuredly die (Genesis 2:17). The Hebrew construction of the phrase is an infinitive absolute, literally translated "you shall die dead." It conveyed to Adam the certainty that he would die if he disobeyed. God always keeps His promises. His holiness makes it impossible for Him to do otherwise. And so, death, first spiritual and then physical, entered the creation as judgments just as God promised.

Romans 8 expands our understanding, explaining that judgment had to come upon the entire creation, upon the entire universe. That first sin was that bad! God promised in Genesis 3:15 that He would deal with sin not only ultimately but throughout world history. You could consider Genesis 3:15 a summary of the rest of world history. Sin always has negative consequences though sometimes we are able to convince ourselves we got away with it. God never sleeps. He always knows when we sin, even in our thinking. Just as with Adam and Eve, His justice and righteousness demand that He judge our sin.

SCIENCE INSIGHTS IN ROMANS 8:18-21

- REASONED CONCLUSION → Use of logizomai (v. 18)
- PREDICTIVE MODELING → "eager anticipation" (v. 19)
- HISTORICAL EVENT → use of aorist tense (v. 20)
- OUTSIDE CAUSE → use of passive voice (v. 20)
- CHANGED NATURE → "subjected to futility" (v. 20)
- DIVINE ACTION → "Him who subjected it" (v. 20)
- DIVINE PURPOSE → "in hope" (v. 20)
- FIXED PRINCIPLE → "slavery to corruption" (v. 21)
- RADICAL FUTURE CHANGE → "will be set free" (v. 21)

THE TAKE AWAY: The scientific statements made in these verses are profound. The most likely mechanism for God's promised subjection is His implementation of a principle of degeneration and death, described as the curse. This foundational principle of modern science is the universal tendency of structured systems to move from order to disorder. Entropy is the quantification of this negative movement. Death is the state of maximum entropy.

The portion of Romans 8 that I call the anticipation of glory (8:19-21) has one of the clearest descriptions of this drastic change found anywhere in Scripture. Paul referred to it as a basis for suffering. The last part of the passage (8:21) anticipates a still future release from this curse, an anticipated glory that is certain because God, the Creator, promises it.

Simply put, the law of decay is the tendency in all systems to unwind, unravel, or wear out. It is the movement from organization to disorganization in the natural realm. Every woman knows her home does not clean itself but instead gets dusty and chaotic just sitting there. Waiting a few years doesn't improve it. Every man knows his brand-new car begins to deteriorate the moment he drives it off the car lot. Driving a car more does not improve it. Both show this law in operation. The passage of time inevitably brings deterioration with it. Paul described it as a law of "futility" (8:20), "slavery to corruption" (8:21) and a cause of suffering.

I hesitate to mention that you see this law at work every time you look in a mirror. This tendency is undeniably the result of an ever-increasing number of genetic mutations in our germ and somatic cells as we age. Visible aging is hard to ignore, isn't it, especially the older we get? Our twelve-year-old doesn't understand yet. All too soon, he will.

THE TAKE AWAY: Secular science excludes divinely revealed data and supernatural explanations and, therefore, cannot discover the origin of scientific truths like the law of decay. It cannot understand the radical changes that first sin caused to the pre-Fall universe. The law of decay invalidates the assumption of uniformitarianism, the fundamental belief of secular science that the universe today is the same as it has been for hundreds of thousands of years, that the natural realm has not changed in any significant way, that the study of science today is the key to understanding science in the past. We discussed and rebutted this fallacy earlier. The law of decay is an example of a law of nature that has changed from one dispensation to another and that refutes evolution and all its ancillary tenets.

> Every formula which expresses a law of nature is a hymn of praise to God.
>
> **MARIA MITCHELL, FIRST AMERICAN PROFESSIONAL FEMALE ASTRONOMER**

The details of Romans 8:20-21 give a concise description of this principle of degeneration and death:

> For the creation was subjected to futility, not willingly, but because of Him who subjected it, in hope that the creation itself also will be set free from its slavery to corruption into the freedom of the glory of the children of God.

In the first descriptive term, the word "futility" has the sense of emptiness, purposelessness, and transitoriness.[22] Isn't that a good description of degeneration and death?

A second phrase, "slavery to corruption" (8:21), indicates that the operation of this principle is an irreversible and unchanging constant. It truly is a fixed principle because God made it that way after the Fall. Death will continue to operate and prevail until He again changes the Earth and does away with sin and its consequences.

The term "corruption" means ruin, destruction, dissolution, and deterioration.[23] It describes the constant decay, degeneration, and principle of death so readily observable all around us.

"...the whole creation groans and suffers the pains of childbirth" (8:22). What a graphic description of sin's destruction of the "very good" creation! Notice that the whole creation, the whole universe, was devastated and continues to be devastated by the consequences of sin. The personification verb "groans" indicates ongoing effects, a good description of the shadow of death and the sorrow it includes. The personification phrase "pains of childbirth" foreshadows the inevitability of suffering and pain in a sinful creation. Could there be a clearer

22. Arndt, William F. and Gingrich, F. Wilbur, *A Greek-English Lexicon of the New Testament*, The University of Chicago Press, 1957, 496.
23. Ibid, 865.

description of the negative effects specified by scientific principles, without using technical, textbook language?

Romans is not the only book in the Bible to contain a description of the principle of degeneration and death. See if you can find this scientific principle described in the following verses:

> Of old You founded the earth, and the heavens are the work of Your hands. Even they will perish, but You endure; and all of them will wear out like a garment; like clothing You will change them and they will be changed.
>
> Psalm 102:25-26
> (also quoted in Hebrews 1:10-12)

> Lift up your eyes to the sky. Then look to the earth beneath; for the sky will vanish like smoke, and the earth will wear out like a garment and its inhabitants will die in like manner; but My salvation will be forever, and My righteousness will not wane.
>
> Isaiah 51:6

> A voice says, "Call out." Then he answered, "What shall I call out?" All flesh is grass, and all its loveliness is like the flower of the field. The grass withers, the flower fades, when the breath of the Lord blows upon it; surely the people are grass. The grass withers, the flower fades, but the word of our God stands forever.
>
> Isaiah 40:6-8

The biblical description of this principle is the opposite of the naturalistic view. Evolution teaches that the principle of innovation is inherent in life, that every aspect of life must and will move from disorder to increasing complexity and order. In case I have not yet convinced you, I could accurately state that the law of decay by itself, as described in Scripture, disproves evolution.

> A major issue in all this is the judgment of God on the whole creation at the Fall, which is totally undermined by the traditional vast-ages viewpoint.
>
> **DR. BRIAN STONE, ENGINEERING**

WHAT IS MEANT BY THE PRINCIPLE OF DEATH?

This leads us to significant questions in creation science. Is the fundamental change that took place at the Fall of mankind related to the second law of thermodynamics? More specifically, is the curse described in Genesis 3 and Romans 8 related to the second law? Did the second law begin operation at the Fall when God imposed the judgment for the sin of Adam and Eve or before?

You are probably asking about now, what is this impressive sounding second law of thermodynamics? Technically speaking, the second law of thermodynamics applies to energy and energy transfers (thermodynamics) though it is also applied to biological and other energy systems. The law describes positive, negative, and even neutral effects on creation. As you have probably already guessed, the law of decay is a negative aspect of the second law.

From the perspective of creation science, there seem to be three possible answers to the question of the relationship of the second law of thermodynamics to the "curse" of Genesis 3 as described in Romans 8. Remember, no one was there to observe these characteristics of the pre-Fall world nor are we able to take a time-machine back to do experiments. But let us explore some possibilities.

First, the second law may be directly related to the curse. In fact, God may have instituted the law as the judgment at the Fall of mankind. That would mean that the pre-Fall "very good" world was free of the second law. Many creationists of the past and today have taken this view.

A second possible answer is that some positive aspects of the second law existed before the Fall of mankind while detrimental ones were initiated at the Fall. The law in some form did in fact exist in the pre-Fall world because some processes in the "very good" world probably required it.

God Himself may have activated the detrimental aspects of the second law as a consequence of His judgment. The decay, degeneration, and eventual death of biological life systems, the disintegration of effectiveness in mechanical systems, and the destructiveness of geophysical and astrophysical events like hurricanes, earthquakes, and exploding stars are all examples.

Most scientists broaden the second law to include virtually all systems, including biological. I found this good description of the second law in one of my engineering textbooks:

> Thus, heat always flows spontaneously from a hotter to a colder body; gases always seep thru an opening spontaneously from a region of high pressure to a region of low pressure; gases and liquids left by themselves always tend to mix, not unmix...Rocks weather and crumble; iron rusts; people grow old. These are all examples of irreversible processes that take place naturally in only one direction and, by their one-sidedness, express the second law of thermodynamics.[24]

The author included "people grow old" as part of his broad description before going into his specific thermodynamic definition. Many creationists take this broad understanding of the second law.

The third possible answer is that the entire second law of thermodynamics existed as a part of the original creation and existed before the Fall of mankind. In personal conversations, John Baumgardner called my attention to the problem of claiming that the second law began operation at the Fall of mankind in Genesis 3. He pointed out that some beneficial processes involving the second law would have been necessary in the "very good," pre-Fall world.

Other creationists also give specific aspects of the law that were probably present before the Fall.[25] The transfer of heat from a hotter object to a colder one, say from the sun to the earth, is basic to thermodynamics. Adam and Eve would have needed heavy, heavy coats all the time had they not enjoyed this transfer of heat. In another thermodynamic energy transfer, gases move from high pressure to lower pressures as seen in human and mammal respiration. I believe we are safe in assuming Adam and Eve enjoyed being able to breathe. Another beneficial operation of the law occurs during digestion of food when our bodies breakdown complex food molecules into simple components that fuel and build our cells.

---

24. Sears, F.W., and Zemansky, M.W., *University Physics*, Addison-Wesley Publishing Co, 1964, 431.
25. Sarfati, 395. https://creation.com/the-second-law-of-thermodynamics-answers-to-critics#-fall (Answer #3).

Sarfati suggests that:

> …rather than postulating that the Second Law began at the Fall, it is more likely that God withdrew some of His sustaining power at that time. He still sustains the universe (Colossians 1:17) otherwise it would cease to exist. But most of the time He doesn't sustain it in the way that He did when He prevented the Israelites' shoes and clothes from wearing out during the 40 years in the wilderness (Deuteronomy 29:5). But this special case may have been the normal state of affairs before the Fall.[26]

The third answer to the question seems to be the best. Our Creator probably built the second law of thermodynamics into the "very good" creation, but something drastic occurred as a result of that first sin. God introduced another comparable law that the Bible calls the curse as the judgment for sin upon man and the natural realm, mainly for biological systems. This curse, which included the corruption of the information in the genetic make-up of all living things, had profound and far reaching effects.

I do not equate the curse of Genesis 3 with the second law. My purpose is to stress that the drastic change in the entire natural realm as a result of the Fall of mankind was as law-like as the second law is in physics. A fundamental change took place in all the natural realm that affected everything and can be observed in all the sciences today. In Genesis 3, Moses described it as a "curse" while Paul described it as "subjected to futility" in Romans 8.

> Surely all of us are familiar with the everyday expression of this [second] law [of thermodynamics] in open systems. (The humorous popular version of the second law is Murphy's Law: "Whatever can go wrong will go wrong.") Metals corrode, machines break down, our bodies deteriorate, and we die. Constant maintenance and planning against contingencies are required if life is to be sustained for even a transitory period, such as the lifetime of the individual. Ultimately, the second law takes over, and our bodies return to dust and our automobiles to the junkyard. By the application of our minds, we can resist the demands of the second law temporarily. General evolution collapses around this concept, however, because at the initiation of the evolutionary process in antiquity, there was no mind available to construct purposive "machines" to temporarily obviate the second law's demands. The idea that the second law can be confined to closed systems is a piece of confusion on the part of the proponent of such a concept.
>
> **DR. KER C. THOMSON, GEOPHYSICS**

26. Sarfati, pp. 395-396.

WILL THE PRINCIPLE OF DEATH EVER BE REVERSED?

In Romans 8, Paul began by contrasting our present suffering on this sin-cursed Earth with our glorious future with our Creator in His heavenly home, which will be sin-free, of course:

> For I consider that the sufferings of this present time are not worthy to be compared with the glory that is to be revealed to us.
>
> Romans 8:18

For those of you who haven't believed that Christ took the punishment for your sins on the cross, thereby conquering sin, and was resurrected from the dead, thereby conquering death, you need to reevaluate your worldview. Believing that you don't need Christ's substitutionary work on your behalf is even more dangerous and stupid than believing that you evolved from ooze and will eventually reach perfection. You won't.

THE TAKE AWAY: Continued decay is an unending problem only for those who don't have the assurance of an eternal life through Jesus Christ. Believe now that Jesus Christ died for you and was resurrected. God ensures that simple faith in His Son results in a glorious future forevermore. For the believer in Christ, physical death brings the future transformation that life on a grief-filled Earth can never hope to bring.

Paul encouraged all believers by reminding us of God's promise to radically transform all creation again. Just as God changed all creation after Adam's sin, so He will change it in the future, this time back into perfection. At that future time, He will remove the curse of the law of death. If we have believed in Christ's substitutionary death and resurrection, our bodies will die, but we will receive new, perfect bodies that will live on in a perfect environment similar to the Garden of Eden, one unaffected by the corruption of sin.

Try to imagine a world in which the principle of death has been lifted. Hard, isn't it? But you can be sure that Creator God always keeps His promises. Such a world awaits all who accept Christ's gift of eternal life. Salvation from the curse of sin comes only through Christ.

"For the anxious longing of the creation waits eagerly for the revealing of the sons of God." In Romans 8:19, Paul began describing the future that awaits the creation. Using personification again, he described the creation as "waiting eagerly" for that future day. At the end of 8:20 ("in hope") and then into 8:21, he presented a hopeful picture, a future that will reflect creation in its pre-Fall perfection.

This passage makes it clear that the creation "will be set free from its slavery to corruption." This freedom involves the lifting of the law of death, a future time that will coincide with every believer being set free from all suffering. Just as the creation will be radically transformed to a condition without death and sorrow so will our new bodies display no vestige of mortality or disease.

Only the God of the Bible can restore not only mankind but all creation through His re-creating power. Not only can He, but Romans 8 assures us that He will. What a magnificent promise!

> God's nature embodies justice and holiness, but is also a personal and loving God who cares for each creature (so the name 'father' is indeed appropriate).
>
> **DR. GEORGE ELLIS, ASTROPHYSICS**

THE TAKE AWAY: While Paul wanted us to understand the devastating effects of sin, even more he wanted us to appreciate God's grace in providing blessings through suffering—yes, through suffering. The key, he taught, is in holding to the eternal perspective while suffering the problems caused by living with a sin nature inside us and a sin-cursed world outside us. For those who trust Christ, the future is always hopeful. Contrary to the teachings of evolution, we truly will find unending joy and peace when our bodies perish, and we are face to face with our Lord.

WHAT CAN WE CONCLUDE?

Romans 8:19-22 tells us that God caused radical changes in the past and will initiate radical changes again in the future. He was the One who "subjected" the creation to "futility." Nothing in the natural realm caused those changes. They were not natural, evolving changes but supernatural ones. God cursed both mankind and the rest of creation because Adam chose to rebel against Him. Sounds grim, doesn't it, but God loves us and promises to one day remove those curses through another display of His absolute sovereignty over the creation.

Francis Schaeffer's words on the Fall of man are worth remembering:

> What does a historic space-time Fall involve? It means that there was a period before man fell; that, if you had been there, you could have seen Adam before he fell; that, at the point when he revolted against God by making a free choice to disobey God's commandment, there was a tick of the clock. Take away the first three chapters of Genesis, and you cannot maintain a true Christian position nor give Christianity's answers.[27]

You will be pleased to know we will discuss God's plan in greater depth so you can be even more confident in your knowledge of the truth about both creation and science.

> The Big Bang says nothing created everything, versus the Bible that teaches that everything was created by the Word of God; the earth, sun, moon and stars formed by gradual accretion over billions of years rather than in a few days by the Word of God; life occurring by accident rather than by holy fiat; humans evolving over millions of years from an ape-like ancestor rather than being created in the image of God in a single day; death being an essential part of the world from the start rather than being an unintended but necessary consequence of mankind's original sin.
>
> **DR. JIM MASON, EXPERIMENTAL NUCLEAR PHYSICS**

27. Schaeffer, Francis A., *The God Who Is There: Speaking Historic Christianity into the Twentieth Century*, Inter-Varsity Press, 1968, 104.

To me, the very least of all saints, this grace was given, to preach to the Gentiles the unfathomable riches of Christ, and to bring to light what is the administration of the mystery which for ages has been hidden in God who created all things.

EPHESIANS 3:8-9

For by Him all things were created, both in the heavens and on earth, visible and invisible, whether thrones or dominions or rulers or authorities—all things have been created through Him and for Him. He is before all things, and in Him all things hold together.

COLOSSIANS 1:16-17

PAUL, APOSTLE AND JEWISH SCHOLAR

CHAPTER 14

SCIENTIFIC FOUNDATION STONE #6

CREATION'S REVELATION OF GOD NOT PURPOSELESSNESS

The sixth foundation stone for accurate and enduring scientific conclusions centers around another of God's purposes for creating the universe as He did. We already saw that He created the heavens and earth to supply our every need because we are the crown jewel of His creation. He wants us to succeed. Now we will learn He built information about Himself into the creation so we can't miss it. Yes, you are right. He designed creation to SHOUT of His magnificence and perfection! He yearns for everyone to know Him.

Naturalism and evolutionism see no ultimate purpose for the natural realm. For them, all creation is random with no purposeful direction. Through evolution, creation has supposedly progressed from nothing to man all on its own with no outside guidance. All has happened by chance with no predetermined plan. Doesn't that sound like placing the pieces of a car on the ground and expecting them to come together on their own as a gleaming, fine-tuned racing machine? It won't happen. In fact, it can't happen.

> Life cannot have had a random beginning....The trouble is that there are about 2000 enzymes, and the chance of obtaining them all in a random trial is only one part in 10^40,000, an outrageously small probability that could not be faced even if the whole universe consisted of organic soup.
>
> **SIR FRED HOYLE, ASTRONOMER**

Sadly, most Christians today believe creation somehow turned out fine-tuned, all possible because of the passage of billions of years. As if the passage of time by itself can produce order out of chaos! Believing such ideas requires tremendous faith, far more than believing in God, because faith in evolution contradicts all the evidence of our senses and science. Believing in evolution is the essence of blind faith.

THE TAKE AWAY: I'm sure you have heard skeptics accuse the believer in God and His Word of having blind faith. They refuse to understand our faith is not blind at all but, rather, based on sound evidence. We put our faith in objective, observable, and, therefore, scientific facts. Faith in the God of Scripture and His declarations in it have passed the test of time, not just of our time but down through every century since the six days of creation. All secular faith falls apart on objective examination. Theirs is the essence of blind faith.

> God is a mathematician of a very high order and He used advanced mathematics in constructing the universe.
> DR. PAUL A. M. DIRAC, 1933 NOBEL PRIZE IN PHYSICS

Our brief study of Genesis 3 and Romans 8 alone prove God not only exists but is Who He says He is, perfect in every way. Our faith is not blind but absolutely objective. We shouldn't let anyone convince us otherwise.

HOW DO WE KNOW CREATION HAS A PURPOSE?

Scripture clearly establishes the purpose for all things, including the creation itself. All mankind is to love, honor, and obey God and, thereby, glorify Him.[1] The rest of creation is to glorify Him as well.[2]

Mankind tends to ignore God's purpose for him. Every sin we commit proves it, but the rest of creation always fulfills His desires. In fact, creation serves to introduce every member of the human race to Him.

1. Westminster Shorter Catechism, question 1: "What is the chief end of man? Man's chief end is to glorify God, and to enjoy him forever." Ps. 22:23; Matt. 5:16; 1 Cor. 6:20; 1 Pet. 2:12; 4:16.
2. Isaiah 43:20.

> Because that which is known about God is evident within them; for God made it evident to them. For since the creation of the world His invisible attributes, His eternal power and divine nature, have been clearly seen, being understood through what has been made, so that they are without excuse.
>
> Romans 1:19-20

Notice all the words in these two verses that relate to knowing: "known," "evident" (twice), "clearly seen," and "understood."[3] Endeavoring to know is science at its finest. It involves theorizing, investigating, observing, and learning from the creation, not only information about the creation itself but also about the Creator of it. In these verses, the apostle Paul spoke of both an internal revelation through conscience ("evident within them") and an external, observable revelation through the creation ("understood through what has been made"). This divinely orchestrated creation, which God made to be known and used, is the object of all scientific inquiry.

TERMS ON THE CLARITY OF GOD'S REVELATION

"to reveal" ἀποκαλύπτω (Romans 1:18)

"known" γνωστός (v. 19)

"evident" φανερός (v. 19)

"to be evident" φανερόω (v. 19)

"to be seen" καθοράω (v. 20)

"to understand" νοέω (v. 20)

"to know" γινώσκω (v. 21)

God's revelation through creation has been available to mankind "since the creation of the world." He has made it evident to every human being on the face of the earth throughout history. It is "clearly evident" to all mankind, not just to scientists, though, by definition, science at its finest is the search through observation for knowledge about the Creator and the workings of His creation. Scientists should be the first to recognize His revelation through its countless manifestations. If you have a science background, meditate on that for a while.

> There is for a free man no occupation more worthy and delightful than to contemplate the beauteous works of nature and honor the infinite wisdom and goodness of God.
>
> **JOHN RAY, FOUNDER OF BIOLOGY, NATURAL HISTORY**

3. For a detailed exposition of Romans 1:18-34 refer to my Romans series (#14-25) on my website at the link, http://www.forhisglorynm.com/index.php/exposition/romans?start=140.

HOW DOES GOD REVEAL HIMSELF TO US?

Theologians classify God's revelation of Himself to mankind as either special revelation (revelation through inspired knowledge) or general revelation (revelation through creation). What is the difference?

- God recorded His special revelation in Scripture. He reveals His general revelation every time we observe His creation.
- Special revelation is the most important and comprehensive way in which we learn about God because it is direct, propositional, and objective. General revelation is not as clear and can be interpreted accurately only when studied through the special revelation of Scripture, a point I have emphasized throughout this book.
- The general revelation obtained by observing creation is the only information about God easily obtained by all people regardless of location.
- General revelation through creation clearly reveals God's existence and His "invisible attributes, His eternal power and divine nature." This revelation is clear enough to hold us accountable to God, "so that [unbelievers] are without excuse."
- General revelation can convict us of our sins and reveal our need for the further revelation of the only means of obtaining eternal life, belief in Jesus Christ alone as Savior.
- For those unbelievers who respond to the message of the creation and desire to know the Creator, the special revelation of Scripture makes clear they are separated from God through sin and in need of the salvation only Jesus Christ can provide.
- The special revelation of Scripture can be understood by those who have believed the general revelation of creation and chosen to embrace the Creator Christ as Savior.

- Only believers can understand any special revelation of Scripture because only they have the Holy Spirit's power to guide and direct their study of it.
- While the special revelation of Scripture is essential for every believer's spiritual growth, he can gain greater appreciation of God's magnificence by studying His creation.

THE TAKE AWAY: The general revelation of creation introduces an unbeliever to God and enhances a believer's appreciation of Him. Only the special revelation of the Bible can fill in the historical and spiritual details.

As we've seen, Scripture presents objective and observable truth. Scriptural truth is propositional truth, meaning it can be demonstrated to be true. We can accept Scripture's promise of salvation as true, for example, because it records eye-witness accounts of Jesus' resurrection as observed by believers and unbelievers alike. Scripture never contradicts itself, and every new archeological and scientific discovery supports its accuracy. Further, history teaches that God always keeps His promises. He fulfills every prophesy literally and completely at the right time.

> It's difficult to see how science will point to a Creator who wants a personal relationship with us, who loves us, who wants us to love each other, who has expectations for us that come to us by the wisdom of Scriptures.
>
> DR. WILLIAM D. PHILLIPS, 1997 NOBEL PRIZE IN PHYSICS

We can trust God to reveal only truth because He is absolute truth. Since He inspired all sixty-six books of the Bible, we can be sure of their accuracy in the original languages and any literal translations. We can trust His written Word, which is the source of all special revelation, because God, the ultimate Author of Scripture, is perfect and infallible.

Let's make it simple. God's revelation is like those gifts that have wrapped box inside wrapped box with the gift itself inside the last box. God, too, wraps box after box for us, but He puts a wonderful gift inside each box, not just in the final one. The more we study creation (general revelation) and the more we study Scripture (special revelation), the more wonderful insights about God we receive. The more we focus on both general and special revelation, the more we recognize God's majesty and His love for us. God reveals gift after amazing gift at just the right time in ways that astound us if we choose to pay attention to them.

> So I hold out my arms to my Redeemer, who, having been foretold for four thousand years, has come to suffer and to die for me on earth, at the time and under all the circumstances foretold. By His grace, I await death in peace, in the hope of being eternally united to Him. Yet I live with joy, whether in the prosperity which it pleases Him to bestow upon me, or in the adversity which He sends for my good, and which He has taught me to bear by His example.
>
> **BLAISE PASCAL, MATHEMATICS, PHYSICS**

WHAT IS SPECIAL REVELATION?

Though every generation in history has had easy access to the general revelation of creation, in our dispensation, the Church Age, the Bible is our only source of special revelation. God inspired every word of Scripture and uses them to reveal all we need to know about life and eternity.

As we've seen, when Scripture mentions a historical event in which God revealed Himself directly, He expects us to appreciate its historical significance but also its application to our lives today. Through these events, we learn about Him, about living in a relationship with Him, and about serving Him.

> There can be no real conflict between the two Books of the Great Author. Both are revelations made by Him to man–the earlier telling of God-made harmonies coming up from the deep past, and rising to their height when man appeared, the later teaching man's relations to his Maker, and speaking of loftier harmonies in the eternal future.
>
> **JAMES DWIGHT DANA, GEOLOGY, MINERALOGY, VOLCANOLOGY**

Before the Church Age, God's visible appearances in history as recorded in Scripture were not only obvious but often spectacular. We have already marveled at His direct appearances to Adam and Eve. Genesis 2:16-17 records His instructions to Adam about the only test of his obedience, not eating the fruit. In Genesis 2:18-20, God appeared again to instruct him about the yet-to-be-created woman and the naming of the animals.

After the Fall, God initiated a confrontation with Adam and Eve about their sin (Genesis 3:9-19). History doesn't record a more dreadful or consequential conversation. As we have seen, God's special revelation at that time included curses on all creation, information that science must believe if it is to understand the universe.

THE TAKE AWAY: This direct special revelation through actual conversations with God continued down through the ages until the Apostle John recorded Jesus' words in Revelation, the last book of the Bible. At that time, all special revelation ceased because Scripture contained all God wanted us Church Age believers to know about Himself and His plan as well as all we need to know. We can anticipate with joy receiving more new special revelation during the future Tribulation period and the Millennial Kingdom.[4]

God warned against adding to or taking away from the book of Revelation:

> I testify to everyone who hears the words of the prophecy of this book: if anyone adds to them, God will add to him the plagues which are written in this book; and if anyone takes away from the words of the book of this prophecy, God will take away his part from the tree of life and from the holy city, which are written in this book.
>
> Revelation 22:18-19

That passage closes the entire canon of Scripture. God no longer needs to reveal Himself through additional special revelation because His written Word already gives us all He wants us to know at this time in history.

Special revelation has been progressive down through the ages. With each new book added to the Bible, God revealed additional information about Himself and His plan that believers were now capable of understanding. Even though the information was new to the reader, because God is immutable, neither He nor His plan has ever changed.

Remember those boxes of gifts? Before all Scripture was revealed, each new book of Scripture was like opening a huge new box. Each contained

4. God will raise up two prophets at the beginning of the Tribulation (Revelation 11:3), and it appears dreams and visions will occur then (Joel 2:28-29). In the Kingdom, we will receive special revelation from the Logos Himself, Jesus Christ.

new information about God's person and plan at exactly the time its readers needed to know more and had the capacity to understand more.

The most important revelation of God to mankind occurred when Jesus Christ, the Second Person of the Trinity, was born of the virgin Mary and lived on earth.[5] The term incarnation describes this concept of God appearing on Earth as a true human. Not only was Christ's first appearance prophesied throughout the Old Testament, but His life fulfilled every one of those prophecies exactly. When He said, "It is finished" on the cross, He referred to His fulfillment of every prophecy related to His mighty work as Messiah and especially His work of redemption of the entire human race.

The special revelation of Scripture declares emphatically that in Jesus' humanity dwells all the fullness of the deity of God.[6] In the words of Jesus Himself:

> All things have been handed over to Me by My Father, and no one knows who the Son is except the Father, and who the Father is except the Son, and anyone to whom the Son wills to reveal Him.
>
> Luke 10:22

Scripture is the revelation of "who the Son is;" therefore, it is the revelation of who God is.

WHAT ABOUT GENERAL REVELATION?

Not only did God choose to reveal Himself through the special revelation of Scripture and spectacularly in the incarnation of Jesus, but He also chose to do so through the general revelation of creation. In fact, an important purpose of the creation is to introduce us to our Creator. Though God built into the creation evidence of His person and plan, we can't accurately understand it without studying His Word. In other words, we can't know Him personally unless we read and study the Bible, but we can enjoy hints of Him any time we notice His creation.

THE TAKE AWAY: General revelation through creation gives us a broad picture of God as Creator, but only Scripture can fill in the

5. John 1:1-18; 14:9; Hebrew 1:1-2.
6. Isaiah 9:6; John 1:1; 8:58-59; Titus 2:13; Colossians 2:9; Hebrew 1:8; Revelation 1:8.

essential details of His being that allow us to know Him personally. If we are wise, we will continue studying both creation and Scripture until our souls leave our bodies.

Using that box analogy again, the outer boxes we open first contain information about God we receive from creation. We may need to open many boxes before creation convinces us of God's presence, and we want to know more about Him. After we make that decision, we find that the next box we open contains special revelation about salvation from sin, about having a personal relationship with our gift-Giver through faith in Christ's work on the cross. We may need to open many more boxes before we finally believe the good news that Jesus Christ died as our Substitute, took the punishment for our sins, and was resurrected three days later.

After we believe in Christ as Savior and begin reading the Bible, we eagerly open box after box of special revelation from Scripture that help us grow in our understanding of Him and His plan for us. Interspersed in those special revelation boxes, we continue opening boxes of general revelation from creation that excite us to learn more about the Creator. God has just the right box inside the last one so that we continue adding to our love and appreciation of Him. God designed general and special revelation to work together perfectly. Neat, isn't it!

> If there are a bunch of fruit trees, one can say that whoever created these fruit trees wanted some apples. In other words, by looking at the order in the world, we can infer purpose and from purpose we begin to get some knowledge of the Creator, the Planner of all this. This is, then, how I look at God. I look at God through the works of God's hands and from those works imply intentions. From these intentions, I receive an impression of the Almighty.
>
> **DR. ARNO PENZIAS, 1978 NOBEL PRIZE IN PHYSICS**

WHAT OTHER PURPOSE FOR CREATION SHOULD WE CONSIDER?

A secondary purpose of the creation as general revelation is to make mankind accountable to his Creator.

> For since the creation of the world His invisible attributes, His eternal power and divine nature, have been clearly seen, being understood through what has been made, so that they are without excuse.
>
> Romans 1:20

Note the concluding phrase, "so that they are without excuse." Since God is "clearly seen" in creation and since He Himself "made it evident" to every member of the human race, every one of us deep down knows enough about Creator God to respond to Him. After seeing the proof of His existence in the creation, we must make a decision. Do we want to know more about Him, or do we prefer to ignore His shouts?

No one is really an atheist. Anyone who claims to be an atheist has merely "suppressed" the truth about God to such an extent that he has convinced himself of God's non-existence (Romans 1:18). That is the epitome of self-deception, isn't it? After learning of the existence of God through his interaction with the creation, he has decided that he doesn't want to know more about his Creator.

> Atheism is so senseless. When I look at the solar system, I see the earth at the right distance from the sun to receive the proper amount of heat and light. This did not happen by chance.
>
> **ISAAC NEWTON, PHYSICS, MATHEMATICS**

Any unbeliever who continues to suppress the truth that God made evident to him in the creation will remain dead in his "trespasses and sins."[7] Nothing could be worse both in this life and in eternity. Every unbeliever deserves eternal judgment because he has chosen to reject God's introduction of Himself through the general revelation of the creation. He has chosen to ignore the truth. He has closed the boxes on further revelation though at any time until his death he can resume opening them.

THE TAKE AWAY: God wants all to believe in Him;[8] therefore, the creation continually shouts His existence to the world. Every time we see the sky painted with a sunset, touch a velvety rose petal, have a newborn baby smile at us, feel the licks of a faithful pet, see lightning streak across

7. Ephesians 2:1.
8. 1 Timothy 2:3-4.

the sky, hear a bird's song, marvel at the intricacies of the human eye, or notice a scab turning into healthy skin, we are experiencing God's shouts.

In grace, He continues shouting. Our responsibility is to respond with faith to those shouts. When we ignore or reject them, we suffer. If we continue rejecting them until death, we will end up suffering excruciating pain in the Lake of Fire forever. Remember, please, that forever is a really long time. Listen to His shouts!

> God's nature is revealed most perfectly in the life and teachings of Jesus of Nazareth, as recorded in the New Testament of the Bible, who was sent by God to reveal the divine nature, summarized in "God is Love."
>
> **DR. GEORGE ELLIS, ASTROPHYSICS**

All us believers need to help unbelievers open their ears to hear His shouts. What could be easier than mentioning God's kindness or creativity as you point to the sunset or a flower? Easy, yes, and by doing so you are helping a God-suppressor awaken to the One lovingly shouting to a lost and dying world through His creation.

Romans 1:21-32 records the results of mankind's rejection of the general revelation of the creation. We will focus on Romans 1:21-23:

> For even though they knew God [through seeing Him revealed by His creation], they did not honor Him as God or give thanks, but they became futile in their speculations, and their foolish heart was darkened. Professing to be wise, they became fools, and exchanged the glory of the incorruptible God for an image in the form of corruptible man and of birds and four-footed animals and crawling creatures.

Even though God adequately, abundantly, and continually reveals Himself through creation so that everyone should want to know more about Him, and even though everyone observes and understands that revelation (1:19-20), most reject it (1:21). They rationalize God's revelation by speculating on anti-God theories (1:21). Do evolution, naturalism, or secularism fit into this category of speculation? Such foolishness produces a hardening of heart, a resistance to God's revelation of truth. We could call that a dismissal of God and truth reprobation (1:22).

> There is a God shaped vacuum in the heart of every man which cannot be filled by any created thing, but only by God, the Creator, made known through Jesus.
>
> **BLAISE PASCAL, MATHEMATICS, PHYSICS**

Because God created man with an inward need for Him and a desire to worship Him, when anyone rejects the truth of general revelation, a God-shaped void remains in his heart. To fill that void, he replaces the One true God, the Creator, with parts of the creation (1:23). Statues of Buddha and Zeus are examples. So is animism. The list of absurd God-substitutes man creates in his foolish speculation is extensive. Our twelve-year-old might point to the cobra Egyptian pharaohs worshipped. A deadly serpent? Really?

Or what about crystals that are supposed to have spiritual powers? Or the money or power some expect to bring them happiness? Some even worship trees, cows, or men and women who have died and remained dead. How sad anyone would reject the Creator and worship some crazy part of the creation incapable of saving him!

Are you noticing the alliterative pattern here, revelation, realization, rejection, rationalization, reprobation, and replacement? Can you think of an "R" sounding word to complete our alliteration? Try the word wrath. Because of man's tendency to reject the Creator and worship the creation, God is forced to act against him.

> For the wrath of God is revealed from heaven against all ungodliness and unrighteousness of men who suppress the truth in unrighteousness.
>
> Romans 1:18

"The wrath of God," His righteous judgment, "is revealed" against all suppressors of revealed truth. Why? Because they have chosen to believe foolishness rather than their Creator. Every person who has ever lived has had equal access to the creation. When anyone rejects it's impossible-to-ignore information about God, he also chooses to reject God's plan for his eternal existence. He deserves God's wrath because he himself chose to not trust the witness of his senses about creation shouting praises to the Creator.

Notice that the verb "is revealed" is in the present tense. Paul was not referring to the wrath of God that had already taken place at the Fall and the Flood or even at Sodom and Gomorrah. He wasn't discussing future displays of wrath, including the final judgments of the Tribulation period and the Lake of Fire. He was warning against a form of divine wrath in present, on-going time. We can see the consequences of this wrath throughout human history. Wars, famine, pestilence, depression, decay, destruction, and death come to mind.

Paul described this present tense, on-going manifestation of God's wrath in detail at the conclusion of this passage in Romans 1:24-32. In His wrath, God allows mankind to commit the sins of depravity that are clearly observable all around us today.

After someone views God's revelation in creation and chooses to ignore it, which is in essence choosing to reject his Creator, God allows him to follow his own inclinations. The phrase "gave them over," repeated three times (Romans 1:24, 26, 28), means just that. God is a gentleman. He wants us to know Him; He makes His presence and person impossible to ignore in the creation, but if we choose to reject Him, He respects our decision. Our foolishness in rejecting Him invariably leads to a life of misery, self-destruction, and eventual eternal separation from Him in the Lake of Fire.

The suffering described in 1:24-32 applies to all individuals, cultures, and nations that reject Him. A culture in decline experiences the wrath of God in a present tense sense, just as do the individuals in it. In His wrath, God allows those who continually reject creation's shouts to commit sodomy, adultery, pedophilia, gender confusion, bestiality, the physical mutilation even of children, pornography, all the sins of depravity that are clearly observable all around us today.

THE TAKE AWAY: Every day, the wrath of God is revealed in this on-going sense of allowing us to reap what we sow. Our rejection of God sows His rejection of us and His acquiescence to our desire to self-destruct. That is God's wrath worked out in present, on-going time. A sobering thought, isn't it? We need to give the Gospel boldly and often. You could even choose to become a scientist and point unbelievers to the God revealed so thoroughly and beautifully in the creation. God has made us His ambassadors. We need to act like it.

The general revelation of creation has been available to every human being who has ever lived.

> The heavens are telling of the glory of God, and their expanse is declaring the work of His hands.
>
> Psalm 19:1

The heavens surround everyone. Even the blind can feel rain and the heat of the sun. The glories of the moon and stars affirm God's love even in the darkness. The rainbow declares His faithfulness. The birds that fill the skies literally sing out the existence of their Creator.

> God always reveals Himself. Again I think as Psalm 19, "the heavens proclaim the glory of God," that is, God reveals Himself in all there is. All reality, to a greater or lesser extent, reveals the purpose of God. There is some connection to the purpose and order of the world in all aspects of human experience."
>
> **DR. ARNO PENZIAS, 1978 NOBEL PRIZE IN PHYSICS**

The text says that God has made revelation of Himself available "since the creation of the world" (Romans 1:20), even before He created mankind. Does God want everyone to know Him and accept His gift of salvation? Yes, so much so that He put that revelation in place even before He created the first human being!

The general revelation of creation is adequate to make all mankind accountable to the Creator—and to condemn all humanity. Those who reject God "are without excuse." That phrase literally means without any defense. No unbeliever will have any defense when he stands before holy God on judgment day because creation declared his need for the Savior, and he chose to replace the true God with something useless to save him.

I should not leave you without hope and depressed. When anyone responds positively to God's self-revelation in creation, God makes sure that person, even if he lives in the remotest jungle, receives the special revelation of Scripture about salvation through the Lord Jesus Christ. Maybe you will have the privilege of taking the Gospel to him. What joy that would bring you both!

THE TAKE AWAY: Salvation is not based on any effort or work on our part. Good thing, isn't it, since we are creatures, not the Creator? We are helpless to save ourselves and incapable of adding anything to our Creator's perfect salvation work. Salvation depends solely on Jesus Christ's sacrifice for our sins on the cross. Trusting His finished work for all mankind is the only means of entering into a real relationship with Him.[9] God made Christ's sacrificial salvation work the focus of all special revelation in Scripture. Nothing is more important.

You'll be relieved to know that you can't lose your salvation either, no matter how bad your sins, not even if you later reject or even curse God.[10] The moment you believe in Christ, you become a member of His family. You are an heir of your heavenly Father, the King of kings. Your eternal destiny in heaven with Him is forever secure. Nothing can change that accomplished fact. That absolute truth deserves a huge Hallelujah!

See how important special revelation is? We meet God through the general revelation of creation, but we gain His joy and peace only through studying His Word, learning about His great love for us, and accepting it as true.

THE TAKE AWAY: Every member of the human race is accountable to God. Imagine the sorrow of scientists who spent their lives studying the creation and still rejected the Creator. Secularists, naturalists, evolutionists, and unbelieving scientists choose to reject and suppress the divine view of the creation because "they exchanged the truth of God for a lie and worshiped and served the creature rather than the Creator, who is blessed forever. Amen" (Romans 1:29).

Every Christian scientist should be motivated to seek meaning and purpose behind the aspect of the creation that is the focus of his study. He has the special privilege of minutely examining a significant part of the creation for evidence of the Creator's essence, of uncovering a little more of His genius that will benefit all mankind. Here is another place you may want to set this book down for a moment to worship our gracious Creator.

---

9. John 3:16; Romans 3:20-24; Ephesians 2:1-10; John 14:6; Act 4:12.
10. John 10:27-29; Romans 8:31-39.

> Whoever applies himself to medicine should seriously weigh the following considerations: First, that he will one day have to render an account to the supreme Judge of the lives of sick persons committed to his care. Next, whatever skill or knowledge he may, by divine favour, become possessed of, should be devoted above all things to the glory of God and the welfare of the human race. Thirdly, he must remember that is no mean or ignoble creature that he deals with. We may ascertain the worth of the human race since for its sake God's only begotten Son became man and thereby ennobled the nature that he took upon him. Finally the physician should bear in mind that he himself is not exempt from the common lot but is subject to the same laws of mortality and disease as his fellows and he will care for the sick with more diligence and tenderness if he remembers that he himself is their fellow sufferer.
>
> **DR. THOMAS SYDENHAM, PIONEER OF MODERN MEDICINE**

WHAT CAN WE CONCLUDE?

God has embedded into the creation information that can and should motivate us to know Him. The sciences help illuminate the wonders of His creativity. Remember those gift boxes inside gift boxes? Scientists have the joy of unwrapping them day after day as they seek to uncover the mechanics behind the creation. Though most of us aren't scientists by profession, we, too, have the pleasure of seeing the wonders of His creation unwrapped before our eyes every day. None of us can miss the vastness, variety, and beauty of the creation if only we look with believing eyes.

Through the Bible, His special revelation, God encourages us to have a personal relationship with Him and to grow in our knowledge of His person and plan. That knowledge also comes in gift box wrapped in gift box, each helping us understand one divine truth after another. Each truth brings with it joy and peace beyond comprehension. Those gifts make us richer than any amount of money can.

THE TAKE AWAY: The creation serves a purpose for both mankind and the Creator. For mankind, it introduces us to the Creator and makes us aware of our need for greater knowledge of Him. For God, the creation makes us accountable to Him and thus becomes the basis for divine judgment. Acceptance of God as Creator leads to great rewards. Rejection leads to judgment. Our twelve-year-old would roll his eyes and wonder what kind of idiot would choose judgment.

> We take the highest and best of human faculties, and, exalting them in our imagination to an unlimited extent, endeavour to attain an imperfect conception of that Infinite Power which created everything around us.
>
> CHARLES BABBAGE, PIONEER OF MODERN COMPUTERS

...and saying, "Where is the promise of His coming? For ever since the fathers fell asleep, all continues just as it was from the beginning of creation." For when they maintain this, it escapes their notice that by the word of God the heavens existed long ago and the earth was formed out of water and by water.

2 PETER 3:4-7

PETER, APOSTLE AND ELDER

CHAPTER 15

SCIENTIFIC FOUNDATION STONE #7

GOD'S SOVEREIGN CONTROL OF CREATION NOT CHANCE

The seventh foundation stone, God's sovereign control of creation, should be a confidence booster. How could we not have increased confidence in life and eternity when we understand our Creator's complete control over all creation? What better hands could all heaven and earth be in? Who better than the Creator to guide our use of all He has given us? Who better to trust with our future? I hope this chapter blows you away, especially you believers. Knowing how extensively our Lord is involved in His creation should do just that.

The Creator, because He is the Creator, is not bound by natural laws, physical constants, or anything else in the creation. Likewise, the creation cannot operate independent of its Creator though naturalists try to pretend otherwise. Events in the natural realm are not the random effects of chance as evolution supposes. On the contrary, Creator God is sovereign over the entire creation. He designed it to benefit those created in His image. He freely uses it to accomplish His will and purpose.

While the Bible teaches that God is sovereign over all things, theologians generally concentrate on His sovereignty over people and history. Though not as commonly taught, Scripture is equally clear about His absolute authority over the natural realm. Hundreds of passages speak in relation to science.

I am about to take you on an exciting exploration of some of the many Bible passages that mention scientific topics. Most also describe God's direct intervention in history. In this chapter, I will summarize His sovereign control using major categories of science. If you wish to study these passages in chronological order as they occur in the various periods of history, please turn to Appendix B.

My hope is that this chapter totally obliterates any credibility the religion of naturalism still has in your mind. Even more, I hope you are filled with awe by the Creator's abundant and varied shouts of love through His creation. It seems to me that He has been shouting at the top of His voice since the moment He created the first man. See if the truths in this chapter don't confirm that for you.

> This now tells how precise the Creator's aim must have been, namely to an accuracy of one part in 10, to the power of 10, to the power of 123. This is an extraordinary figure. One could not possibly even write the number down in full in the ordinary denary notation: it would be 1 followed by 10 to the power of 123 successive 0's. Even if we were to write a 0 on each separate proton and on each separate neutron in the entire universe—and we could throw in all the other particles for good measure—we should fall far short of writing down the figure needed.
>
> **DR. ROGER PENROSE, MATHEMATICAL PHYSICIST, MATHEMATICIAN**

IS GOD SOVEREIGN OVER INANIMATE NATURE?

Is He ever! We already discussed His control over inanimate nature, that aspect of nature that does not have life, during the six days of creation in Genesis 1 (e.g., light, water, land, sun, stars) and the acts of judgment in Genesis 3 (e.g., childbirth for women, toil for men, thorns and thistles). Certainly, we could discuss the Genesis Flood here, but I will include it later under God's judgments.

After the Flood, God instituted the Noahic Covenant in which He legally bound Himself to never bring another universal Flood, an act of kindness over inanimate nature we don't appreciate nearly enough. In the Bible, a covenant was the same as a legally binding contract in our culture. It had parties to the covenant, stipulations, and behavior specified that was legally binding. As we have seen, God could guarantee His promise because He alone is able to control the entire natural realm, including all sciences. By instituting an unconditional covenant, God gave more than double assurance to the parties to the covenants. Why hasn't the earth been inundated by another universal flood? Simply put: God honors His covenants and promises. The Noahic Covenant guaranteed a stable universe that has predictability, functions in law-like fashion, and can be studied scientifically. Without the Noahic Covenant there may be no assurance of any constants or laws of science to study in the era after the Flood. And so, God continues to keep that contract by controlling all creation. History shows that God has acted in accordance with this legal contract.

The second book of the Bible, the Book of Exodus, records the Creator directly orchestrating inanimate physical events during the plagues of Egypt. He turned the Nile River and all other water sources into blood (chemistry, physics) (Exodus 7:14-21), changed soot into boils on mankind (chemistry) (9:8-11), assaulted the earth with hail mixed with fire without harming the Israelites (climatology, astrophysics) (9:18-26, 29-33), and dropped a supernatural, palpable darkness on every Egyptian for three days and nights (physics, optics) (10:21-23).

Many miraculous events also surrounded the children of Israel during their exodus from Egypt. The Old Testament references God's deliverance of the Jews at the Red Sea as the supreme example of His sovereignty over inanimate creation.[1] Eventually, it was superseded in importance and magnitude by God's sovereignty over sin and death at the resurrection of Christ, the greatest ever expression of His sovereignty over all things in heaven and on Earth.

Read with awe the historical record penned by Moses, the one who made these scientific observations:

> So it came between the camp of Egypt and the camp of Israel; and there was the cloud along with the darkness, yet it gave light at night. Thus, the one did not come near the other all night. Then Moses stretched out his hand over the sea; and the Lord swept the sea back by a strong east wind all night and turned the sea into dry land, so the waters were divided. The sons of Israel went through the

1. Deut. 4:32-38; Josh. 2:8-11; 1 Sam. 10:17-18; 2 Sam. 7:22-23; Neh. 9:9-11.

> midst of the sea on the dry land, and the waters were like a wall to them on their right hand and on their left.
>
> Exodus 14:20-22

A supernatural cloud gave the Jews light in the night and protected them from their enemies (climatology, optics) (14:20). In parting of the Red Sea, God controlled at least hydrodynamics, hydraulics, physics, and climatology (14:21-22, 29-31). At exactly the right time, He closed the waters to trap the Egyptians and save the Jews (14:26-28), a stunning example of His love, justice, and righteousness expressed through His sovereignty. Every miraculous aspect of this encounter between good and evil showed God's mastery of His creation and His desire to use it to care for those who choose to obey Him.

At Mount Sinai, God used thunder, lightning, clouds, smoke, and the quaking of the mountain to announce and warn the Israelites of His presence (climatology, chemistry, tectonics, geophysics) (19:16-19; 20:18, 22).

> Now Mount Sinai was all in smoke because the Lord descended upon it in fire; and its smoke ascended like the smoke of a furnace, and the whole mountain quaked violently. When the sound of the trumpet grew louder and louder, Moses spoke, and God answered him with thunder.
>
> Exodus 19:18-19

Later, an event similar to the Red Sea crossing but on a much smaller scale initiated the Jews' conquest of the Promised Land. The Lord announced His conquest plan in advance to Joshua because He wanted all Israel to realize He was sovereign over them as well as over their enemies. The conquest began when He stopped the flow of the waters of the Jordan River in one heap (hydrodynamics, hydrology) (Joshua 3:13-17; 4:23-24), and the Israelites were able to cross into the Promised Land, again on dry land (hydrology) (5:1).

Consider that miracle for a moment. Both at the Red Sea with Moses and the Jordan River with Joshua, God dried land that had been under water for thousands of years so the Israelites could cross with ease. After they all crossed safely, He commanded the waters to return, and they obeyed (4:18). You can bet that the land under the water was immediately muddy and impassable!

God asked Job questions related to many areas of science, including the inanimate and animal worlds.[2] Each question implied God is sovereign over all these areas and that Job, because he was a creature and not the Creator, had minimal control over them. Even that limited control had to be given to him by the Creator. God Himself framed these questions, showing He orchestrates these actions that still continue four thousand years later.

> Thousands of years ago Job saw the futility of finite man's attempting to define God when he cried, "Can man with searching find out God?" Similarly, wise men ever since have always looked in amazement at the wonderful orderliness of nature and then recognized their own ignorance and finiteness and have been content to stand in silence and in reverence before the Being who is immanent in Nature, repeating with the psalmist, "The fool hath said in his heart, there is no God."
>
> **DR. ROBERT MILLIKAN, 1923 NOBEL PRIZE IN PHYSICS**

About the inanimate, God asked Job:

> "Where were you when I laid the foundation of the earth? Tell Me, if you have understanding, Who set its measurements? Since you know. Or who stretched the line on it? On what were its bases sunk? Or who laid its cornerstone... (geophysics) Or who enclosed the sea with doors when, bursting forth, it went out from the womb; When I made a cloud its garment and thick darkness its swaddling band, and I placed boundaries on it and set a bolt and doors, and I said, "Thus far you shall come, but no farther; and here shall your proud waves stop?" (oceanography) Have you ever in your life commanded the morning, and caused the dawn to know its place, that it might take hold of the ends of the earth, and the wicked be shaken out of it?
>
> Job 38:4-13

I highly recommend you stop for a moment and read God's entire speech to Job. You will be enlightened and encouraged, glad to be able to depend on your Creator and assured that you can.

In response to the prophet Elijah's faith, God caused a three-year drought (climatology). Later, He expanded the food supply for a

2. Refer to Appendix B for the details.

widow (nutrition) (1 Kings 17:15-16), sent fire to consume Elijah's sacrifice to prove His superiority to Baal (physics) (18:21, 38-39), brought prophesied rain after the prophesied drought (climatology) (18:43-4; James 5:17), sent fire to consume two delegations of soldiers (physics) (2 Kings 1:10-14), and split the waters of the Jordan River so Elijah and Elisha could cross on dry land (physics, hydraulics) (2:8).

God's miracles during the prophet Elisha's ministry were equally impressive. Even though we are not told that God performed them, the clear implication is they were displays of His sovereignty performed through the "man of God," meaning God used Elisha as His instrument. God again split the waters of the Jordan River, this time so Elisha could re-cross after Elijah's miraculous disappearance (physics, hydraulics) (2 Kings 2:14), used bears to tear up forty-two boys mocking His prophet Elisha (zoology) (2:23-24), purified bad water near Jericho (chemistry) (2:19-22), increased a widow's oil supply (chemistry, nutrition) (4:1-6), purified a poisonous stew (toxicology, chemistry) (4:38-41), and made an iron axe head float in the Jordan River (physics, hydraulics, metallurgy) (6:5-6).

God's absolute authority over nature, one of the reasons we worship Him, is recorded beautifully in Psalm 104, a poetic parallel to Genesis 1. In verses 3-4, we see that God orchestrates climatology:

> He lays the beams of His upper chambers in the waters; He makes the clouds His chariot; He walks upon the wings of the wind; He makes the winds His messengers, flaming fire His ministers.
>
> Psalm 104:3-4

These verses do not simply describe the independent but interconnected working of hydrology, winds, and atmospheric forces but also His sovereign direction of those processes.

More specifically, Psalm 104 declares that the whole hydrological cycle is under God's control:

> He sends forth springs in the valleys; they flow between the mountains...He waters the mountains from His upper chambers; the earth is satisfied with the fruit of His works.
>
> Psalm 104:10-13

Botany is also described as dependent on God's care:

> He causes the grass to grow for the cattle, and vegetation for the labor of man, so that he may bring forth food from the earth, and wine which makes man's heart glad, so that he may make his face glisten with oil, and food which sustains man's heart. The trees of the Lord drink their fill, the cedars of Lebanon which He planted.
>
> Psalm 104:14-16

God also controls the area of physics called optics:

> You appoint darkness and it becomes night, in which all the beasts of the forest prowl about.
>
> Psalm 104:20

In His first recorded miracle, Christ demonstrated His omnipotence by transforming, in an instant, a simple compound of two elements (water) into one of the most complex carbon-based compounds known to man (wine) (chemistry, nuclear physics, nutrition, oenology) (John 2:6-10). This miracle showed that our Creator is sovereign over every subatomic particle in the universe. Had you thought about His turning water into wine in scientific terms? If you enjoy chemistry or nuclear physics, you should have!

Later, Christ demonstrated His sovereignty over all creation by stilling dangerous storms instantly (climatology, geophysical fluid dynamics) (Matthew 8:23-27; 14:25-26, 29) and walking on water with Peter (physics, climatology) (Matthew 14:24-33). Think about the omnipotence needed to perform those miracles. Feel free to worship at any time. You don't need to wait until Sunday.

In the Book of Acts, Luke recorded miracles that verified the credentials of early church leaders. On one occasion when Peter was in prison, an angel, acting as God's instrument, opened the prison gate to release him (mechanics, physics) (Acts 5:19-20).

Scripture gives many more examples, but these suffice in showing that God is sovereign over inanimate creation. What about animals and other living creatures? Is He sovereign there, too? You guessed it!

> To know the mighty works of God, to comprehend His wisdom and majesty and power; to appreciate, in degree, the wonderful workings of His laws, surely all this must be a pleasing and acceptable mode of worship to the Most High, to whom ignorance cannot be more grateful than knowledge.
>
> NICOLAUS COPERNICUS, MATHEMATICIAN, ASTRONOMER

IS GOD SOVEREIGN OVER ANIMALS, INSECTS, BIRDS, AQUATIC CREATURES, AND REPTILES?

Scripture contains more examples of God's sovereignty over living creatures than over the inanimate creation. I will mention only a few. Check out Appendix B for more.

In the Book of Job, God's questions made His complete control over living creatures incontrovertible. The rhetorical answer to each of God's questions is: "No, but I can do what you cannot."

- "Can you hunt the prey for the *lion* or satisfy the appetite of the young *lions*?" (38:39)
- "Do you know the time the *mountain goats* give birth? Do you observe the calving of the *deer*?" (39:1)
- "Who sent the *wild donkey* out free? And who loosed the bonds of the swift *donkey*?" (39:5)
- "Will the *wild ox* consent to serve you, or will he spend the night at your manger?" (39:9)
- "The *ostriches'* wings flap joyously with the pinion and plumage I love...Because *God* has made her forget wisdom and has not given her a share of understanding." (39:13, 17)
- "Do you give the *horse* his might? Do you clothe his neck with a mane?" (39:19)
- "Is it by your understanding that the *hawk* soars, stretching his wings toward the south?" (39:26)

- "Is it at your command that the *eagle* mounts up and makes his nest on high?" (39:27)
- "Behold now, *Behemoth*,[3] which I made as well as you; He eats grass like an ox." (40:15)

Notice the variety of animals. Isn't it reasonable to conclude that God is sovereign over every other creature as well, even over viruses like Covid-19, monkeypox, and bird flu?

> When confronted with the order and beauty of the universe and the strange coincidences of nature, it's very tempting to take the leap of faith from science into religion. I am sure many physicists want to. I only wish they would admit it.
>
> **DR. TONY ROTHMAN, THEORETICAL PHYSICIST**

Later, in the time of Moses when God brought the children of Israel out of Egypt, He performed many miraculous acts that impacted living creatures. Even before the ten plagues, He turned Moses' wooden staff into a snake (physics, herpetology) and his hand leprous before restoring it to health (pathology, medicine) (Exodus 4:6-8). These two miracles gave Moses confidence that God could and would perform the equally miraculous plagues as promised (4:1-5).

Plagues number two through five and number eight involved animals and insects, including the infliction of a super-abundance of frogs (herpetology) (Exodus 8:2-6, 10-14), an infestation of gnats (entomology) (8:16-19), a huge swarm of insects (entomology) (8:20-24, 31), the destruction of only Egyptian livestock (zoology) (9:1-7), and an invasion of locusts (entomology) (10:4-6, 12-15, 19). The timing of the afflictions and their removal were obviously supernatural because God announced both in advance. They were also selective and precise, often sparing the Israelites from among the Egyptians.

After the exodus, while the children of Israel were in the wilderness, God supernaturally brought them quail for food (zoology, nutrition) (Exodus 16:12-30, 35; Numbers 11:6-32; John 6:31-33). He made the quail available every day at exactly the same time at exactly the same

3. The translation Behemoth comes from earlier translations before dinosaur fossils were discovered. The description in the text fits a sauropod dinosaur.

distance from the camp. He wanted them to recognize His grace to them and learn to trust Him faithfully.

An interesting incident late in Israel's time in the wilderness involved a Moabite prophet named Balaam whom God used as His instrument. Balaam's donkey was not only redirected by an angel of God but also given the ability to speak. The donkey and Balaam had an intelligible conversation (zoology, linguistics) (Numbers 22:21-30). Notice that the donkey not only gave revelatory insights but used proper grammar. I wonder where in Moab it went to school! Ask your twelve-year-old if his grammar is as good as that donkey's.

> How came the bodies of animals to be contrived with so much art, and for what ends were their several parts? Was the eye contrived without skill in Opticks, and the ear without knowledge of sounds?... and these things being rightly dispatch'd, does it not appear from phænomena that there is a Being incorporeal, living, intelligent...?
>
> **ISAAC NEWTON**

In one of my favorite miracles, after Elijah confronted King Ahab on God's behalf, God fulfilled his needs in exile by having a raven bring bread and meat to his hiding place on time every day (ornithology, nutrition) (1 Kings 17:2-6).

We already saw that Psalm 104 discusses the inanimate creation. That psalm and others also describe God's sovereignty over living creatures. He quenches the thirst of the beasts of the field and the wild donkeys (zoology) (104:10-11). He feeds cattle (104:14), lions (zoology) (104:21), and sea creatures (ichthyology) (104:25-28). He causes the birth of deer (zoology) (29:9) and provides homes for birds (ornithology) (84:3).

In Jesus' teaching in the Sermon on the Mount, He indicated His sovereign care of birds (ornithology) (Mathew 6:26; Luke 12:24). At the beginning of the Church Age, He delivered the Apostle Paul from the mouth of a lion (zoology) (2 Timothy 4:17) and from the venom of a deadly viper (herpetology) (Acts 28:1-6).

Clearly, God is sovereign over every creature He gives life. Are you overwhelmed yet by His majesty? Can you hear His shouts clearly? We have just begun!

> Since we are assured that the all-wise Creator has observed the most exact proportions of number, weight and measure in the make of all things, the most likely way therefore to get any insight into the nature of those parts of the Creation which come within our observation must in all reason be to number, weigh and measure. And we have much encouragement to pursue this method, of searching into the nature of things, from the great success that has attended any attempts of this kind.
>
> **STEPHAN HALES, PIONEER IN BOTANY, CHEMISTRY, PHYSIOLOGY**

IS GOD SOVEREIGN OVER HUMANITY?

Of course, He is! How could we expect less? A striking example occurred before the Genesis Flood. Genesis 5:24 and Hebrews 11:5 record that believer Enoch, who probably literally "walked with God," the preincarnate Jesus Christ, was taken up to heaven without dying (medicine, physics). The text describes this as a divine work, "for God took him." Only one other time does Scripture record God's taking someone into eternity before death. Over two-thousand years after Enoch, God took the prophet Elijah into heaven in a whirlwind (medicine, physics, chemistry) (2 Kings 2:8). Both were obviously divine miracles. Who but God has control over life and death?

The Tower of Babel event was unique in the history of mankind (Genesis 11:1-9). The account, brief though it is, describes an event so transformative it changed all mankind for the rest of world history. Before Babel, everyone spoke one language with a single vocabulary (11:1). After God's intervention, they spoke many different and formerly unknown languages. All the languages we speak today, except for the original one spoken before Babel, had their origins here. The Lord Himself "confused the languages" (11:7-9), and so He is demonstrably sovereign over languages, linguistics, and communication. He is also sovereign over our ability to form thoughts mentally and communicate them to others both by mouth and in writing.

Because those who spoke the same language left Babel and banded together in different parts of the world, the Babel event divided people into nations. Yes, God is sovereign over nations and their leaders, too.

> He makes the nations great, then destroys them. He enlarges the nations, then leads them away.
>
> Job 12:23

Later, in Egypt, God hardened pharaoh's heart but only after he had hardened his own heart to the point of no return in his rejection of God (Exodus 2-4).

During the ministries of Elijah and Elisha, God intervened miraculously many times. Most dramatically, He raised from the dead the son of the Gentile widow who helped Elijah (medicine, physics) (1 Kings 17:17-24). During Elisha's time, He gave fertility to a Shunammite widow for the birth of a son whom He later raised from the dead (medicine, anthropology) (2 Kings 4:13-17, 31-37), healed the leprosy of Naaman, the commander of the army of the King of Aram (bacteriology, medicine) (5:8-14), cursed Gehazi and his descendants with leprosy (medicine) (5:26-29), opened Elisha's servant's eyes to see an angelic army (optics, physics) (6:17), struck the Aramean army with blindness (medicine, pathology) (6:18), and finally, opened eyes to see in Samaria (medicine, pathology) (6:20). If you think that list is long, remember these are merely the interventions God chose to record in the Scriptures. Elijah and Elisha probably performed many more.

Jesus confirmed the historicity of Elisha's miracles, specifically the healing of Naaman (Luke 4:27). Did Jesus see the Father as sovereign over the creation and the affairs of mankind? Certainly. Since Jesus and the Holy Spirit are equal with the Father, are they also sovereign? Certainly.

I hope you never read Scripture again without seeing God's control of all creation. Yet, there is more!

> The first-ever study on the skin colour of Africans has now been published, and the results are not what most people expected. It turns out that the genes that control both light and dark skin colours are found across the world. In other words, these variants were in the original human population before we spread across the globe. This is exciting news! It means that all people on earth really are descended from a single source population, which we believe was at Babel.
>
> **DR. ROBERT CARTER, MARINE BIOLOGIST**

WAS GOD SOVEREIGN OVER THE CREATION IN THE HISTORY OF ISRAEL?

God sovereignly worked throughout the history of Israel, His chosen nation, by miraculously using the natural realm. Psalm 105 summarizes the formation of the nation Israel while praising God's work on behalf of them. This praise psalm recounts the various stages of God's supernatural intervention. I recommend you read the psalm in its entirety and marvel at God's kindness and control.

The Psalm begins with praise to the Lord (105:1-7), then follows with a summary of Israel's origin:

- Abraham and the Abrahamic Covenant (105:8-15)
- Joseph and the Egyptians (105:16-24)
- The plagues on the Egyptians (105:25-36)
- Israel's exodus from Egypt (105:37-38)
- The wilderness experience (105:39-43)
- The conquest of the land under Joshua (105:44-45)

God clearly intervened in His dealings with individuals but also notice the references to His affecting the creation so He could preserve them as a nation set apart to Him:

- Physical protection from their enemies (105:14-15) but the judgment of a famine (agriculture, geophysics) (105:16)
- Fruitful childbearing (anthropology) (105:24)
- Miraculous plagues (zoology, physics, geophysics, herpetology, entomology, climatology, botany, anthropology) (105:27-36)
- Illumination at night (optics, physics) (105:39)
- Provision of quail and manna (zoology, botany, nutrition) (105:40)
- Provision of water from rocks (chemistry, physics) (105:41)

God continually demonstrated His compassion for Israel through His sovereign control of their physical circumstances. He even kept their sandals and clothing from wearing out, another example of His kind attention to all our needs (Deuteronomy 29:5).

Israel's patriarchs, Abraham, Isaac, and Jacob, experienced God's sovereignty over nature in an unusual way. God brought about the first promised son, Isaac, by altering the physiology of both Abraham and his wife Sarah. Abraham at ninety-nine years old and Sarah at ninety were too old to bring about conception by natural means (Genesis 11:30; 16:1; 17:17). The text emphasizes Sarah's barrenness, making it clear that God was the One who produced their son, not natural means (fertilization, obstetrics) (17:16-19).

Sarah was not the only wife of a patriarch to experience a supernatural conception. Her son Isaac's wife, Rebekah, was also barren but conceived by a work of God (fertilization, obstetrics) (25:21-22). In the next generation, Jacob's wife, Rachel, had a similar experience, years of barrenness followed by conception (fertilization, obstetrics) (30:22-24). We can conclude that the descendants of the patriarchs originated through supernatural interventions, not natural processes. God reversed the aging process so that the pregnancies were natural and normal, but the conceptions were supernatural.

The first four-hundred years of the nation Israel were filled with God's omnipotent and sovereign control over the creation. As already noted, the plagues God used to free Israel from Egyptian tyranny involved sovereignty over many aspects of creation, including hydrology, chemistry, and biology (the Nile River turned to blood); zoology (frogs, gnats, flies, and locusts); climatology and hydrology (hail); and optics and physics (darkness); medicine and anthropology (diseases on beasts and man, and the death of Pharaoh's son). The exodus itself was a display of power over the hydrodynamics of the Red Sea.

When God was preparing Moses for the major historical events in the formation of Israel, He often displayed His power and sovereignty over mankind, including in hardening pharaoh's heart after he had reached the point of no return in his rejection of God (Exodus 2-4).

During Israel's forty-years in the wilderness, God often intervened directly to preserve and protect them. At Marah, He made bitter water drinkable so they could survive, not only as individuals but as a nation (chemistry) (Exodus 15:23-25). He conditionally prevented them from getting Egyptian diseases (medicine) (15:26). He provided nutrition through manna for forty years (nutrition) (16:12-31, 35; 1 Corinthians 10:3). Even Jesus calls attention to this supernatural food delivery service by making clear that His Father orchestrated the event (John 6:31-33). At Meribah, He again provided water (Exodus 17:5-7; 1 Corinthians 10:4-5) and then quail (Numbers 11:6-32). For forty years, He worked supernaturally to preserve millions of Israelites so they could leave the wilderness and enter the Promised Land (nutrition, medicine).

Numbers 26:51 records that 601,730 Jewish men survived to enter the Land. Because that number does not include women and children, it is reasonable to believe God preserved at least 2.5 to 3 million Jews. Nothing is impossible with God.

During the time of the conquest of the Promised Land under Joshua's leadership, God intervened often. Some events implied supernatural intervention while others explicitly stated it. The text implies God won the nation's first battle, Jericho, through supernatural forces.

Preceding the battle, Joshua had a vision of an angelic Being who identified Himself as the "captain of the host of the Lord," implying He was the Leader of an army of angels (Joshua 5:13-15). Doesn't it make sense this invisible angelic army led by its Captain crumbled the walls? The Captain was probably the Lord Himself since all appearances of "the Angel of the Lord" in the Old Testament were pre-incarnate appearances of Christ. Often, the description of this Angel indicated deity (John 1:18). Further, Joshua bowed in worship before the Angel and was not rebuked as was the Apostle John when he tried to worship a mere angel (Revelation 19:10; 22:8-9).

The text clearly states God orchestrated all events. God told Joshua, "See, I have given Jericho into your hand, with its king and the valiant warriors" (Joshua 6:2). After the Israelites followed God's instructions, which included marching around the city walls and blowing horns (not your normal battle plan), the city walls "fell down flat" exactly as He promised (6:20).

Archaeological discoveries, not only in Jericho but in many surrounding cities,[4] show that their walls were massive in thickness (eighteen feet at the base) and in height (thirty feet). Some scholars have tried to make the case the walls of Jericho crumbled because of an earthquake or some other unrecorded physical phenomenon,[5] but the context of the passage gives a supernatural explanation (Joshua 6).

In a display of divine precision, God preserved the small section of the wall that housed believer Rahab and her family, just as He had promised. Talk about control down to the smallest detail! Putting all these facts together, we can accurately conclude Jericho fell when divine power acted on the massive city walls. With the walls destroyed by God, Joshua and the Israelites finished the annihilation of this barbaric and habitually unrepentant city.

Another unusual incident occurred during the conquest of Canaan that again demonstrated God's sovereign control over Jewish history. The Israelites were deceived into entering into a covenant with a Canaanite people, the Gibeonites. When five Canaanite armies attacked the Gibeonites because of their alliance with Israel (Joshua 10:1-5), the Gibeonites requested Israel's help (10:6). The Lord assured Joshua He would give him victory. "I have given them into your hand; not one of them shall stand before you" (10:8).

While God used the instrumentality of the Israelite army to win the battle, notice that He Himself got directly involved by slaying many, followed by their complete destruction when He let loose some of the hailstones He stores in the heavenlies:

> And the Lord confounded them before Israel, and He slew them with a great slaughter at Gibeon, and pursued them by the way of the ascent of Beth-horon and struck them as far as Azekah and Makkedah. As they fled from before Israel, while they were at the descent of Beth-horon, the LORD threw large stones from heaven on them as far as Azekah,

---

4. According to Steve Collins, the head archaeologist at the site, who is also director of the School of Archaeology, Veritas International University, an archaeological excavation at Tall el-Hammam has been underway since before 2005. This site is just across the Jordan River from Jericho and is probably the ancient site of Sodom.
5. John J. Davis lists explanations of natural causes from other scholars but supports a supernatural cause in Conquest and Crisis: Studies in Joshua, Judges, and Ruth, Baker Book House, 1969, 46-47.

> and they died; there were more who died from the hailstones than those whom the sons of Israel killed with the sword.
>
> Joshua 10:10-11

God demonstrated sovereignty over physics, military tactics, and climatology in the first part of the battle, but these were not the only displays of His sovereign power. The next passage reveals an even more spectacular one. He stopped the earth in its rotation! We know Earth, not the sun, stopped since the description is from the perspective of Earth and because the sun appeared stationary.

Again, God showed He is sovereign over astrophysics. The passage says it all:

> Then Joshua spoke to the Lord in the day when the Lord delivered up the Amorites before the sons of Israel, and he said in the sight of Israel, "O sun, stand still at Gibeon, And O moon in the valley of Aijalon." So the sun stood still, and the moon stopped, Until the nation avenged themselves of their enemies. Is it not written in the book of Jashar? And the sun stopped in the middle of the sky and did not hasten to go down for about a whole day. There was no day like that before it or after it, when the Lord listened to the voice of a man; for the Lord fought for Israel.
>
> Joshua 10:12-14

God's promises to the nation Israel concerning its future success or failure in the land of Canaan, the land He had promised to the patriarchs, preceded the nation's entering the land and were presented in the form of blessings for obedience to the Mosaic Covenant and curses (judgments and discipline) for disobedience. Old Testament prophets tied every curse on Israel to its disobedience to the Mosaic Covenant.

Both the blessings and curses were primarily physical, including in agriculture (Deuteronomy 28:4, 18, 38-40), climatology (vv. 12, 23-24), the productivity of farm animals (vv. 4, 11, 30), and medical and health issues (vv. 4, 21-22, 27-29, 35, 59). Because every blessing and curse was literally fulfilled in the history of the nation, culminating in its expulsion from the Promised Land, we know God has sovereign control over every aspect of nature and uses His omnipotence to fulfill all covenants, plans, and promises.

When Solomon dedicated the first temple to the Lord in the kingdom age of Israel, the priests sacrificed massive numbers of animals (1 Kings 8:62-63). The writer of 2 Chronicles recorded that God revealed His pleasure in a spectacular, visible manifestation (optics).

> Now when Solomon had finished praying, fire came down from heaven and consumed the burnt offering and the sacrifices, and the glory of the Lord filled the house. The priests could not enter into the house of the Lord because the glory of the Lord filled the Lord's house. All the sons of Israel, seeing the fire come down and the glory of the Lord upon the house, bowed down on the pavement with their faces to the ground, and they worshiped and gave praise to the Lord, saying, "Truly He is good, truly His lovingkindness is everlasting."
>
> 2 Chronicles 7:1-3

As you meditate on this historical event and visualize the display of God's glory, you may want to imitate the Jews who "bowed down on the pavement with their faces to the ground."

Sadly, after King Solomon's death, the Kingdom of Israel divided, and the two separate kingdoms continually defied God to the point He intervened, not to bless them but to destroy them as a nation but not as His people. Even during that time of great apostasy, God intervened in the physical realm, especially, as we have seen, during the ministries of the prophets Elijah and Elisha.

Though we have given only a few examples of God's sovereign control over His chosen nation Israel, you can see He intervened for their benefit by using the natural realm. He is equally compassionate to us as individuals in the Church Age, another reason to worship Him and view the world as He does.

> Amazing fine tuning occurs in the laws that make this complexity possible. Realization of the complexity of what is accomplished makes it very difficult not to use the word "miraculous" without taking a stand as to the ontological status of the word.
>
> DR. GEORGE F.R. ELLIS, MATHEMATICIAN

WAS GOD SOVEREIGN OVER PAST CATASTROPHES?

The Scriptures are filled with examples of God using the natural realm to inflict His judgments on mankind. These supernatural events involved the creation and thus impacted scientific laws. The most striking is the worldwide flood in the time of Noah (Genesis 6-8). The Genesis text attributes the Flood to God directly by quoting His own words:

> The Lord said, "*I will blot out* man whom I have created from the face of the land..."
>
> Genesis 6:7

> "Behold, *I, even I am bringing* the flood of water upon the earth…"
>
> Genesis 6:17

This judgment on mankind radically transformed the physical earth. God burst open the "fountains of the great deep" (geophysics, tectonics, volcanism, oceanography) (7:11). He sent rain in torrents for forty straight days (climatology, hydrology, meteorology) (7:11). He allowed tsunami flood waters to rise for one-hundred-fifty days (physics, hydraulics, oceanography) (7:24). He sent winds to subside the waters (climatology, oceanography, hydraulics) (8:1). He closed the fountains and floodgates and restrained the rain (geophysics, hydraulics) (8:2) so that the waters receded for one-hundred-fifty days (oceanography, hydraulics) (8:3). The description of the destruction is so massive it is difficult if not impossible to imagine. Remember it was a worldwide cataclysm (geophysics).

Animals were affected by the Flood. Those animals that entered the ark two by two were not brought by Noah but instead confounded their natural instincts and "went into the ark to Noah" and "entered as God had commanded him" (zoology) (7:15-16). The text notes that "the Lord closed it [the door of the ark] behind him" (mechanics, physics) (7:16), thereby ensuring the safety of all that entered.

As we have seen, another divine judgment that transformed creation occurred at the Tower of Babel (11:1-9). The effects of that judgment on mankind's intellect and speech persist to this day (anthropology, linguistics) (11:7, 9). As a result of that judgment, the people scattered all over the world (11:8-9), which was God's intention from the beginning (1:28; 9:1). This scattering according to languages produced nations. All languages and all nations can trace their origins to the languages and peoples from Babel as recorded in the table of nations (10). The Tower of Babel event showed God's sovereignty not only over mental processes and languages but over nations, politics, and governments.

In the days of Abraham, God sent a localized judgment on some Canaanites when He caused the destruction of the cities of the plain, including Sodom and Gomorrah (Genesis 18-19). A cosmic airburst of a meteor or asteroid entered the atmosphere near the cities, annihilating them (astrophysics, geophysics, physics, meteoritics).[6] Scripture records many other natural disaster judgments, but are any predicted in the future? The answer is yes, some just as severe involving the natural realm.

God promises that many of His judgments in the future Great Tribulation will be in the form of natural disasters that directly affect all mankind. In the first Trumpet Judgment, hail and fire will burn one-third of all vegetation on Earth (astrophysics, climatology, botany, chemistry) (Revelation 8:7). In the second Trumpet Judgment, a volcanic eruption will turn one-third of all seas into blood (volcanism, orogeny, oceanography) (8:8-9). In the third Trumpet Judgment, an astrophysical collision with earth will pollute one-third of all spring waters (astrophysics, geophysics, chemistry) (8:10-11). The fourth Trumpet Judgment will darken one-third of the sun, moon, and stars (astrophysics) (8:12). The incredible geophysical and astrophysical disasters at the end of the tribulation period, described as bowl judgments will clearly show that God is sovereign over His judgments and thus sovereign over the natural means He uses to effect them (16:12-17, 17-21).

THE TAKE AWAY: Scripture declares that God's judgments catastrophically affect not only mankind but the rest of creation as well. The reality of God's sovereign intervention in history is incontrovertible to the honest student of history and science. And Scripture gives us even more proof!

---

6. An archaeological team based in Albuquerque led by Steve Collins is presently excavating what is believed to be the city of Sodom. They are in the process of publishing many of their findings. One of the published works concludes that evidence points to the cause of the destruction of the cities as a cosmic airburst of a meteor or asteroid. See https://www.nature.com/articles/s41598-021-97778-3.pdf.

> There are many ways in which people are made aware of their power to believe in the supremacy of Divine guidance and power: through music or visual art, some event or experience decisively influencing their life, looking through a microscope or telescope, or just by looking at the miraculous manifestations or purposefulness of Nature.
>
> SIR ERNST CHAIN, 1945 NOBEL PRIZE IN MEDICINE

WHAT DOES THE NEW TESTAMENT SAY ABOUT GOD'S SOVEREIGN INTERVENTIONS?

As amazing as these Old Testament miracles were, the New Testament records God's most important interventions into the creation. The conception and birth of Jesus were the miracles of all miracles, God's ultimate interventions into the creation and the supreme events of all world history. We should consider the Incarnation not only a historical event but also a scientific miracle. The concept of God Himself becoming a physical human being is beyond human comprehension, which is probably why you will not find it discussed anywhere in secular science.

A biblical worldview is essential to understand the Incarnation. Key passages will help us comprehend the profound effect of this event on all mankind for all human history:

> In the beginning was the Word, and the Word was with God, and the Word was God. He was in the beginning with God. All things came into being through Him, and apart from Him nothing came into being that has come into being. In Him was life, and the life was the Light of men...And the Word became flesh, and dwelt among us, and we saw His glory, glory as of the only begotten from the Father, full of grace and truth.
>
> John 1:1-4, 14

> Have this attitude in yourselves which was also in Christ Jesus, who, although He existed in the form of God, did not regard equality with God a thing to be grasped, but emptied Himself, taking the form of a bond-servant,

> and being made in the likeness of men. Being found in appearance as a man, He humbled Himself by becoming obedient to the point of death, even death on a cross. For this reason also, God highly exalted Him, and bestowed on Him the name which is above every name, so that at the name of Jesus EVERY KNEE WILL BOW, of those who are in heaven and on earth and under the earth, and that every tongue will confess that Jesus Christ is Lord, to the glory of God the Father.
>
> Philippians 2:5-11

We could expound on the truth in these verses in detail, but they are clear enough that if you study and meditate on them, you will not only understand them but also bow your knee and confess that Jesus is God. Eventually "at the name of Jesus EVERY KNEE WILL BOW." He is the omnipotent and sovereign Creator who graciously took on human form to save us from our sins.

> If God annihilates or deflects or creates a unit of matter, He has created a new situation at that point. Immediately nature domiciles this new situation, makes it at home in her realm, adapts all other events to it. It finds itself conforming to all the laws. If God creates a miraculous spermatozoon in the body of a virgin, it does not proceed to break any laws. The laws at once take over. Nature is ready. Pregnancy follows, according to all the normal laws, and nine months later a child is born.
>
> C.S. LEWIS, *MIRACLES*

Because God took the form of a human baby, the biological conception of Jesus needed to be miraculous. Though called the virgin birth by theologians, Jesus' conception was the real miracle. Clearly, Mary became pregnant through the supernatural work of the Holy Spirit, not because of Joseph or any other man (fertilization, medicine, anthropology) (Matthew 1:18-23).

In the miracle of miracles, God Himself became man! Think about that in relation to biology. How could the self-existent and transcendent Creator become part of the conception of a human by the name of Mary? How could that God-man embryo grow in a normal pregnancy in the womb of a human sinner? How could that God-baby be born into the fallen world without the contamination of sin? Such thoughts are beyond our finite understanding. They are certainly beyond any scientific investigation.

THE TAKE AWAY: Theologically and scientifically, God's taking on the form of man is known as the hypostatic union. This union did not mix two natures to form a new third nature but was the personal union of One who is 100% human and 100% God. In this union, the attributes of one nature were not transferred to the other nature; therefore, neither nature changed. The priority of the divine nature means we should describe this union as the God-man and not the man-God.

In the hypostatic union, which will continue forever, Jesus remains fully human but without any sin. His divine essence didn't change, but He chose to limit His access to the divine prerogatives. Philippians 2:6-11 summarizes what theologians call the doctrine of kenosis:

> Christ Jesus, who, as He already existed in the form of God, did not consider equality with God something to be grasped, but emptied Himself by taking the form of a bond-servant and being born in the likeness of men. And being found in appearance as a man, He humbled Himself by becoming obedient to the point of death: death on a cross. For this reason also God highly exalted Him, and bestowed on Him the name which is above every name, so that at the name of Jesus every knee will bow, of those who are in heaven and on earth and under the earth, and that every tongue will confess that Jesus Christ is Lord, to the glory of God the Father.

Is your head starting to hurt or at least spin? There is certainly much mystery here. We can't begin to fully grasp this miraculous event. At least I can't.

> I believe in God. It makes no sense to me to assume that the Universe and our existence is just a cosmic accident, that life emerged due to random physical processes in an environment which simply happened to have the right properties. As a Christian I begin to comprehend what life is all about through belief in a Creator, some of whose nature was revealed by a man born about 2000 years ago.
>
> **DR. ANTONY HEWISH, 1974 NOBEL PRIZE IN PHYSICS**

Other events associated with the birth of Jesus were miraculous as well. Zecharias, the father of John the Baptizer, lost his ability to speak because he refused to trust the angel sent to him by God (medicine) (Luke 1:20). When, after nine months, he showed his trust in his Creator, God restored his speech (audiology) (1:64). Consider, too, the appearance and reappearance of the star guiding the magi to Bethlehem (astrophysics) (Matthew 2:2, 7-10) and Simeon knowing that an eight-day-old boy was the promised Messiah (Luke 2:25-35) (neuroscience).

Jesus' miracles were a necessary and prominent part of His authentication as the Son of God, the promised Messiah. They confirmed His deity. His miracles were interventions into the physical realm and often involved individuals. Most were medical miracles and included His healing of leprosy, paralysis, fever, demon possession, hemorrhaging, blindness, muteness, deafness, deformities, lunacy, and dropsy (pathology, medicine). He turned water into wine (chemistry) and stilled storms (climatology), fed over five-thousand and then over four-thousand from a few pieces of bread and fish (nutrition) (Matthew 14:19-21; 15:35-38). He raised from the dead the synagogue official's daughter (Matthew 9:18, 24-25), Lazarus (John 11:40-44), and the widow of Nain's son (Luke 7:11-15) (pathology, medicine, physics). He caused a fig tree to wither (botany) (Matthew 21:19-20).

Jesus often used a part of creation to teach His disciples. We will concentrate on His fish lessons. One time, several disciples had fished all night but caught nothing (Luke 5:1-5). Jesus ordered Peter to lower the nets again. This time they had such a large catch that their nets threatened to break (ichthyology) (5:6-7). Was this merely coincidence or supernatural? The passage indicates that Peter recognized the latter (5:8-9). He knew from experience that fish are less likely to be caught in the early morning.

In a similar occasion after Jesus' resurrection and after another fruitless night fishing, Jesus had the disciples cast their nets on the opposite side of the boat, the right side (John 21:1-5). Again, they had such a large catch that they could not haul in all the fish (ichthyology) (21:6). Were these conservative fish that preferred to remain on the right side of the boat?

In another fish miracle, Jesus taught a lesson on paying taxes. He told His disciples they would find the exact amount needed to pay a temple tax in the first fish they caught (ichthyology) (Matthew 17:27). What was a Roman coin doing in the mouth of that fish? Why hadn't the fish swallowed it or spit it out? How did Jesus know a coin would be in that fish? Is not Jesus' sovereignty over fish apparent? Could He also be sovereign over every other living creature? Silly question, our twelve-year-old would say.

We should also include a unique event in Christ's life, an event secular science is incapable of evaluating. Theologians describe it as the transfiguration of Christ (Matthew 17:2-5) during which He revealed part of His glory to three of His disciples, Peter, James, and John. Elijah and Moses appeared with Him. Remember that these two prophets had been gone for centuries. No camera had recorded them; no contemporary painter had memorialized their countenances, but the disciples recognized them.

Though Scripture gives few details about the transfiguration, we surmise that various branches of science were involved, including physics, optics, audiology, and anthropology. The Lord chose not to reveal most details of this miracle, but what little we know is awe-inspiring. Who but a sovereign God could have accomplished them? Who, indeed! Wasn't that at least one purpose for that miraculous event?

Miraculous physical phenomena also accompanied Christ's death on the cross. From noon to three, the sky was impenetrably dark as Christ took on Himself the punishment for every sin ever committed by every person who will ever live (physics, optics, and possibly astrophysics) (Matthew 27:45). The veil of the temple tore from top to bottom—and it was thirty feet high and several inches thick (physics) (27:51). As Christ died, an earthquake shook the earth, and many dead rose alive from their graves (medicine, geophysics, tectonics, seismology) (27:51-53).

The second miracle of miracles was Christ's resurrection from the dead (physics). Because of its importance to all mankind, all four gospels conclude with the record of this historical event. They also all record His appearances in His resurrection body to a wide variety of disciples, including five hundred at one time.

> The miracles of human history are those in which God has spoken to men. The supreme miracle for Christians is the Resurrection. Something happened to those few men who know Jesus, which led them to believe that Jesus yet lived, with such intensity and conviction that this belief remains the basis of the Christian Church two thousand years later.
>
> **DR. NEVILL MOTT, 1977 NOBEL PRIZE IN PHYSICS**

We should not neglect God's supernatural interventions through the disciples in the early church. We referenced some as general miracles, but a variety of specific miracles filled that time period. During the evening before His crucifixion, Jesus promised His eleven remaining disciples He would send the Holy Spirit in a special way. Did He ever!

> When the day of Pentecost had come, they were all together in one place. And suddenly there came from heaven a noise like a violent rushing wind, and it filled the whole house where they were sitting. And there appeared to them tongues as of fire distributing themselves, and they rested on each one of them. And they were all filled with the Holy Spirit and began to speak with other tongues, as the Spirit was giving them utterance.
>
> Acts 2:1-4

Notice the physical phenomenon that called attention to the indwelling presence of the Holy Spirit. "A noise" (audiology) and "a violent rushing wind" (physics, dynamics) filled the room. "Tongues as of fire" (physics, combustion) appeared on their heads, an image impossible to comprehend. The most important miracle was the ability of these believers to speak languages they had never learned so their Jewish audience, who had come from around the world for Passover, could understand them (linguistics, mental processes). This ability was a gift from the Holy Spirit to authenticate that God had given them authority to speak for Him.

In the Book of Acts, Luke recorded miracles that also verified the credentials of early church leaders. On one occasion when Peter was in prison, an angel, acting as God's instrument, opened the prison gate to release him (physics) (Acts 5:19-20). Another time, the Holy Spirit instantly transported Philip to a far distant location (dynamics, physics) (8:39-40).

Other representative miracles in the early church include Peter's healing of a lame man in Jerusalem and a dead woman in Joppa (medicine) (Acts 3:2-10, 12; 9:35-41), Philip casting out demons and healing paralytics and the lame in Samaria (medicine) (8:7, 13), and Paul healing a lame man at Lystra (14:8-10) and restoring Eutychus to life at Troas (medicine) (20:9-12).

If these miracles were not so awe-inspiring, they would be unbelievably repetitive, wouldn't they? Can we ever praise and thank God enough? I think not, but we should try.

> I think only an idiot can be an atheist. We must admit that there exists an incomprehensible power or force with limitless foresight and knowledge that started the whole universe going in the first place.
>
> DR. CHRISTIAN ANFINSEN, 1972 NOBEL PRIZE IN CHEMISTRY

DO YOU MARVEL YET AT THE VOLUME OF GOD'S SHOUTS?

Since all three members of the Trinity were involved in creating the heavens and the earth and all that is in them,[7] we can be sure all three continue to use the creation for Their own glory.

THE TAKE AWAY: We should apply the principles we covered in Chapters 7 and 8 here. Because Scripture is a revelation from the Creator Himself, in it He makes known that He is controlling the universe, that chance plays no part. Because He inspired the Scriptures, we can be confident His descriptions of His sovereignty are true. Because Scripture is inerrant, we have truth that is absolute and far beyond anything secular science can ever discover. Such an understanding is the basis for the biblical worldview.

Did I mention that literally hundreds of passages describe the Creator's sovereignty over the creation? Go back and count the number included in this chapter alone. Is God not shouting through His interventions in His creation? Does not each of these shouts refute the religion of naturalism? Does evolutionary thinking not show itself to be ridiculous?

7. God the Father took part in creation (Genesis 1:1) as did the Holy Spirit (Genesis 1:2) and Jesus Christ (Colossians 1:16-17).

THE TAKE AWAY: Chance, coincidence, and fate are figments of man's imagination, idols they worship. They reflect a worldview that denies God and His control over both history and creation. Any scientist who ignores or underestimates God's sovereignty is doomed to make faulty and uninformed scientific claims. He needs to realize that God alone is God and that every bit of the creation shouts that truth. May we all learn to depend on Him alone in every area of life, even in the study of science.

> The case against evolution is summed up by Berkeley University law professor Philip Johnson, who makes the following points:
>
> 1. evolution is grounded not on scientific fact, but on a philosophical belief called naturalism;
> 2. the belief that a large body of empirical evidence supports evolution is an illusion;
> 3. evolution is itself a religion;
> 4. if evolution were a scientific hypothesis based on a rigorous study of the evidence, it would have been abandoned long ago.
>
> STEPHEN TAYLOR, ELECTRICAL ENGINEERING

"You, Lord, in the beginning
laid the foundation of the earth,
And the heavens are the works of Your hands."

HEBREWS 1:10

For it was fitting for Him,
for whom are all things,
and through whom are all things,
in bringing many sons to glory,
to perfect the author of their salvation
through sufferings.

HEBREWS 2:10

FIRST CENTURY JEWISH WRITER

CHAPTER 16

SCIENTIFIC FOUNDATION STONE #8

RADICAL CHANGES IN HISTORY NOT UNIFORMITARIANISM

As you have probably realized, God used two worldwide catastrophes to radically change the creation. Science can't be practiced successfully apart from an understanding of those events. At the same time He was reworking creation, God instituted new dispensations that define history from the divine viewpoint. History can't be understood apart from an understanding of dispensations.

Theologically, a dispensation is a time period in which God changes the way in which He deals with mankind according to His broader, overall plan. During each past dispensation, God interacted with mankind differently, both spiritually and physically.

I'm going to let you in on a little secret. This is my favorite chapter of this book, and I have not even written it yet! By pointing out the Creator's two extreme interventions in His creation, I hope to prove the inadequacy of secular science to explain ancient events. If I present the information correctly, this chapter will refute uniformitarianism and even evolution. It will show that, because the universe experienced changes that drastically affected many of the laws of nature and even many physical constants, uniformitarianism is a fraudulent idea, as is the idea of evolution that depends on it.

It will also refute the teaching of secular science that the universe is billions of years old. Unfortunately, the evolutionary time frame of billions of years is accepted by most of the Christian community, including many old earth creationists. The young universe view I present is by far the minority view, but I will show you it is the only worldview validated by Scripture.

> The rate of change of scientific hypothesis is naturally much more rapid than that of Biblical interpretations, so that if an interpretation is founded on such an hypothesis, it may help to keep the hypothesis above ground long after it ought to be buried and forgotten. At the same time I think that each individual man should do all he can to impress his own mind with the extent, the order, and the unity of the universe.
>
> **JAMES CLERK MAXWELL, PHYSIC**

Historically, biblical dispensationalism has stressed the theological distinctions and changes from one God-designed era to the next. Little consideration has been given to the extreme physical changes that occurred along with them. The implications of these creational changes are far-reaching. We need to remember that the present is not the key to understanding the past as uniformitarianism has been telling us for years. The present is the key to understanding our present age only. The Bible is the key to understanding the past. We live in a post-Flood world and age. What does that mean, and how does it affect science? Keep reading.

If your mind was blown away by the enumeration of God's sovereign control over creation in the last chapter, just wait until you finish this chapter. I hope you never again see science in the same way. I hope God's shouts will reverberate in your mind for as long as you are on this earth and that you will continue to marvel as you see His handiwork revealed throughout His creation. I hope these truths will stick with you until your physical body experiences the next radical transformation scheduled on God's timetable. Have I encouraged you to keep reading?

> The more man inquires into the laws which regulate the material universe, the more he is convinced that all its varied forms arise from the action of a few simple principles...The works of the Creator, ever present to our senses, give a living and perpetual testimony of his power and goodness far surpassing any evidence transmitted through human testimony. The testimony of man becomes fainter at every stage of transmission, whilst each new inquiry into the works of the Almighty gives to us more exalted views of his wisdom, his goodness, and his power.
>
> **CHARLES BABBAGE, MATHEMATICIAN AND INVENTOR**

HOW CAN WE BE CERTAIN THESE RADICAL CHANGES REALLY HAPPENED?

Only the Bible describes the radical changes God instituted between past eras. For a New Testament passage that gives an inspired interpretation of ancient events, turn to 2 Peter 3. The context of the passage is Peter's disapproval of a common first century skeptic's view of Christ's second coming. The skeptic's argument that Christ would not return was based on his assumption of uniformitarianism. Bet you didn't know this pseudo-scientific idea that no radical changes have occurred in history was confusing men as far back as the first century. Interesting how unoriginal secular man's thinking is, isn't it? Because Satan isn't omniscient, he can't come up with original ideas. He's stuck with repeating the same lies over and over. Too bad most people are ignorant of history and find his lies new, intriguing, and even scientific.

THE TAKE AWAY: Peter knew his readers needed to be spiritually prepared for the coming onslaught of false teachers, and so he used past divine judgments to warn them of future judgments if they were not vigilant. Today, we need to be equally prepared to recognize and repel the arguments of the many wolves in sheep's clothing who attempt to take us away from the faith.

God's direct intervention in the form of judgments caused every one of these transformative events. None were natural but rather God's direct, sovereign work. We need to examine these radical transformations in the physical realm so we can better understand their impact on science and history.

The book of 2 Peter can be divided into three parts:

Ch. 1: The Motivation to Maturity as protection against false teachers
Ch. 2: The Malevolent Makeup of false teachers
Ch. 3: The Mocking Message of false teachers refuted

In 2 Peter 1, Peter encouraged his reader to grow to maturity (Motivation to Maturity) as a protection against the blasphemy of false teachers. Notice the following verse, which contrasts "the true knowledge of our Lord Jesus Christ" with false teaching:

> For if these qualities are yours and are increasing, they render you neither useless nor unfruitful in the true knowledge of our Lord Jesus Christ.
>
> 2 Peter 1:8

In 2 Peter 2, Peter described the characteristics of false prophets and teachers that make them beguiling and dangerous. Notice especially:

> But false prophets also arose among the people, just as there will also be false teachers among you, who will secretly introduce destructive heresies, even denying the Master who bought them, bringing swift destruction upon themselves.
>
> 2 Peter 2:1

In 2 Peter 3, Peter refuted not only all skepticism about Christ's second coming but also the unscientific assumption of uniformitarianism that supports it. These skeptics assumed that since they themselves had observed no supernatural interventions, none had occurred in the past, and none should be expected in the future. No one had observed a supernatural beginning for the world, had they? Why, then, should anyone expect a supernatural second coming in the future? Uniformitarianism in the first century underscored false assumptions about world history just as it continues to do today.

The final chapter focuses on refuting the main false doctrine (Mocking Message) of false teachers:

> Know this first of all, that in the last days mockers will come with their mocking, following after their own lusts,

> and saying, "Where is the promise of His coming? For ever since the fathers fell asleep, all continues just as it was from the beginning of creation."
>
> 2 Peter 3:3-4

In these two sentences, Peter summarized the mocker's uniformitarian assumption by quoting their claims. In doing so, he alludes to different eras of human history, dividing it into four distinct dispensational ages:

1. the present Church Age, "the last days"
2. the past Age of Israel, "the fathers fell asleep"
3. the future Kingdom Age, "His coming"
4. the time of the original creation, "the beginning of creation"

Radical changes to natural laws and physical constants between some of these ages underlie the text. All refute the assumption of uniformitarianism.

Peter countered the skeptics' false belief and underlying assumption by giving two examples of transformative past events that must have been the Creator's supernatural interventions:

> For when they maintain this, it escapes their notice that by the word of God the heavens existed long ago and the earth was formed out of water and by water, through which the world at that time was destroyed, being flooded with water.
>
> 2 Peter 3:5-6

Peter referred to the original creation with the words, "the heavens existed long ago, and the earth was formed out of water and by water." He was reminding his readers of Genesis 1:1-2 concerning the creation of the heavens and earth. The phrase in verse six, "the world at that time" alludes to the time between the Fall of mankind and the Flood. That world was the fallen world God judged with the Flood. As we have learned, the Fall of mankind in the persons of Adam and Eve radically changed the original "very good" creation into the sinful mess it has been ever since (Chapters 14).

Next, Peter described the Genesis Flood, writing that the "world at that time was destroyed." Later in this chapter, we will learn that the Flood was the second historical event that radically affected the universe. Keep reading.

Peter then predicted the future catastrophic event that will end the universe as we know it:

> But by His word the present heavens and earth are being reserved for fire, kept for the day of judgment and destruction of ungodly men.
>
> 2 Peter 3:7

Notice that Scripture expands this idea in verse 10, thereby adding emphasis:

> But the day of the Lord will come like a thief, in which the heavens will pass away with a roar and the elements will be destroyed with intense heat, and the earth and its works will be burned up.
>
> 2 Peter 3:10

This judgment and the destruction associated with it will happen after Christ's second coming. Peter was not giving a detailed prophetic exposition. His focus was on exposing the error of their uniformitarian assumption. How he must have hated their mocking of Christ's second coming!

CATASTROPHIC EVENTS MENTIONED IN 2 PETER 3

CREATION
"earth formed out of water" (v. 5)

THE FALL
"world at that time" (v. 6)

THE FLOOD
"present heavens and earth" (v. 7a)

NEW HEAVENS AND NEW EARTH
"judgment and destruction" (v. 7b)

Many additional prophecies add to our understanding of Christ's return to earth in glory. A horrible seven-year tribulation period (Revelation 6-18) will precede it. By the end of that time of divine judgment, often called Daniel's Seventieth Week (Daniel 9:24-27), the nation Israel will have turned to God in faith (Revelation 7:1-9; Romans 11:25-27). Christ will then establish the thousand-year Millennial Kingdom (Revelation 20). At the end of it, He will preside over the Great White Throne Judgment during which He will consign all unbelievers to the Lake of Fire forever (Revelation 20:11-15). Then, He will destroy the universe with fire and create the new heavens and new earth to last eternally (2 Peter 3:7, 10-13; Revelation 21:1-22:5).

Peter understood the eternal importance of refuting the skeptics' mocking of the second coming. So should we! Why? Because anyone who denies Christ's second coming also denies all the associated eschatological events. How can anyone trust God for salvation if he doesn't trust God's timeline for history and His omnipotence that makes it happen? Peter wanted us to know the truth so we will not be fooled by skeptics and scoffers. Instead, we will await Christ's return with confident expectation, secure in our knowledge He will fulfill these prophecies just as He has all others.

Since radical changes occurred in the past, Peter reasoned, scoffers are foolish to reject a second coming of Christ that will bring new and even more radical changes (3:8-13). He went on to describe a future event when "the heavens will be destroyed by burning, and the elements will melt with intense heat" (3:12). He anticipated a glorious future age that he described as the "new heavens and a new earth" (3:13).

THE TAKE AWAY: In this apologetic, Peter alluded to four eras/dispensations of world history and three radical transformative events, all instituted by God. The four eras are the original creation of the heavens and earth, the pre-flood world that perished with the Flood, our current post-Flood world, and the future new heavens and new earth with righteousness dwelling therein.[1] The three transformative events are the original creation in six days, the universal Flood of Noah's time, and the cataclysmic judgments by fire that will end the universe as we know it.

These four eras are part of God's design for world history. An eternal state that we commonly call heaven will extend beyond the history of the physical universe. 2 Peter 3 and Revelation 21-22 give hints about this eternal state, but Scripture does not give us many details, maybe because our finite minds cannot comprehend such magnificence.

> I believe in God. In fact, I believe in a personal God who acts in and interacts with the creation. I believe that the observations about the orderliness of the physical universe, and the apparently exceptional fine-tuning of the conditions of the universe for the development of life suggest that an intelligent Creator is responsible.
>
> **DR. WILLIAM D. PHILLIPS, 1997 NOBEL PRIZE IN PHYSICS**

1. Traditional dispensationalists usually include seven dispensations during which God manages mankind in very different ways. They see three dispensations between the Fall of mankind and the Church Age.

WHAT HAPPENED IN HISTORY THAT NECESSITATED SUCH RADICAL CHANGES?

Sovereign God orchestrates world history and omnipotently transforms the natural realm. Throughout history, He has used different dispensational governance programs to correspond with the fundamentally different physical conditions He initiated in each era.

The three past dispensations with the most radical changes were:

1. The pre-Fall world (original creation), ending with God's judgment for that first sin.
2. The time after the Fall, ending in God's Flood judgment for universal sinfulness.
3. The time from the Flood to God's Second Coming judgment (the time in which we live).

THE TAKE AWAY: God's judgments on mankind's rebellion against Him filled the transition time after the Fall and Flood before the next era finally stabilized. God used changes in natural laws and physical constants to execute these judgments. The Tribulation era before the Millennium, which Christ's Second Coming will usher in, will be equally catastrophic, filled with major changes in scientific laws. Sadly, secular science disagrees with the historical timeline presented in God's inerrant Word. Satan has sown skepticism about God' sovereign control of creation and history, using scientific lies to fool many into believing uniformitarianism and other absurd theories.

God's judgments of the world come in cycles. Each of the past two dispensational cycles began well with God accomplishing a work of grace. Scripture describes recurring cycles of sin that precipitated these dispensational changes, beginning after Adam's first sin. Let's use the cycle of sin between Adam's sin and the Flood of Noah to illustrate the five stages of each cycle that give the pattern for all world history.

In the first cycle after Adam's first sin, God graciously intervened in Adam and Eve's lives (Genesis 3:6-7). He initiated contact with them (3:8-9), confronted their sin to illuminate it to them and convict them of it (3:8-13), made shocking promises that involved the potential salvation of their descendants (3:15), and implied He was giving them a new opportunity for restoration to fellowship with Him.

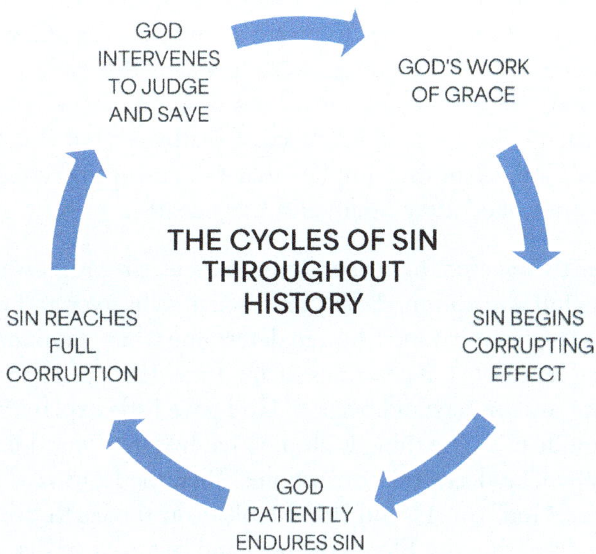

Adam expressed faith in God's promises by giving his wife the name Eve, meaning life (3:20) because he believed God's promise in Genesis 3:15 that she would be the mother of future generations. When they accepted the animal skins God provided (3:21), they also accepted that only through a blood sacrifice could their sin be covered and they be redeemed. Their acceptance of God's salvation provision indicated their faith in Him as their Savior/Deliverer.

The second stage of this first dispensational cycle began when the corrupting effects of sin became visible in that first post-Fall society. Genesis 4 describes the multiple sins of Cain, the first-born son of Adam and Eve. First, he offered a less acceptable sacrifice than his brother Abel because he did not know God well enough to understand the importance of a blood offering for sin (4:3-5; Hebrews 11:4). His attitude in giving his offering was all wrong because He focused on pleasing himself, not God (Genesis 4:5). When, in another act of grace, God confronted him, Cain was unrepentant (4:6-7). In anger, he murdered his brother Abel (4:8). Divine judgment followed swiftly when God removed him from fellowship with his family and Himself (4:9-14).

During the third stage of this first dispensational cycle of sin, God patiently endured mankind's sinful rebellion against Him. The Bible often describes God as slow to anger. In most cycles, God endures many generations of increasing wickedness. In the cycle leading up to the Flood, the two generational lines of Cain (4:16-24) and Seth (4:25-5:32), the third son of Adam and Eve, thrived but slowly moved away from God. Genesis 6 implies that the corrupting effects of sin gradually intensified after Adam and Eve's deaths.

In the fourth stage of the dispensational cycle, sin progresses until it reaches its full corruption. As the omniscient Observer and righteous Judge, God is the only One who can determine when mankind reaches this stage. Genesis 6:1-8 describes such a time. This passage is difficult to interpret and understand because God gave little explanatory revelation to guide us.[2] One thing is clear. Whatever man was doing, it was extremely evil. God saw the corruption ("Then the Lord saw" (6:5), "in the sight of God" (6:11), and "God looked on the earth" (6:12)) and determined to bring the Flood, the next and last cycle of discipline for the post-Fall dispensation.

In the fifth stage of the dispensational cycle of sin, God intervenes in history to bring judgment and salvation. He brought the Flood to stop the corrupting effects of sin and to save and protect His image bearers whom He loves, a work of grace in itself. Thankfully, every divine judgment of unrepentant, unrestrained evil includes a story of salvation.

THE TAKE AWAY: After the original creation, all historic transformations have resulted from God's direct intervention in the form of judgments. Sadly, mankind tends to ignore the lessons of history, and so the cycles inevitably begin again. Like Cain, most people refuse to respond to God's grace and gradually descend deeper into sin. God brings judgment to slow that descent. Though some may respond to His grace for a time, soon sin escalates among most of mankind, eventually necessitating a destructive judgment and God's institution of a new dispensation with its accompanying changes in the creation.

---

2. Competent scholars interpret this passage in three ways: 1. The descendants of Seth ("sons of God") merged with the descendants of Cain (daughters on men), corrupting the godly line. 2. Powerful demon-possessed despots married the daughters of men, producing evil descendants. 3. Fallen angels appeared as humans and married human women producing strange Nephilim. For a good explanation of these views, read Tim Chaffey's *Fallen: The Sons of God and the Nephilim*, Rise Books, 2019.

If you are alert, you can see the cycles playing out in your own history and adjust your lifestyle, so sin loses its power and God takes control of your life. You should study history in part so you can reject in your own life the mistakes of past generations.

> My greatest discovery was that I needed God, and that I was nothing without him and that he loved me and showed his love by sending Jesus to save me.
>
> **DR. ALEXANDER FLEMING, 1945 NOBEL PRIZE FOR MEDICINE**

WHAT RADICAL CHANGES OCCURRED IN THE ORIGINAL CREATION AFTER THE FALL OF MAN?

The first radical transformation of the entire creation took place when sin entered the newly created universe. We learned in Chapter 13 that sometime after God declared His original creation "very good," the first man chose to listen to his wife and disobey his Creator, bringing judgment on himself and all creation. The changes that ensued, perhaps the most radical in all world history, were neither naturalistic nor evolutionary but supernatural ones imposed by God. The specific examples recorded in Genesis 3 imply that virtually every area of science was affected. Zoology, anthropology, geophysics, botany, and physics changed most dramatically.

WHAT RADICAL CHANGES TOOK PLACE BETWEEN THE PRE-FLOOD AND POST-FLOOD WORLD?

The second radical transformation of the entire creation took place through God's Flood judgment. Some changes were direct divine interventions while others were natural processes God orchestrated to accomplish His purpose. We'll look at them in some detail.

Let's begin, as we should, with the biblical data about the Flood before we look at the particular physical changes it initiated. The biblical narrative describes a Flood that was worldwide. Though many dispute the universality of the event, the information supporting its magnitude is overwhelming. We can group the data that corroborates the universal nature of the Flood into seven broad categories.

1. UNIVERSAL CORRUPTION

The sinful corruption that brought about God's Flood judgment was universal, practiced by all but eight people. The biblical text says all mankind had declined spiritually to such an extent that the Lord had to intervene to preserve the human race.

> Then the Lord saw that the wickedness of man was great on the earth, and that every intent of the thoughts of his heart was only evil continually. The Lord was sorry that He had made man on the earth, and He was grieved in His heart..."I am sorry that I have made them."
>
> Genesis 6:5-7

God evaluated the pre-Flood culture as totally wicked. Their unbridled evil sorrowed and grieved Him (6:6-7). Notice His emphasis on the worldwide corruption that included all humanity: "great on the earth," "every intent," and "only evil continually." Their blatant and pervasive unrighteousness was about to destroy the Earth. He needed to judge them so those made in His image could be preserved.

The text goes on to tell us that while the Lord had determined to bring the Flood judgment, He was equally intent on preserving the one righteous (believing) family that remained (6:8-10).

> Now the earth was corrupt in the sight of God, and the earth was filled with violence. God looked on the earth, and behold, it was corrupt; for all flesh had corrupted their way upon the earth. Then God said to Noah, "The end of all flesh has come before Me; for the earth is filled with violence because of them; and behold, I am about to destroy them with the earth."
>
> Genesis 6:11-13

Notice that the text again mentions the depravity of the entire world, "corrupt" (6:11, 12) and "filled with violence" (6:11, 13), emphasizing that God's judgment was both righteous and just. His patience is evident in that He waited to bring worldwide judgment until only eight righteous people (believers) remained.

2. THE DETERMINATION OF GOD

God determined, predicted, initiated, and orchestrated the Flood, making it primarily a supernatural event. No naturalism there. Genesis 6:7-8 makes His sovereign control clear:

> The Lord said, "*I* will blot out man whom I have created from the face of the land, from man to animals to creeping things and to birds of the sky; for I am sorry that I have made them." But Noah found favor in the eyes of the Lord.
>
> Genesis 6:7-8

A later verse emphasizes that God Himself ("I, even I") brought about the catastrophic Flood:

> Behold, I, even I am bringing the flood of water upon the earth, to destroy all flesh in which is the breath of life, from under heaven; everything that is on the earth shall perish.
>
> Genesis 6:17

Notice the universal wording of "all flesh... under heaven," and "everything on earth shall perish." The Lord was the One who determined to bring a worldwide (universal) Flood on the whole earth as judgment for worldwide rejection of His righteousness.

In another passage, God again announced His determination to bring the Flood:

> For after seven more days, I will send rain on the earth forty days and forty nights; and I will blot out from the face of the land every living thing that I have made.
>
> Genesis 7:4

Mankind's evil was universal, covering the whole earth, and so God's judgment had to be equally comprehensive.

3. THE EXISTENCE OF THE ARK

The ark (chest or box) that carried Noah, his family, and the animals during the Flood was massive, three-hundred cubits in length (~450 feet), fifty cubits in width (~75 feet), and thirty cubits in height (~45 feet).[3] Building such a mammoth ship would have taken many years. Why did God instruct Noah to put so much effort and time into building something so large? If the Flood had not been worldwide, He could have instructed Noah to take a week-long, month-long, or even a years-long journey to a place the flood would not reach. We know God later instructed Lot and his family to leave the immediate vicinity of Sodom before He destroyed the cities of the plain (Genesis 19). Why didn't He do the same with Noah? Have I mentioned the Flood was worldwide? God knew Noah couldn't find safety apart from entering the ark.

4. THE UNIVERSAL DESTRUCTION OF LIFE

The text describes the total annihilation of all animals (Genesis 6:5-7, 11-13, 17; 7:4) and all mankind (6:7, 12-13, 17; 7:4, 8, 21-23) through phrases such as the "end of all flesh" (6:13) and "every living thing that I have made" (7:4). The only exceptions to this universal judgment were the animals in the ark and Noah and the seven members of his family (6:8-9, 17-18, 22; 7:1, 23, 24; 8:1). This clear description of the universality of God's destruction of life means the Flood encompassed the entire Earth (6:5-6, 13, 17). Hence, the Flood was worldwide, not localized merely to one area as many would have us believe.

5. THE UNIVERSALITY OF THE HEBREW WORDS

The Hebrew word *kol* is translated in the English text as "all" or "every." The word occurs thirty-eight times in the Flood description (Genesis 6-8). In the five verses of Genesis 7:19-23 alone, the term "all" occurs a remarkable eight times.

> The water prevailed more and more upon the earth, so that all (*kol*) the high mountains everywhere (*kol*) under

3. Today, we can gain an appreciation for the size of the ark by visiting the built-to-scale ark at the Ark Encounter exhibit built by Answers-in-Genesis in northern Kentucky. https://arkencounter.com. The original ark would have been at least the size of 1.5 football fields in length and half its size in width with a total capacity of more than 570 standard railroad container cars.

> the heavens were covered. The water prevailed fifteen cubits higher, and the mountains were covered. All (*kol*) flesh that moved on the earth perished, birds and cattle and beasts and every (*kol*) swarming thing that swarms upon the earth, and all (*kol*) mankind; of all (*kol*) that was on the dry land, all (*kol*) in whose nostrils was the breath of the spirit of life, died. Thus He blotted out every (*kol*) living thing that was upon the face of the land, from man to animals to creeping things and to birds of the sky, and they were blotted out from the earth; and only Noah was left, together with those that were with him in the ark.
>
> Genesis 7:19-23

Only someone who wanted to discredit the Flood account and, therefore, discredit God would try to explain away the thirty-eight uses of the word "all." All means all. Our twelve-year-old would sigh in exasperation at anyone who tried to say otherwise.

Notice the variety of objects the Hebrew word for "all" modifies:

all flesh - 6:12, 13, 17, 19; 7:15-16, 21-22; 9:11, 15-17
all the earth - 7:3, 8:9
all fountains of the deep - 7:11
all mountains - 7:19
all mankind - 7:21
all fish - 9:2
everything - 6:17; 8:22
every living thing - 6:19; 7:4, 23; 8:17
every kind - 6:20-21
every creeping thing - 6:20; 7:8; 9:2
everywhere - 7:19
every swarming thing - 7:21
every beast - 9:2, 10
every bird - 9:2
every living creature - 9:10, 12, 15-16

Other Hebrew words also indicate the universality of the Flood. The Hebrew word *mabul*, which means flood, was used twelve times in the Old Testament, all in reference to the Genesis Flood, never about localized floods. Other Hebrew words that indicate local floods, flooding, or moving waters were never used about the Genesis Flood. Such words are translated torrent (*nachal*; Psalm 74:15) and rivers (*yeaor* and *nehar*). The Genesis Flood was unique because it was universal. Even the Hebrew word used to describe it is unique, used only to describe the worldwide Genesis Flood.

Similarly, five times in the New Testament, the Greek word *kataklusmos* refers only to the Genesis Flood. The exclusivity of this usage confirms the Genesis Flood was universal. As with the Hebrew in the Old Testament, the New Testament used other Greek words to refer to small floods or moving waters. These include words translated flood (*potamos* in Revelation 12:15 and *plemmura* in Luke 6:48). Those words were never used to describe the Genesis Flood, another indication the Flood was unique, unlike any localized flood.

In both the Old and New Testaments, God inspired the writers to use words that distinguish the worldwide Flood from all local floods. Again, we see His attention to detail and His desire for us to understand His Word clearly. He knew many would dispute the universality of the Flood, so He bombarded us, even grammatically, with evidence to prove the truth.

6. THE UNIVERSAL GEOLOGICAL UPHEAVAL

The biblical text describes a destruction so massive that only a worldwide flood can explain it. We will study this upheaval in more detail later, but for now here is a summary. The text describes not only a 40-day rain event but the bursting of the "fountains of the deep" (Genesis 7:11). The flood "waters prevailed" 150 days (7:24) until they were fifteen cubits (~23 feet) above "all" the mountains (7:18-20). The waters were so great another 150 days were needed for them to recede (8:3). The flood event lasted 370 days, indicating a flood of unprecedented proportions. Even the largest floods recorded in history lasted only a few days, not weeks, much less over a year as did the Genesis Flood. The world has never since and never again will see such a flood.

7. GOD'S COVENANT WITH NOAH

A powerful argument for a universal Flood is the covenant instituted by God with Noah and all creation (Genesis 9). In what theology calls the Noahic Covenant, God legally bound Himself to never again bring a universal Flood on the world (9:8-17). The Flood had to have been universal because history has recorded many local floods all over the world since that time. If the Genesis Flood was merely local, God has failed to keep His promise and violated His covenant many times. That is unthinkable for our sovereign and faithful Creator.

THE TAKE AWAY: Does the biblical description show that the Flood was worldwide and, therefore, radically changed the entire earth? Peter said that the "world at that time was destroyed" (2 Peter 3:6), emphasizing that the post-Flood world was entirely different from the pre-Flood world. Just to be clear, if God destroyed the world, it was destroyed. Destruction is a radical change. Such massive, worldwide destruction disproves the satanic claims of uniformitarianism.

BIBLICAL SUMMARY FOR A WORLDWIDE FLOOD

1. Universal Corruption
2. The Determination of God
3. The Existence of the Ark
4. The Universal Destruction of Life
5. The Universality of the Hebrew Words
6. The Universal Geological Upheaval
7. God's Covenant with Noah

DOES SCIENTIFIC EVIDENCE SHOW THE FLOOD CAUSED RADICAL CHANGES?

The Bible clearly describes a global flood of unimaginable scale. No other flood or natural disaster has come close. Historical science has abundant, even overwhelming traces of the Flood event available for analysis. Why then do so many deny the historicity of the Flood? Mainly because they trust secular science more than God's Word. The official position of secular science is that geology has uncovered no evidence whatsoever to support a worldwide flood but much to support evolution. Both are lies.

THE TAKE AWAY: Did I mention that science has found overwhelming evidence for the Flood? The problem is not the amount of evidence but the interpretation of it. Secularists choose to apply their flawed worldview to their examination of the geological data. Their biases blind them against God and His judgment at the Flood. They need new worldview glasses. The conflict once again is with differing worldviews, naturalistic versus biblical, not with insufficient data.

> Nature is our kindest friend and best critic in experimental science if we only allow her intimations to fall unbiased on our minds.
> **MICHAEL FARADAY, PHYSICIST, INVENTED THE FIRST ELECTRIC MOTOR**

We are left with a dilemma. How can we successfully and boldly answer such wildly inaccurate but universally accepted assertions and do so with grace and knowledge? A growing movement among geologists has set out to help us. They are called flood geologists, and they specialize, as you would expect, in flood geology.[4] They have examined the abundance of trace materials and presented their findings in an easy to understand but also intensely scientific and thorough manner.[5] When you've finished reading the following summary of their main findings, you will have plenty of ammunition to use against those who ridicule a worldwide and unbelievably destructive Flood as described in the Scriptures.

A quick overview of their findings will help us see that no matter how we look at the data, the pre-Flood world was much different from the post-Flood world that we live in today. We will focus on the evidence presented by several scientific specialties. Put on your lab coat and biblical worldview safety glasses, and let's get into the science.

GEOPHYSICS AND PLATE TECTONICS

Plate tectonics, a relatively new specialty within the broader science of geophysics, studies the movement and composition of the continental plates that compose landmasses.[6] Tectonic plates during the Flood moved much more rapidly than today and would have transformed Earth's crust just as Scripture describes. These tectonic actions, called catastrophic plate tectonics, would have affected the composition of the

4. Andrew Snelling detailed the current young earth, Flood geology evidence in his two volume *Earth's Catastrophic Past - Geology, Creation, and The Flood*, Institute for Creation Research, 2009.
5. Details on much of this research can be found at globalflood.org, icr.org, answersingenesis.org, and creation.com.
6. In his first volume of *Earth's Catastrophic Past,* Snelling gives a brief history (363-371) and an excellent description (373-378) of this young science.

entire natural realm, meaning the seven continents today are radically different from the one continent before the Flood.

We will start this scientific discussion where we should always start, with the biblical text:

> In the six hundredth year of Noah's life, in the second month, on the seventeenth day of the month, on the same day all the fountains of the great deep burst open, and the floodgates of the sky were opened. The rain fell upon the earth for forty days and forty nights.
>
> Genesis 7:11-12

First, this passage emphasizes the historicity of the Flood by giving its precise beginning to the very day of Noah's life, the *seventeenth day of the month*. The book of Genesis gives a complete chronology of most of the events described in it. This chronology can be derived from the time notes associated with the ages of the patriarchs in the genealogies. By using a correlating date with Genesis and a well-established date from world history, we can give actual dates to major events in Genesis. Using a conservative dating approach, we come up with a date of about 2500 BC for the Flood.[7]

> Historian Bill Cooper's research in After the Flood provides dates from several ancient cultures. The first is that of the Anglo-Saxons, whose history has 5,200 years from creation to Christ, according to the Laud and Parker Chronicles. Cooper's research also indicated that Nennius' record of the ancient British history has 5,228 years from creation to Christ. The Irish chronology has a date of about 4000 B.C. for creation, which is surprisingly close to Ussher and Jones! Even the Mayans had a date for the Flood of 3113 B.C.
>
> **KEN HAM, *A FLOOD OF EVIDENCE: 40 REASONS NOAH AND THE ARK STILL MATTER***

The phrase "the fountains of the great deep burst open" describes in simple terms one awe-inspiring physical cause of the Flood. The wording is brief, so we can't envision all the details. In the past, creationists interpreted it as the bursting of underground water reservoirs from the pre-Flood world. For obvious reasons, the existence of underground reservoirs has never been verified, so it has remained a hypothesis only.

7. The late Dr. Herold Hoehner has done much work on biblical chronology. Using his chronological work we can calculate a date for the Flood of 2487 BC using the Masoretic text numbers and about 3200 BC using the Septuagint numbers.

A better explanation of the phrase is offered by geophysicist Dr. John Baumgardner,[8] who has done extensive research on the Flood for over forty years. While employed at Los Alamos National Laboratory, he developed a numerical Flood model that not only effectively accounts for the data in Genesis 6-8 but also for the data from the natural realm.[9] Using his 3-D, finite element program for modeling the earth's mantle and lithosphere (its outer shell), he has ascertained that the Flood was not only a hydrological cataclysm but also, and even more so, a tectonic one.

To understand his model, I need to briefly explain the science of plate tectonics. Extensive slabs of rock, called tectonic plates, form Earth's outer shell. Today, Earth has about a dozen major plates. They are typically fifty to one-hundred miles thick and several thousand miles wide. A layer of lower density granitic rock about twenty miles thick covers portions of some of these plates. These areas form the continents. The portions of the plates without the granitic layer form deep ocean floors.

Just below the plates is a layer of much hotter and, therefore, much weaker rock known as the asthenosphere. Because this area is roughly a thousand times weaker than the tectonic plates, the plates can slide across the Earth's near surface with almost no bottom drag. According to GPS measurements, plates today move only a few inches per year.

What do these plates do? The plates on opposite sides of the mid-ocean ridge system pull apart from each other and move in opposite directions. In other places in the ocean basins, one plate plunges beneath another and slides into the 1,800-foot-thick layer of crystalline rock known as the mantle (which includes the asthenosphere). This process of a plate sliding into the mantle is known as subduction. In yet other places, such as along the San Andreas Fault in California, two plates slide past each other gently, like ships passing in the night but with earthquakes and tremors felt on the surface.

Most of the time when an oceanic plate is subducting beneath an overriding plate, the two plates are locked together, and the overriding plate is increasingly bowed upward along its leading edge as the subducting plate moves continuously downward into the mantle. When

8. Most of the following description of Baumgardner's research in geophysics and plate tectonics was reviewed by him with corrections and many additions made by him.
9. You can find most of Dr. Baumgardner's research and papers at his website: globalflood.org.

the stress on the fault separating the two plates exceeds the strength of the rocks locking the fault, the rocks break, there is rapid slip along the fault, and the overriding plate rapidly returns to its un-deformed shape. This sudden slip releases a vast amount of stored elastic energy, producing an earthquake and often a tsunami.

In the 2011 magnitude 9.1 Tōhoku subduction earthquake off the east coast of Japan, some one-hundred-sixty feet of sudden slip occurred between two plates. In this most powerful earthquake ever recorded in Japan, the slip generated a devastating tsunami that killed more than 18,000 people and destroyed some 130,000 buildings.

Evidence is overwhelming that, during the Flood, tectonic plates moved thousands of miles across the face of the Earth at speeds of five miles per hour or more and that the entire pre-Flood oceanic floor was subducted into the Earth's interior. Such a vast amount of subduction, with rapid episodic locking and unlocking of the plates, would have generated at least fifty-thousand mega-tsunamis that eroded and destroyed the continental portion of the Earth. This erosion produced fossil-containing sediment that today averages six-thousand feet in depth across the world's continents. In other words, the Flood cataclysm destroyed the entire surface of the Earth, just as God disclosed in advance to Noah (Genesis 6). The apostle Peter emphasized that change, writing, "the world at that time was destroyed."[10] Destroyed means destroyed!

Because the tectonic plates have a much lower temperature than the hot mantle rock beneath them, they have a higher density and, therefore, a tendency to sink. In a landmark paper in 1986, Baumgardner showed that this condition can lead to catastrophic consequences. In a subsequent work, he pointed out that experiments reveal mantle minerals weaken dramatically under stress—by factors of billions or more. Baumgardner has published computer simulations that validate this catastrophic, runaway behavior. The rapid resurfacing of the Earth during the Flood through runaway subduction of oceanic plates is known as catastrophic plate tectonics.

In addition to the vast number of mega-tsunamis generated in subduction zones, another remarkable process unfolds along the

10. 2 Peter 3:6.

mid-ocean ridge system when two oceanic plates migrate away from each other. As the two plates move apart, rock from below rises to fill the gap. The temperatures needed to melt mantle minerals decrease with decreasing pressure. As mantle rock moves vertically upward to fill the gap created by two plates moving apart, minerals suddenly find themselves above their melting points and, therefore, begin to melt. This process is called decompression melting.

Minerals with the lowest melting temperatures melt first, which means only partial melting occurs in the rising mantle rock. Because of high pressure, even small amounts of melted minerals get squeezed from the un-melted minerals to rise toward the surface. At a mid-ocean ridge, the magma that reaches the surface today forms a layer of basalt three to four miles thick when it cools and hardens. This process also occurred during the Flood.

Dr. Baumgardner points out another spectacular consequence of this process during the Flood. With the plates moving apart at five to ten miles per hour, a broad strip of molten rock was in direct contact with seawater across some 40,000 miles of the ocean bottom. Because the rock temperature was about 1900°F, the seawater in contact with it became supercritical water, which then flashed to high velocity steam and formed a chain of supersonic steam jets above the strip. Those jets entrained liquid seawater as they rose upward through the overlying ocean and became spectacular fountains reaching into the stratosphere. The liquid water from those fountains then fell back to the surface as heavy rain over much of the Earth. Baumgardner suggests that the phrase "fountains of the great deep" (Genesis 7:11) is a fitting description of those water entraining jets.

In summary, the pre-Flood world was destroyed by the Flood and replaced with the surface geology of the Earth that exists today. To better understand, let's go back about sixteen-hundred years to Day Three of creation.[11] Notice the wording:

> Then God said, "Let the waters below the heavens be gathered into one place, and let the dry land appear;" and it was so. God called the dry land earth, and the gathering of the waters He called seas; and God saw that it was good.
>
> Genesis 1:9-10

11. https://answersingenesis.org/bible-timeline/timeline-for-the-flood/.

THE TAKE AWAY: Since the waters were "gathered into one place," they must have formed only one ocean in the beginning. That suggests all land was also in one place, forming one large continent that some early geophysicists named Pangea. Baumgardner's model proposes that this single large continent was broken up during the Flood due to the rapid movements of the tectonic plates. Some have referred to that rapid movement as continental sprint, in contrast to today's continental drift of a few inches per year. After the Flood, the Earth stabilized, leaving behind our present continents. Those tectonic movements not only radically transformed the continents but also all geophysics.

On Day Three, God created only one continent. The rapid and intense tectonic plate movements of the Genesis Flood produced, in less than one year, the seven major continents we have today, North America, South America, Europe, Asia, Africa, Australia, and Antarctica. That is a dramatic change! Absolutely no uniformitarianism there. Keep your lab coat and biblical worldview safety glasses on as we enter the geology lab.

GEOLOGY

The Flood probably affected the sciences of geophysics and geology the most. The Bible does not give many details about pre-Flood geology but does make two inferences. First, as noted above, the earth had only one continental landmass. Second, Noah's ark floated in the flood waters above all the mountains of the earth (Genesis 7:20).

If the Genesis Flood was of the scale described in Genesis 6-8 and in Baumgardner's Flood model, we would expect to find evidence in the geological column, and we do. As we have seen, traditional historical geology interprets the geological column from a naturalistic, evolutionary worldview. It applies uniformitarianism and concludes that each of the geological layers was laid down by slow processes of either wind or water over millions/billions of years. If, instead, we adopt the objective biblical worldview to interpret the geological column down to the Precambrian layers, we see overwhelming evidence of the Genesis Flood.

> As we conquer peak after peak we see in front of us regions full of interest and beauty, but we do not see our goal, we do not see the horizon; in the distance tower still higher peaks, which will yield to those who ascend them still wider prospects, and deepen the feeling, the truth of which is emphasized by every advance in science, that 'Great are the Works of the Lord.
>
> **SIR JOSEPH J. THOMSON, 1906 NOBEL PRIZE IN PHYSICS, FOUNDER OF ATOMIC PHYSICS**

We already discussed fossil evidence when we considered the fallacies of evolution.[12] This time, we'll briefly look at fossils from another perspective. Tectonic movements during the Flood produced huge tsunamis whose turbulent water waves ripped up everything in their path, including soil, rocks, boulders, even entire forests. They destroyed all life and transformed massive volumes of bedrock into sand, silt, and clay particles, transporting this material for hundreds, even thousands of miles.

When the explosive speed of the water slowed, the suspended particles settled to the bottom of the water column, producing, under the pressure of the sediment above, the different rock layers in the geological column. In some locations, animals and plants were trapped and buried, forming fossils.

THE TAKE AWAY: These fossil layers, essentially in the same order, are found today on all seven continents. The fossils are the dead remains of the pre-Flood world and should remind us of God's judgment of out-of-control sin.

> The evolutionist may then grope for evidence in the fossil record. But again he is trotting out another batch of circular reasoning. This is so because evolution is used to interpret the fossil record, so it cannot be used to justify evolution. To do so puts the proponent in the intellectual booby hatch. Whatever the explanation for the fossil record may be, it cannot be one that in effect denies the Second Law of Thermodynamics. In fact, the most obvious feature of the fossil record is not upward synthesis but rather death and decay. We find strong evidence for the steady loss of species within the fossil record. This is more in consonance with the Second Law of Thermodynamics than with the upward growth posited by evolution.
>
> **DR. KER C. THOMSON, GEOPHYSICS**

---

12. For much more detailed geological evidence for the Flood, see *The Genesis Flood: The Biblical Record and Its Scientific Implications*, John C. Whitcomb and Henry M. Morris, Baker Book House, 1961 and *Earth's Catastrophic Past – Geology, Creation, & The Flood*, Andrew Snelling. icr.org, answersingenesis.org, and creation.com also provide many details.

The vast lateral extent of most of these geological layers as well as the small amount or absence of erosional channeling on the boundaries between them indicates a single catastrophic Flood event. In other words, the layers were not down long enough for natural erosion to take place. They were quickly covered by the next layer of flood residue.

The geological layers of rock we see today could not have been laid down over long periods of time. Once again, uniformitarianism is shown to be pseudo-science. You can add hiking boots to your lab coat and worldview glasses for our next observations.

MONTOLOGY/OROGENY

Montology, traditionally known as orogeny, is the study of mountain formation. Because Scripture says Noah's ark floated above all the mountains of the earth (Genesis 7:20), we know the pre-Flood world had mountains. Genesis 7:20 also says floodwaters covered all mountains by approximately twenty-three feet.[13] By calculating the quantity of water on the surface of the earth today, montologists have determined pre-Flood mountains were much lower than today's mountains.

Flood geology and tectonic studies suggest present-day mountains rose to the heights observed today because of landmass pressures produced by the Flood's tectonic after-effects, not by the Flood itself. It also implies that today's high mountain ranges, including the Himalayas, Alps, Andes, and Rockies, were formed at the end of and after the Flood. God's judgment set the Flood in motion and produced the subsequent, associated after-effects. Had God not instituted judgment, these natural processes would not have occurred. And climbing Mount Everest would not be nearly as challenging! This is another example of the pre-Flood world being radically different from the post-Flood world.

> The Hebrew phrase in Psalm 104:8a is the basis for the correct translation of mountains rising and valleys sinking. This shows that mountains and valleys during the Flood were not the same height as they are today. Even today mountains and valleys are changing their height; volcanic mountains, for instance, can grow very quickly, such as Surtsey (a new island) or Paricutin (a volcanic mountain in Mexico that formed in 1943).
>
> **KEN HAM, *A FLOOD OF EVIDENCE: 40 REASONS NOAH AND THE ARK STILL MATTER***

13. The text specifies 15 cubits. Using a cubit of 18 inches results in about 23 feet.

PALEONTOLOGY

The entire science of paleontology depends for its existence on the evidence left by the Flood. I held a creation seminar for mainly high school home schoolers several years ago. While discussing the Flood, I gave a simple explanation of paleontology. A sharp eight-year-old asked why scientists didn't call it fossilology. I did not have a good answer, so for the rest of our seminar we called it fossilology. For those of you who want to know, the name paleontology comes from the Greek words that mean the study (-logy) of ancient things (paleo-). I prefer fossilology.

Every layer of sedimentary rock, from the Cambrian layer to the surface, contains many fossils of dead animals and plants. Because the rocks below the Cambrian contain no such fossils, it is reasonable to conclude the destruction of the Flood reached the top of the Precambrian rocks. The abundance of fossils indicates the worldwide death caused by the Flood, thereby confirming the biblical record.

First, the very existence of fossils argues for catastrophic conditions. Normally, when an animal dies, either predators eat it, or it decomposes quickly through natural decaying processes. Occasionally, fossils are formed today but only under rapid and catastrophic conditions, such as the Mount Saint Helen's volcanic eruption in 1980, which acted much like the Flood but on a significantly smaller scale. Note that the many fossils from this modern event were formed in days, not over billions of years, another nail in evolution's coffin.

Not only are fossils present in every geological layer from the Cambrian layer to the surface, but these layers exist in the same order all over the world. Both facts indicate a worldwide event. Fossils are even found on top of all mountains worldwide. How could they be there apart from a worldwide flood that covered every mountain? The abundance of fossils in every sedimentary stratum on every continent indicates massive death due to worldwide catastrophic flooding and sudden burial. Simply put, these creatures didn't have time to decompose naturally.

Second, consider fossil graveyards, which are accumulations of many, often different creatures in one area. These creatures appear to have been transported and catastrophically buried all at one time. Again, fossil graveyards are found all over the world, including in the United States. Dinosaur National Monument on the border of Colorado and Utah contains an abundance of different species of dinosaurs. What kind of catastrophe could have gathered such huge creatures and buried them rapidly all together? Our twelve-year-old would say a massive flood! Once again, he would be right.

Third, polystrate fossils are found worldwide. Polystrate fossils go through more than one sedimentary layer, sometimes through multiple layers. Many are trees that transverse layers that are supposedly each many millions/billions of years old. Why didn't these trees decompose as they waited to be fossilized during those interminable millennia? It makes far more sense that, because the layers were laid down rapidly and suddenly, trees could be preserved as fossils and span several sedimentary layers. Because dinosaurs were also large, they are often preserved as polystrate fossils as well. If evolution is correct, why didn't they, too, decompose and disappear without a trace before they could be fossilized?

Though much other data in the fossil record supports a worldwide Flood, let's move on to other radical changes between the pre-Flood and post-Flood worlds. We have looked at geophysics, plate tectonics, geology, orogeny, and paleontology, but what about all that water? "Water, water, everywhere and nary a drop to drink."[14] You can exchange your hiking boots and lab coat for a raincoat but continue wearing your worldview glasses for our next observations.

CLIMATOLOGY

The Scriptures suggest that the climate changed after the Flood. The biblical data about climate is limited but does give small indicators. Genesis 2:5-6 refers to the Garden of Eden being watered by a mist before God "sent rain upon the earth." It may well be that the first rain on the planet was the rain of the Flood. If so, the forty days and nights of continuous rain (Genesis 7:11) are significant. The geophysical conditions necessary to produce such a long and comprehensive storm have never been recorded at any other time in world history.

The secular magazine, *Scientific American*, claims that the world's largest storm was Typhoon Tip in the vicinity of southern Japan.[15] It was

---

14. *The Rime of the Ancyent Marinere*, Samuel Taylor Coleridge, published in 1798 in the first edition of Lyrical Ballads.
15. https://www.scientificamerican.com/article/earths-strongest-most-massive-storm-ever/. Downloaded 2020.

huge but nothing like the deluge described in Genesis 7. The article states that the peak winds occurred on October 12, 1979, and that it slowly weakened when it hit landfall in southern Japan on October 19, only seven days later. Typhoon Tip was 1380 miles across, the distance between Dallas and New York. The "world's largest storm," though significant, was tiny in comparison to the storm described in Genesis.

As would be expected, secular climatologists claim that even a super-saturated atmosphere would not be capable of producing such a storm. Is there a scientific way to explain a forty-day storm event? Dr. Baumgardner's Flood model gives a plausible explanation. As discussed above, the separating of the ocean tectonic plates that brought mantle magma into contact with the waters at the bottom of the ocean would have created massive water jets that included steam. That steam would have shot into the atmosphere around the world, producing massive clouds that would have spread throughout the earth, providing a continuous supply of atmospheric water vapor to produce the forty-day rain event. Apparently, after forty days, the tectonic plates stabilized both in terms of movement and temperature, and the storms subsided.

The fossil record gives hints about pre-Flood climatology. The variety of flora and fauna and the many extinct creatures represented in the fossil record indicate a pre-Flood environment far lusher and, therefore, warmer than now. This warmth indicates the composition and proportions of the atmospheric gases changed during the Flood. After the Flood, the environment must have been much harsher since many species of flora and fauna, including dinosaurs, survived only a short time.

Trace data from the ice age that followed the Flood support the existence of a worldwide flood that changed the climate across the earth. Most scientists agree on the existence of at least one ice age because its remnants are observable today. Our twelve-year-old would mention glaciers. He'd probably roll his eyes again, too.

Climatologists do not understand what caused this last and probably only ice age. According to climatologist William Stokes:

> The underlying cause of glaciation remains in doubt...At least 29 "explanations" have been advanced to account for

> widespread glaciations. Most of these had little chance of survival from the first, but others enjoyed some degree of success until they were rendered untenable by subsequently accumulated information.[16]

The number of secular theories continues to increase because each has flaws and must be revised. Atmospheric scientist Michael Oard noted that as of 1968 over sixty theories had been proposed, all of which have serious, unresolved inconsistencies and problems.[17]

> The only solution is to put Neanderthals after the Flood. They were the first people (yes, *Homo sapiens*) to make it up into Eurasia. They struggled to cling to a marginal environment. They never achieved a large population size. They became incredibly inbred and were on their way to extinction, except that another, larger group of people moved into the area. These new people were not hunter-gatherers. They had more food. Thus, they had more children. The Neanderthals were overwhelmed, but not completely because about 60% of their DNA lives on in us.
>
> **DR. ROBERT CARTER, MARINE BIOLOGY**

Climatologists who accept the biblical description of the Flood have concluded a massive flood scenario could certainly have caused the ice age, which began a short time after the receding of the waters described in Genesis 8. Oard presents a convincing argument for the cause of the ice age.[18] In a thumbnail explanation of his theory, worldwide volcanism during and after the Flood, the same phenomenon Dr. Baumgardner found in his Flood model, warmed the world's oceans. Warmer water produced increased evaporation and thus increased clouds and precipitation. Dust particles from the volcanic action of the Flood also clogged the atmosphere. These massive water and dust clouds blocked the sun, producing a many-years-long snowfall, causing the atmosphere and, thus, the earth to cool. This global cooling produced the ice age. Obviously, all that snow and ice changed the science of climatology radically.

Eventually, the earth began to warm as the ice sheets of the ice age began to recede. Have we been experiencing global warming ever

---

16. Stokes, William L., "Another Look at the Ice Age," *Science*, Oct 28, 1955, 815 cited in *The World That Perished*, John C. Whitcomb, Jr., Baker Book House Company, 1973, 85.
17. Oard, Michael J., *An Ice Age Caused by the Genesis Flood*, Institute for Creation Research, 1990, 13.
18. Ibid.

since the Genesis Flood? Is it not interesting that when we look to Scripture for answers in one area, we are given answers in other areas as well? This phenomenon is especially true in science because all areas of creation interconnect.

> Most believe the Flood of Noah triggered the Ice Age. The rising magmas, lavas, and hot waters associated with continental plate movements would have caused ocean temperatures to rise. Also, fine ash from volcanic eruptions probably lingered in the upper atmosphere in post-Flood years, which, unlike a greenhouse effect, would reduce the sunlight for cooler summers. So the mechanism for such a rare event was in place due to Genesis 6-8. But what happens in an ice age? A lot of water is taken out of the ocean and deposited on land, so the ocean level drops. This exposes land bridges. One well-known land bridge was the one that crossed what we call today "the Bering Strait" from Alaska to Russia, so it is easily feasible for animals to have walked from Asia to North and South America.
>
> **KEN HAM, *A FLOOD OF EVIDENCE: 40 REASONS NOAH AND THE ARK STILL MATTER***

OCEANOGRAPHY

According to Baumgardner's model, the Flood affected the oceans through tectonic movements of the ocean floors, subductions of these floors into the earth's mantle, and tsunami movements of ocean waters over the continents. These forces caused a massive mixing of earth materials as tumultuous water scoured everything in its path.

In Genesis 7-8, God, through Moses, tells us that the ocean waters destroyed everything. Notice the emphasis on this fact in 7:18-20 through the use of the phrases "water prevailed" (7:18), "increased greatly upon the earth" (7:18), "more and more upon the earth" (7:19), "all the high mountains everywhere under the heavens were covered" (7:19), and the "water prevailed fifteen cubits higher" than all the mountains (7:20). The text says the waters prevailed (continued) upon the earth for one-hundred-fifty days (7:24). The waters then receded from the earth with another reference to the one-hundred-fifty days (8:3). That is a flood of inconceivable scale. Typhoon Tip of only about a week duration was insignificant in comparison.

From its start (7:11) to its end (8:13-14), the Genesis Flood spanned three-hundred-seventy days, more than a year. Even the most destructive floods of recent times lasted only a few days. The 2017 Houston flood precipitated by Hurricane Harvey lasted only four or five days.[19] Though a major destructive force for that area, it, again, was inconsequential compared to the Flood described in Genesis 6-8.

Since the Flood affected the crust of the earth, oceans today conform to the increased landmass boundaries the Flood created. We should thank God the oceans stay within His boundaries and can't prevail over the whole earth in another worldwide flood.

> He set for the sea its boundary so that the water would not transgress His command.
>
> Proverbs 8:29

The ocean's composition also varies significantly because of the magnitude of the Flood. For example, the dissolved salts and minerals in the oceans' waters would be quite different after the Flood because of the massive rearranging, mixing, and dissolving of plant and animal materials during the Flood.

PHYSICS, SPECIFICALLY OPTICS AND HYDROLOGY

The biblical text indicates the Flood brought the first rain on Earth. If so, and of course it is true since the Bible is inerrant, raindrops would not have existed from creation to the Flood. We can also assume that, after the Flood, God arranged the surface tensions and optics of raindrops to create the rainbows we observe today.

The optics alone show evidence of divine design. Rainbow optics include white (normal) light entering each raindrop, refracting inside the drop and reflecting within, refracting out of the drop, and finally separating and scattering out to observers. The cumulative effect is a rainbow with a wide range of colors. Note that the order of those colors is always the same. Our God is the God of order and controls every detail so we receive the greatest benefit. He wants us to recognize His hand in every aspect of creation—another of the ways He shouts His lovingkindness to us.

> The order, the symmetry, the harmony enchant us...God is pure order. He is the originator of universal harmony.
>
> **GOTTFRIED WILHELM LEIBNIZ, MATHEMATICS, INVENTOR OF CALCULUS**

19. https://en.wikipedia.org/wiki/Hurricane_Harvey.

THE TAKE AWAY: Rainbows are more than beautiful combinations of physics, chemistry, hydrology, climatology, and other associated physical phenomena. Scripture tells us that God produces each rainbow as a reminder that He will faithfully keep His promises of the Noahic Covenant, including His promise to never again bring another worldwide flood. Every time we see a rainbow, we should rejoice that God is sovereign over all His creation and faithful to fulfill every promise. Be reminded of our discussion on our Lord's sovereignty over all the creation. Don't let our unbelieving culture hijack God's sign! Think of a rainbow as a giant and stunningly beautiful billboard that God places in the sky to shout His authority and love for us.

Are there even more sciences that were radically changed as a result of the Genesis Flood? You will not need a Geiger counter but continue wearing your worldview glasses for our next observations.

NUCLEAR PHYSICS

For over one hundred years, secular science has dominated the conversation on the age of the earth with its evolutionary timescale. In 1997, a group of PhD creation scientists began an eight-year multimillion-dollar research project to study the chronology of earth history and the assumptions of a long evolutionary process.[20] This research project was identified with the acronym RATE for Radioisotopes and the Age of The Earth.[21] Both John Baumgardner and Russ Humphreys were main researchers on the team. The project was privately funded, mainly by contributions from Christians, and is the most extensive and thus most expensive and significant creation research project ever.

Why have you never heard of it? You already know the answer. This research project exposed the false ideas at the heart of secular science. It showed that the major assumptions of radiometric dating are bogus and the evolutionary timeframe based on those dating methods has no validity at all (*Thousands...Not Billions*). It showed that the universally accepted idea of uniformitarianism is false and the evolutionary

20. DeYoung, Don, *Thousands...Not Billions,* Master Books, 2005, 17-19 gives a simple description of the project.

21. The technical results of the RATE research project are contained in the two volumes of *Radioisotopes and the Age of The Earth: A Young Earth Creationist Research Initiative,* Larry Vardiman, Andrew Snelling, and Eugene Chaffin, editors, Institute for Creation Research, 2000.

idea based on deep time and uniformitarianism is unscientific, more pseudo-science. To be blunt, it showed that secular science with evolution at its heart has dangerous problems that jeopardize many scientific conclusions.

Do you think the science establishment ran to microphones with a news flash to announce that much of their work is deeply flawed? Do you see that my book will also be ignored by most, especially in science and the media? I expect my book to be crushed by the momentum of naturalism, which has no intention of changing course. It is up to you to get the word out. Get an extra copy of this book and give it to your favorite scientist—or to anyone, especially young people wondering about evolution or the scientific process. Each of us must do our part in spreading the truth. Think of it as a ministry opportunity.

The RATE project made a huge contribution to science itself (though ignored), especially to biblical worldview science. It produced much scientific data to support a young earth/universe that is about six-thousand-years old. In other words, it showed we can trust the Bible to reveal nothing but truth.

We could spend pages, even volumes, on the issue of the age of the universe. Many books do just that. For our purposes, all I want to say is that the young earth/universe view (about six-thousand-years) fits the biblical worldview of science best. Even though Scripture clearly states this fact, the majority of Christians accept the evolutionary time frame of a 4.5-billion-year-old earth and a 14 to 20 billion-year-old universe without question. Their old earth/universe view supports the secular science worldview but undermines the Bible. Thankfully, we can turn to the scientific evidence produced by RATE scientists and others to refute this inaccurate dating.

Radioisotope decay rates are the basis for all radiometric dating methods that give the billions of years necessary for evolution. One of secular science's main assumptions is that radioactive decay rates, measured in half-lives, are constant and have always been the same. It assumes that these physical constants can give accurate projections into the distant past based on current half-lives rates. We are talking about a distant past of many billions of years! I'll bet you have already realized that secular science applied uniformitarianism to radiometric dating to come to this speculative and now-suspect conclusion.

Several of RATE's experiments indicated that radioisotopes underwent accelerated nuclear decay in the past. If these findings are true, and no one has given scientific evidence that disputes them, conventional secular wisdom on the long age of the earth is flawed as well as unscientific.

The RATE project demonstrated many problems with the entire radiometric dating method. First, half-life decay rates are not constant. Second, the initial conditions (purity) of rock samples cannot be accurately determined. Third, the open or closed condition of the samples is also impossible to determine.[22] We cannot depend on secular science's declarations about the age of the earth and fossils because it depends on a faulty and highly speculative technique to determine it. In fact, dating would be most inaccurate the farther back in time we go, making evolutionary dates for the beginning of the Earth the most inaccurate.

Because secular science did not take into account accelerated decay rates, all secular radiometric dating figures are greatly inflated, thus eliminating the billions of years proposed by the evolutionary time frame. According to RATE:

> Many recent, post-Flood rock formations also hold a record of substantial nuclear decay. This again implies decay rates somewhat shorter than those presently observed. This suggests that during the immediate post-Flood period there was a "tapering off" of accelerated decay to the half-life values which we measure today. In summary, there are three time periods when the acceleration of nuclear decay likely occurred: creation, Flood, and immediate post-Flood times. We conclude the vast majority occurred during the early part of the creation week and during the Genesis flood.[23]

In one of many examples, RATE scientists found evidence of accelerated decay in zircon crystals. The radioactive uranium-238 in the crystals decays into lead over time, producing helium. Helium escapes from the crystals at a known, measurable rate. If the rocks containing zircon crystals are billions of years old, as evolutionists claim, all helium

22. *Thousands…Not Billions*, 177-78.
23. Ibid., 151.

should be long gone. Instead, the presence of much helium supports the biblical view of a young earth.[24]

Coal and diamonds give similar evidence. Carbon-14, which has a half-life of 5,730 years and would have disappeared over billions of years, is still found in them.[25]

> In cases where the actual age of the rock is known, radiometric dating techniques typically give wildly erroneous ages. For example, rock formed in a lava flow from Mt. St. Helens in 1986 was radiometrically dated as 2.6 million years old! If, every time you read a newspaper report concerning an incident about which you had first-hand knowledge, you found that the newspaper report was totally wrong, how many of these would you read before you began to suspect that all the reporting was wrong?
>
> **DR. JIM MASON, EXPERIMENTAL NUCLEAR PHYSICS**

Believing in secular science's declarations about the age of the earth and of fossils is like believing in magic. We see the outcome while ignoring the sleight of hand behind it. In the case of evolution, the magic happens through the faulty and highly speculative technique of radiometric dating. By focusing us on scientific mumbo jumbo, secular science discourages us from examining accelerated decay rates, dating that would be more inaccurate the farther back in time we go. Only by encouraging us to cling to pseudoscience while ignoring scientific studies to the contrary can evolutionists trick us into believing in the billions of years needed to support their anti-God opinions.

THE TAKE AWAY: Again, true science based on meticulous experimentation indicates the pre-Flood world was radically different from the post-Flood world, including in radioactive decay rates. Thus, we have another example of a scientific constant that is not constant at all because God changed it to usher in a new dispensation. The only true constant is God Himself.

---

24. htpps://answersingenesis.org/age-of-the-earth/dating-methods/.
25. htpps:// answersingenesis.org/age-of-the-earth/dating-methods/.

> Numerous pieces of evidence fit a young earth. To mention a few: the historical records, the population growth, the helium content in this world, the missing neutrinos from the sun...the oscillation period of the sun, the decline of the earth's magnetic field, the limited number of supernovas, radioactive halos, the mitochondrial DNA pointing to one mother, and the increase in genetic diseases, etc.
>
> **DR. JOHN K.G. KRAMER, BIOCHEMISTRY**

ANTHROPOLOGY

The biblical text is specific concerning the patriarchs' longevity of life. Bet you didn't know those long lists of patriarchs, the *begat* lists, help prove that the Flood narrative is, indeed, true. The age-related numbers contained in the Masoretic text of Genesis show patriarchal age patterns were significantly different after the Flood. The Septuagint numbers are even greater.[26] This fact on longevity has given rise to many theories. Naturalists contend that the long lives before the Flood prove we should reject Scripture as mythological. It is that uniformitarianism thing again. A closer look suggests a more plausible explanation that supports the biblical position that physical conditions pre-Flood differed from those post-Flood.

The ages of the patriarchs who lived before the Flood were relatively stable, averaging 912.2 years. The youngest died at 777 years old and the oldest at 969. (Our twelve-year-old might tell us that Methuselah was the oldest man to ever live. He would be right.) This average does not include Enoch since, as we learned, he did not live a full, natural life. One of the most dramatic verses in the Bible records that Enoch, at the age of 365, walked with God and was taken up without dying (Genesis 5:21-24). Is that ever an incentive to walk with God!

The Patriarchs' ages after the Flood decreased steadily. By plotting this second set of ages, we see an approximate but recognizable curve, an exponential decay curve. A wide variety of data sets from many different physical phenomena, including physics, nuclear physics, electricity,

26. Many question whether the age numbers in the Masoretic text or the Septuagint are more authentic. In my opinion, the evidence seems to favor the Septuagint. This discussion is beyond the scope of this book. I will use the Masoretic text numbers since they are more widely accepted. Both sets of numbers make the same point.

and chemistry, show when one steady state changes to another steady state, they often follow an exponential decay curve.

THE TAKE AWAY: The longevity data in Scripture supports the position that something, in fact many somethings as we have seen, changed after the Flood. In all likelihood, pre-Flood humans enjoyed a relatively stable and pleasant environment. Before enough time passed after the Flood for another stable environment to be established, a transition period showed steady decline in longevity until one-hundred years became an extra-long life. We should not consider that change either accidental or coincidental but rather a manifestation of the effects of the Flood. As we saw in 2 Peter 3, God destroyed the pre-Flood world (*kosmos*) and replaced it with a new, entirely different environment after the Flood. That change caused decreased longevity. It also resulted in the extinction of some creatures like dinosaurs, another indication the post-Flood environment was much harsher.

> The truth of God's Word is on the line, and ultimately, salvation is on the line. The Bible makes it clear that God created everything in six literal days some 6,000 years ago. He put the genealogies in the Bible for a reason. We must believe He did what He said. Sadly, evolution and long-age thinking are turning people away from God. We need to show people that they can trust God and His written Word.
>
> **DR. TIMOTHY CLAREY, HYDROGEOLOGIST**

Another related bit of data involves the diet God permitted after the Flood. Before the Flood, humans were to eat only plants. After the Flood, God instituted the eating of meat.[27] Why the change? The harsher environment after the Flood required more protein-based food for survival. Even with the addition of protein, lives are now dramatically shorter than before the Flood.

God produced not only spiritual dispensational changes with the Flood but also far-reaching physical changes. Those changes affected some of the laws of nature and physical constants. Except for the resurrection of Christ and the internal regenerating work of God in believers' hearts, no radical transformations have occurred or will occur in the era between the Flood and the Second Coming

27. Genesis 9:3-4.

of Christ. God's Noahic Covenant ensures a stable creation during our post-Flood time. This stability permits the repeatability of phenomena we observe in the creation and allows for today's scientific observations and experimentation.

SOME RADICAL CHANGES CAUSED BY THE FLOOD

- GEOPHYSICS → Geologic Column
- TECTONICS → Continents
- OROGENY → High Mountains
- PALEONTOLOGY → Abundance of Fossils
- CLIMATOLOGY → Seasons, Ice Age
- OCEANOGRAPHY → Landmass Boundaries
- PHYSICS → Constants Changed
- ANTHROPOLOGY → Life Longevity

WHAT RADICAL CHANGES WILL OCCUR BETWEEN OUR PRESENT WORLD AND THE MILLENNIAL KINGDOM?

Only Scripture reveals that the creation will experience another radical transformation because only God knows the future. This third transformation will occur with God's judgment at Christ's Second Coming to inaugurate the Millennial Kingdom. Try to imagine a change as extreme as after the Genesis Flood but with only perfect results. And all Church Age believers, including us, will get to witness it!

Science cannot give us any data about the future, so it cannot describe future conditions or changes. We are dependent on divine revelation for all insights. Thankfully, our Lord has revealed some of His future plans, including some of the changes He will implement in the physical realm. The biblical worldview is unique in being able to describe the future with accuracy. Science will have an exciting future whether today's scientists acknowledge it or not.

The environment after the Flood has remained stable and predictable for over four thousand years.[28] Why? Because of the Noahic Covenant

28. If the age numbers of the Masoretic text are correct, the Flood took place approximately 4,500

of course. As we could have predicted, God has honored the legalities of His covenant and kept His promise. As we could have predicted, too, many, if not most, humans continue their downward spiral away from God. The spiritual fight between those who love God and those who don't, between good and evil, escalates daily.

> As I saw the human wreckage made by Darwinism, as at one of Hitler's ghastly facilities for weeding (by execution) physically and mentally unfit humans out of the race he wanted to purify (just helping out natural selection a bit), I began to realize deep down that the idea of evolution is truly evil...Darwinism is a religion of death. It claims that to improve some, others must die. How could any sane person promote such a religion?...[T]hese men know deep in their hearts that they are wrong. But out of a fierce desire to maintain their independence in this world, these fully-aware intellectual leaders will fight God even as they fall into the lake of fire. It apparently gives them comfort to try to bring others with them by deception, misleading millions into worshipping "an image in the form of corruptible man and of birds and four-footed animals and crawling creatures" (Romans 1:23).
>
> **DR. D. RUSSELL HUMPHREYS, PHYSICIST**

Scripture paints a picture of drastic changes after the conclusion of the Church Age, including a catastrophe-filled seven-year period called the Tribulation, which will transition from our present Church Age to the Millennium. As with other transition periods, this will be a time of extensive and horrific judgments on a degenerate human race (Revelation 4-19). It will conclude with the Second Coming of Christ when He will initiate a final bloody judgment before establishing His one-thousand-year Millennial Kingdom.[29]

Theologians and Bible teachers stress the theological and spiritual aspects of the Millennial Kingdom. In this book we need to stress the physical and scientific aspects related to the physical realm.

Physical descriptions of the future Millennial Kingdom are found primarily in the Old Testament. There, God reveals He will in large measure lift the curse He imposed on creation after the Fall though death

years ago. If the Septuagint numbers are the authentic ones, the Flood took place about 5,200 years ago. This discussion is beyond the scope of this book. See the following for support for the Septuagint numbers as basis of chronology, https://biblearchaeology.org/research/topics/biblical-chronologies/4355-setting-the-record-straight-on-the-primeval-chronology-of-the-septuagint-part-1.

29. At my website (ForHisGloryNM.com), I have extensive teaching on the details of biblical eschatology, including entire seminary courses.

will still be a possibility for those who rebel against the Messiah's rule (Isaiah 65:20). In Acts 3:21, Peter described the Millennial Kingdom as "the times of restoration of all things, which God has spoken by the mouth of all His holy prophets since the world began." Jesus called it a "renewal of all things" (Matthew 19:28). The Millennial Kingdom will inaugurate a return to Garden-of-Eden-type perfection.

The Apostle John, the author of the Book of Revelation, assumed his readers were familiar with all Old Testament descriptions; therefore, rather than citing those same physical descriptions, he revealed new information, the exact length of this dispensation, one thousand years (Revelation 20:2-6). He noted this number six time in the passage. He definitely did not want us to misunderstand, did he?

God will radically change the natural realm in the Millennial Kingdom. The Old Testament records changes in climatology with increased and consistent rainfall (Isaiah 30:23; 35:7), in agriculture with increased productivity (29:17; 41:18-19), in zoology with predatory animals not devouring prey (11:6-9; 65:26), in medicine with increased health and healing (29:18-19; 33:24), and in anthropology with increased longevity (29:18-19; 65:20). Living over one-thousand years is a monster of an increase, isn't it?

Scripture also promises a future resurrection for all mankind. At the end of the Millennium, those who have trusted in Christ for salvation will receive a resurrection to eternal life with God (John 5:25-29; 11:25-26; Revelation 20:4-6) while all unbelievers will receive the judgment of an existence of eternal separation from God in the torment of the Lake of Fire (John 5:29; Revelation 20:11-15).

At the Rapture of the Church before the Tribulation, God will give resurrection bodies to all Church Age believers, which includes all of us living now. We will have these perfect bodies throughout the Millennial Kingdom and all eternity. The Gospels' descriptions of Christ's resurrection body hint at the characteristics of ours. His body is no longer subject to the second law of thermodynamics because He doesn't experience death or degeneration (1 Corinthians 15:20-26). Gravity does not apply to His resurrection body as seen in His awe-inspiring ascension into Heaven (Acts 1:9-11). (He flew supernaturally without wings. For those of you who wonder, we will never have wings.

We were created in God's image and, therefore, will always be set apart from and above birds and angels.)

> When our last hour sounds, we will have the great and ineffable joy of seeing the one whom we could only glimpse in all our work.
> CARL FRIEDRICH GAUSS, THE "PRINCE OF MATHEMATICIANS"

After His resurrection, Christ had the ability to either be recognizable (John 20:19-20, 29; 21:1) or incognito (Luke 24:13). His physical, resurrected body manifested itself as an ordinary man when He was with the Emmaus travelers (24:13-31). The two travelers thought He was an ordinary man when He was easily examinable in bright, outside light (24:16-18) but later recognized Him as Messiah in limited indoor lighting (24:29-31). He even ate a meal with them and later with His disciples (Mark 16:14; Luke 24:29). Does a resurrection body need nourishment? Need? Probably not since Christ's resurrection body is perfect, needing nothing. Will eating be a special treat in Heaven? If so, I would expect no one will have a weight problem, another change we can anticipate with joy! What will perfect chocolate taste like? I can't even imagine.

Anyone investigating the science of resurrection must ask many questions. How will Newton's Laws of motion affect a resurrection body? What will be its molecular properties and chemical composition? What about its optical and biological cell properties? Are there even cells? What is the nature of immortality?

Scriptural data suggests, at a minimum, that resurrection life will be significantly better than our present lives. We won't have Adam's sin nature for one, a change I eagerly await. The changes could be even more radical than those post-Flood and similar in degree to those after the Fall. Only the sovereign Creator of all things can establish such changes, all of them reflections of His perfection. There will be absolutely no Darwinian evolution discussed here!

THE TAKE AWAY: These radical physical changes will bring about significantly different scientific conclusions that will affect many laws of science and physical constants, once again refuting the idea of uniformitarianism. Try to imagine the advances a one-thousand-plus-year-old scientist will be able to make in this almost idyllic environment. The advances we point to with such pride today will seem infantile by comparison.

All of today's science textbooks will be totally out of date and in need of substantial revisions during the Millennium. Why will we even need them when the Creator Himself will teach us? I would advise scientists to adopt the biblical worldview now to get a head start on understanding the future age.

WHAT RADICAL CHANGES WILL OCCUR BETWEEN THE MILLENNIAL KINGDOM AND THE ETERNAL STATE?

The final transformation of the physical realm will come after God's final judgment at the end of the Millennial Kingdom when He introduces the Eternal State, commonly referred to as heaven. The Millennium will end world history. All will then be perfect as well as eternal. The creation will never again undergo substantial change. The laws of decay will never again function. There will no longer be any death or anything associated with it.[30]

As I've said before and love repeating, I believe in a "big bang" but not at the beginning of all things. World history will end with a 'big bang" as described in 2 Peter:

> But the day of the Lord will come like a thief, in which the heavens will pass away with a roar and the elements will be destroyed with intense heat, and the earth and its works will be burned up.
>
> 2 Peter 3:10

The apostle John had a vision of the eternal state he described as a new heavens and new earth with a new Jerusalem:

> Then I saw a new heaven and a new earth; for the first heaven and the first earth passed away, and there is no longer any sea. And I saw the holy city, new Jerusalem, coming down out of heaven from God, made ready as a bride adorned for her husband.
>
> Revelation 21:1-2

30. Revelation 21:3-4.

The Scriptures do not supply much data concerning the physical differences between the Millennial Kingdom and the Eternal State. Revelation 21-22 gives the most extensive description, but eternity will be so different from any other time that the Apostle John seems to have struggled to find words adequate to describe it. His description is brief and full of unfamiliar scenes. We do know Jesus will have already assigned Satan and his demons as well as evil and death to the Lake of Fire. They will no longer be around to plague us.

We can't be sure of the physical properties of the eternal state though, as we've noted, "a river of the water of life, clear as crystal, [will come] from the throne of God and of the Lamb," and "on either side of the river [will be] the tree of life, bearing twelve kinds of fruit, yielding its fruit every month" (Revelation 22:1-2). How these details apply to the rest of the eternal state, we cannot know until we experience it.

We can be certain, though, that God will be the One who transforms the creation at the end of the Millennium to create a "new heavens and new earth" (Revelation 21:1), and God always creates perfection. Why would anyone want to change perfection? I wonder how Adam and Eve would answer that question? Regardless, the Eternal State will remain perfect forever and ever. We will finally, truly, and totally be free of the curse of that first sin.

> And from true lordship it follows that the true God is living, intelligent, and powerful; from the other perfections, that he is supreme, or supremely perfect. He is eternal and infinite, omnipotent and omniscient; that is, he endures from eternity to eternity; and he is present from infinity to infinity; he rules all things, and he knows all things that happen or can happen.
>
> **ISAAC NEWTON**

SURELY GOD'S SHOUTS ARE DEAFENING NOW, RIGHT?

I hope this chapter has convinced you of the total inadequacy of secular science to explain ancient events and of the truth of the biblical worldview and the copious scientific data that support it.

Though much of the church has accepted the lie that the universe is billions of years old, I hope you are no longer one of them. I also hope I have confirmed the need for a paradigm shift in the study of science. This book will probably be ignored today, but its truths will be supported by the Creator Himself in the Millennial Kingdom. That in itself will be a huge change for the better. History and science change, but God is unchanging. We need not fear or worry. Good will triumph because God is absolute good.

> From religion comes a man's purpose; from science, his power to achieve it. Sometimes people ask if religion and science are not opposed to one another. They are: in the sense that the thumb and fingers of my hands are opposed to one another. It is an opposition by means of which anything can be grasped.
>
> **DR. WILLIAM HENRY BRAGG, 1915 NOBEL PRIZE IN PHYSICS**

You may wonder why the ideas of a creation week and a worldwide Flood are denied by scientists. The reason is spiritual, not intellectual. First, if God is Creator, then all mankind stands accountable to Him. Second, mankind will face God as Judge, and the catastrophic Fall and Flood events prove He does not fool around. Most want to avoid God's judgment at all costs, so they deny His existence and the veracity of His account of the Creation and Flood. Their denial can't change the truth. How very sad for them.

On a positive note, if you are a believer in Christ, you can confidently expect an indescribably glorious future. Though few details are available to us, we can be sure that being in a resurrection body face to face with our Lord and Savior can be nothing short of spectacular, indescribable.

Our minds will need transformation for us to be able to comprehend even a small part of that future, won't they? Let us take a moment to thank our Lord for His omnipotent transforming power and His kindness in revealing our glorious future to us. Paul stated it well in one of the passages we discussed earlier:

> For I consider that the sufferings of this present time are not worthy to be compared with the *glory that is to be revealed to us*. For the anxious longing of the creation waits eagerly for the *revealing of the sons of God*. For the creation was subjected to futility, not willingly, but because of Him who subjected it, in hope that the creation itself also *will*

be set free from its slavery to corruption into *the freedom of the glory of the children of God.* For we know that the whole creation groans and suffers the pains of childbirth together until now. And not only this, but also we ourselves, having the first fruits of the Spirit, even we ourselves groan within ourselves, waiting eagerly for our adoption as sons, the *redemption of our body.*

Romans 8:18-23

Each of us takes a unique course through life. I try to cling to the Lord always–even when He offers sharp turns in unforeseen directions, like when I switched from being a university professor to a writer, or when He gave us twins only ten months after our firstborn daughter, or when He opened the door for me to go back to school during my mid-life years. Seek the Lord, stay in the valley as long as He indicates, and step through open doors if and when He directs through circumstances, godly advice–and above all, His Word.

DR. BRIAN THOMAS, PALEOBIOCHEMISTRY

All things came into being through Him,
and apart from Him
nothing came into being
that has come into being.

JOHN 1:3

He was in the world,
and the world was made through Him,
and the world did not know Him.

JOHN 1:10

JOHN, PROPHET AND APOSTLE WHOM JESUS LOVED

CHAPTER 17

SCIENTIFIC FOUNDATION STONE #9

DIVINE MAINTENANCE OF THE CREATION

NOT INDEPENDENT LAWS & CONSTANTS

The ninth foundation stone for accurate and enduring scientific conclusions involves understanding that God actively maintains His creation. He didn't create it, walk away, and leave it to manage itself. He ensures every particle continues to work together as designed. Though we can't observe Him working any more than we can see the wind, Scripture informs us of His quiet, behind-the-scenes efforts on our behalf. Even though scientists can't see Him working either, they observe the stability His maintenance provides in the natural realm.

The secular, naturalistic view ignores His invisible hand upholding creation and defends uniformitarianism, saying that the laws of nature and physical constants are fixed, that they simply exist and are independent of any outside influence, that they are unchanging and timeless. They call them laws of nature and physical constants but do not identify God as the lawgiver and the only constant in our chaotic world.

> My research into the 3D structure of a protein vital to muscle contraction revealed that almost every amino acid, and thus almost every atom, had to take its precise position in sequence for the protein to work—and therefore for the muscles in animal bodies to work. I realized that only a super-genius Creator could have crafted precision instruments down to the atom. I also learned that God takes care of our lives in unseen ways and that His humility keeps Him from complaining about not getting His due credit.
>
> **DR. BRIAN THOMAS, PALEOBIOCHEMISTRY**

HAS GOD ALWAYS PROVIDED STABILITY IN THE CREATION?

Stability in the creation is real and not imagined because God is the embodiment of stability. He is "the same yesterday and today and forever!" (Hebrews 13:8). Though He never changes, He has twice drastically changed the natural realm to suit His purposes. We know He will do so again because He says so in His Word. In the meantime, in this dispensation, we can look to Him for the stability we depend on for science to work.

We live in the post-Flood world, a period that includes the past Age of Israel and the current Church Age. Within each of these two dispensations, the laws of nature and physical constants have been subject to the Creator's sovereign will. While He has kept them constant and predictable as promised, He can and has changed them for a moment. We call these singular divine interventions in the constancy of creation miracles. Scripture records many spectacular ones. Even during these localized, miraculous moments, however, God maintained the worldwide stability associated with our dispensation. Remember Chapter 16?

While stability and predictability characterize the natural realm today, neither is natural, meaning occurring naturally, but each is maintained supernaturally by God. We have stability today because God keeps the promises He made in the Noahic Covenant.

In the ancient world, a covenant was a legally binding agreement, usually between only two parties. We would call it a contract today. The Noahic Covenant was unusual. First, God Himself initiated it and then entered into it. Most other covenants mentioned in the Bible were between individuals,[1] families (tribes),[2] or nations.[3] The Noahic Covenant was between God Himself and every living creature on Earth, including Noah and his family.

Second, this covenant was unconditional. Most covenants contained stipulations that both parties must agree to uphold. In the Noahic Covenant, only God bound Himself to perform the stipulations of the covenant. He did not require man to do anything. He legally bound

1. Genesis 21:27; Malachi 2:14.
2. Joshua 9:7.
3. Deuteronomy 7:2.

Himself to never bring another worldwide Flood like the one that had just ended even if mankind any time in the future chooses to reject Him as they did before the Flood (Genesis 9:8-17).

Scripture records this covenant so that we can marvel at God's grace toward us:

> Then God spoke to Noah and to his sons with him, saying, "Now behold, I Myself do establish My covenant with you, and with your descendants after you; and with every living creature that is with you, the birds, the cattle, and every beast of the earth with you; of all that comes out of the ark, even every beast of the earth. I establish My covenant with you; and all flesh shall never again be cut off by the water of the flood, neither shall there again be a flood to destroy the earth."
>
> Genesis 9:8-11

- Notice that God initiated and then entered into the covenant (9:8).
- The covenant was between God on one side and Noah, all his descendants, and all surviving creatures on the other (9:9-10).
- The covenant included the natural realm and promised specific future outcomes.
- The Creator assured Noah that "all flesh shall *never again* be cut off by the water of the flood, neither shall there again be a flood *to destroy the earth*" (9:11).
- In essence, God promised He would control every atom of creation so no worldwide cataclysmic tectonic and flood event could ever happen again.

Consider God's grace in entering into this legally binding covenant. As the perfect and holy Creator, He had no need to say anything to Noah or his family, who were all sinners (as are we all). None of us deserves anything good from Him, but He continually showers grace and mercy on us, just as He did on Noah so long ago.

Noah and his family would have been shell-shocked after surviving the great destruction of the Flood. Just imagine over a year inside a massive ship being buffeted by unseen forces and assaulted by the ominous sounds of a creation under great stress. And they had no understanding of rain or plate tectonics! They could not have known when or even if the Flood would end!

Because God acted to preserve Noah's family from the catastrophic Flood, they would have assumed He would continue to preserve them under the much more pleasant circumstances after the Flood. They wouldn't have thought a covenant necessary to prove they could rely on Him. He had already demonstrated His great love for them miraculously during the Flood. Certainly, God, the Creator of everything in heaven and on Earth, had absolutely no obligation to enter a legally binding covenant, but He chose to condescend to mankind so He could give double assurance we need not fear another worldwide Flood. Could it also be that God wanted all succeeding generations, especially our scientific age, to have a scientific explanation for the stability observed in the creation? What does our twelve-year-old think?

God made the display of the rainbow a visual reminder of His covenant. He wanted every one of Noah's descendants, which includes you and me, to see its splendor in the sky and remember both His judgment of sin and His incomprehensible grace to those who obey Him (Genesis 9:13-17). The rainbow is a grand celestial signature on the divine contract, an impossible to ignore reminder of His faithfulness.

We have already seen that, as the Creator, God is sovereign over all creation; therefore, we can easily believe He will utilize all plate tectonics, geophysics, geology, climatology, physics, and even nuclear physics for our benefit. And so He does! We don't have to worry about too many glaciers melting and inundating the world. God won't let it happen, political scaremongering to the contrary.

THE TAKE AWAY: God can predict that the world will never again undergo a worldwide flood because He built that predictability into the creation. Nothing can change until He chooses to institute that change, and He has told us in His Word that the next major change in the creation won't happen until the tribulation period at Christ's Second Advent. At that time, Christ will destroy all unbelieving mankind and usher in the Millennium Kingdom filled with all believers—very relieved and thankful believers, I might add.

> As a physicist, that is, a man who had devoted his whole life to a wholly prosaic science, the exploration of matter, no one would surely suspect me of being a fantast. And so, having studied the atom, I am telling you that there is no matter as such! All matter arises and persists only due to a force that causes the atomic particles to vibrate, holding them together in the tiniest of solar systems, the atom. Yet in the whole of the universe there is no force that is either intelligent or eternal, and we must therefore assume that behind this force there is a conscious, intelligent Mind or Spirit. This is the very origin of all matter.
>
> **DR. MAX PLANCK, 1918 NOBEL PRIZE IN PHYSICS**

HOW IS THE NOAHIC COVENANT RELATED TO THE STABILITY OF THE CREATION?

As part of His sovereign control of creation, the Creator maintains a stable creation that keeps every parameter in proper balance for predictability. Science depends on that predictability. For example, because God sovereignly controls astrophysics, He maintains the orbit of the moon in its present precise route. If the moon were only slightly closer to the earth, its gravitational pull would produce tides that would wash over the continents, causing another worldwide flood. The Noahic Covenant assures us of His continued control of the moon's orbit. Next time you admire the moon, thank God it is exactly where it needs to be for our benefit. Thank Him, too, that He maintains the orbit of all heavenly bodies so their effects on the universe are beneficial to those He created in His image.

Several Old Testament passages reference this stability brought on by the Noahic Covenant. When God spoke to that great believer Job, He referred to "ordinances" and "fixed rule."

> "Do you know the *ordinances of the heavens, or fix their rule over the earth*?"
>
> Job 38:33

The answer is an emphatic NO, you don't. Only God has the omniscience, power, and authority to cause the creation to bow before Him in obedience.

Later in history, God referred again to a "fixed order":

> Thus says the Lord, who gives the sun for light by day and the *fixed order* of the moon and stars for light by night, Who stirs up the sea so that its waves roar; The Lord of hosts is His name: "If this *fixed order* departs from before Me," declares the Lord, "Then the offspring of Israel also shall cease from being a nation before Me forever."
>
> Jeremiah 31:35-36

Again, in Jeremiah, God specifically referenced the Noahic Covenant, making the same point about "fixed patterns."

> Thus says the Lord, "If My *covenant* for day and night stand not and the *fixed patterns* of heaven and Earth I have not established, then I would reject the descendants of Jacob and David My servant, not taking from his descendants rulers over the descendants of Abraham, Isaac, and Jacob. But I will restore their fortunes and will have mercy on them."
>
> Jeremiah 33:25-26

Jeremiah recorded these words when the Babylonians were ready to destroy Israel. God wanted His chosen people to understand that His faithfulness to them was as settled as His maintenance of a stable universe. He was reminding them that though they would lose their nation and be exiled from the land, He would fulfill the promises He made to the descendants of Abraham, Isaac and Jacob in another covenant, the Abrahamic Covenant. Just as He had prevented a worldwide flood as promised, so they could trust Him to faithfully keep His word about every covenant. During the suffering of their exile, they should remember His immutability and trust Him to remain faithful to His Word and all His covenants. And so He will continue to do because He is forever stable and immutable.

We can see the truth of God faithfulness throughout history. Has any nation been maintained as has Israel, God's chosen people? Many of those genetically related to Abraham, Isaac, and Jacob are back in the Promised Land thousands of years after being forced from it not only by the Babylonians in BC 586 but also by the Romans in AD 70. Has any worldwide catastrophe like the Flood again destroyed the world? Did the Messiah come as promised? Did He deliver us from sin and death if we believe in Him? God's faithfulness in keeping all past promises proves we can trust Him to be faithful to fulfill every future promise.

THE TAKE AWAY: Naturalists do not understand biblical references to laws and physical constants. They claim to have laws of nature, but they refuse to acknowledge God as the giver and maintainer of those laws. They are blind to the truth that stability does not come from laws of nature independent of God but are derived from God's "ordinances" and "fixed rule." The divine Lawgiver integrated these laws into the creation, just as He did the Noahic Law, the Mosaic Law, and all other spiritual laws. All function as designed because He guarantees it.

> Men became scientific because they expected law in nature and they expected law in nature because they believed in a lawgiver.
>
> **DR. JOHN LENNOX, MATHEMATICS**

The New Testament also attributes the stability and maintenance of the natural realm to God. In Colossians 1:17, the Apostle Paul wrote, "He [Jesus Christ] is before all things, and in Him all things hold together." Notice that Christ holds all things together. That includes keeping electrons and moons in their orbits, doesn't it? Christ as Creator maintains the stability of the entire creation. He did so even as a babe in a manger in Bethlehem. He continues to do so today.

Similarly, Hebrews 1:3 records that Christ, "upholds all things by the word of His power." His word has the power to keep the creation functioning as He designed it. Shouldn't we give thanks that the maintenance of creation is Christ's ongoing work, that the chaos and unpredictability of evolution is, in fact, a lie?

> Life cannot have had a random beginning...The trouble is that there are about 2,000 enzymes, and the chance of obtaining them all in a random trial is only one part in 10^40,000 an outrageously small probability that could not be faced even if the whole universe consisted of organic soup.
>
> **DR. FRED HOYLE, ASTRONOMER**

HOW DO MIRACLES SHOW GOD'S SOVEREIGN MAINTENANCE OF CREATION?

He is sovereign over the creation and can and will use it to accomplish His purposes, even if it interrupts the stability of nature for a short time and in isolated instances. Individual miracles show the precision with which He can interrupt the laws of nature while still maintaining the stability of the creation as a whole.

THE TAKE AWAY: Because God is the Creator, He can change natural laws whenever He desires. He can change them on a broad basis by changing an entire dispensation as He did with the Fall and the Flood, or He can change them on a small-scale with a single miracle. Individual miracles show the precision with which He can interrupt the laws of nature while still maintaining the stability of the creation as a whole.

After the Flood, God allowed abundant miracles only a few times and, then, always for specific purposes. During the life of Moses, He used the miracles of the burning bush, a staff turning into a viper, and an arm becoming leprous to prepare Moses to lead Israel out of bondage and to prove to the Jews that Moses was His chosen leader. He used the ten plagues to show that the Egyptian idols were powerless and to prove He is the only true God. During the exodus of the Israelites from Egypt, He used the Red Sea event to preserve and encourage His people. During Israel's wilderness wanderings, He performed miracle after miracle to remind them of His faithfulness.

These stunning miracles demonstrated that God was transitioning into a new dispensation, the Age of Israel, in His dealings with mankind. He was miraculously forming His own nation as His primary witnessing instrument to the world. Through Israel, He would give the world His Scriptures, both the Old and New Testaments. Through Israel, He would give the world His own Son as Messiah and Savior. These miracles were God shouting about His new direction for mankind and His always abundant works of grace.

The numerous miracles during the ministries of Elijah and Elisha validated them as God's messengers and confirmed that God would bring judgment on Israel for its rampant idolatry, a visible sign of their national rejection of Him. God always shouts when He is about to act either in grace or judgment. Grace began every new dispensation after a judgment ended the previous one.

The next time of plentiful miracles was during Christ's life. The Apostle John called them signs because they pointed to Jesus' deity. These miracles shouted that the promised Messiah had arrived. The volume and diversity of the miracles proved Jesus is superior to and set apart from the earlier prophets. That His miracles fulfilled all Old Testament messianic prophesies further established His authority as the promised Messiah.

The Church Age was the next dispensation and the next period of abundant miracles. Again, God used miracles as validation, this time of the early church leaders as His messengers, that He had chosen them to bring to the world His message of salvation through the risen Jesus Christ. Their miracles shouted the beginning of a new dispensation, our present Church Age.

As we have seen, the Bible predicts a future time filled with miracles. During the Great Tribulation, miracles will alert those living then to prepare for the imminent return of Christ the Messiah in judgment. Those who refuse to understand the message of the miracles will be destroyed by them. These miracles will shout that God is introducing the final dispensation of world history, the Millennial Kingdom Age, and that all mankind should choose to believe in Jesus as Messiah.

> When a miracle takes place, it is the laws of nature that alert us to the fact that it is a miracle. It is important to grasp that Christians do not deny the laws of nature, as Hume implies they do. It is an essential part of the Christian position to believe in the laws of nature as descriptions of those regularities and cause-effect relationships built into the universe by its Creator and according to which it normally operates. If we did not know them, we would never recognize a miracle if we saw one.
>
> **DR. JOHN LENNOX, MATHEMATICS**

WHAT CAN WE CONCLUDE?

Miracles serve several purposes:

1. Fundamentally, miracles remind us God has sovereign authority over His creation.

2. God used miracles to confirm He was intervening in history to establish a new dispensation in His dealings with mankind.

3. God used miracles to inflict His just judgments as with the Flood. He will do so again before the end of our current heaven and earth.

4. God used miracles to validate those chosen to be His dispensational messengers, mainly His prophets, the leaders of the early church, and His Son.

5. In both the Old and New Testaments, miracles demonstrated God's mercy and compassion toward those suffering because of the sin-curse, those who were sick, dying, or even dead (e.g., Lazarus and the widow's son).

Throughout most of history, the Creator has worked moment-by-moment maintaining the stability of the universe. Miracles have been rare. When they have occurred, they have left a record for us of His clear and direct actions so we know we can trust Him to maintain stability in ordinary times. Miracles should alert us that a secular approach to science is inadequate. Uniformitarianism, which dismisses miracles as myths, is a false principle easily disproved by history.

THE TAKE AWAY: God the Father and Jesus the Son maintain the entire physical universe. Secular science can observe Their stabilizing effect and appreciate that scientific work is possible only because of that predictable stability. Even so, secular science refuses to acknowledge that the so-called laws of nature and physical constants are not independent characteristics of nature but under God's direct control. He is not only the Creator but also maintains every process that keeps the creation operating smoothly. We can trust Him to continue His gracious control of our environment until He institutes the next worldwide catastrophe prophesied in Scripture, the seven-year Tribulation.

Praise God that those of us who have believed in Christ as our Savior won't be here to experience the horrendous judgments of the Tribulation. Christ will already have come for us and taken us up to Him in the Rapture of Church Age believers. The Rapture will be a catastrophe for the unbelievers who remain on earth because Jesus will have removed the stabilizing influence of believers.

What a glorious moment it will be for His children, though, as we meet our Savior in the air! Both now and in the future, God is faithful to preserve those who belong to Him. Sometimes, He uses miracles to do just that.

> Without my Saviour I am nothing....The Lord has guided me. He has shown me the way, just as he will show everyone who turns to Him.
>
> GEORGE WASHINGTON CARVER, PIONEER OF AGRICULTURAL CHEMISTRY

Worthy are You, our Lord and our God,
to receive glory and honor and power;
for You created *all things,*
and because of Your will they existed,
and were created.

REVELATION 4:11

...and swore by Him who lives forever and ever,
who created *heaven and the things in it,*
and the earth and the things in it,
and the sea and the things in it,
that there will be delay no longer.

REVELATION 10:6

JOHN, PROPHET AND APOSTLE WHOM JESUS LOVED

CHAPTER 18

SCIENTIFIC FOUNDATION STONE #10

THE INTERPRETATION OF SCRIPTURE NOT RATIONALISM

We have reached the tenth and final foundation stone for accurate and enduring scientific conclusions. Finally. I'm excited. How about you? Though the last, this stone is as critical as all the others. So what is it? By now, you probably can guess. No interpretation of natural phenomena can be accurate unless it rests on God's revelations in Scripture.

The secularist clings to rationalism as his key to evaluating the natural realm, not realizing that his finite intellect and inherent sin nature place unconquerable limits on that ability, no matter his area of study or the weightiness of his degrees.

Because we are creatures, not the Creator, we can only begin to understand the creation, to scratch the surface as we might say, to barely dig into the dust as Adam might say. To accurately understand even that small amount, we must first understand Scripture. Through His Word, the Creator sets us on the right scientific path and then shows us, again through His Word, when we falter off it. We know we have made a mistake if anything we think we have discovered contradicts God's righteous character. We must go back and rethink both our hypotheses and conclusions if we want to achieve true scientific success.

> In questions of science, the authority of a thousand is not worth the humble reasoning of a single individual.
> GALILEO GALILEI, ASTRONOMER, PHYSICIST, ENGINEER

WHAT IS THE RELATIONSHIP BETWEEN SCRIPTURE AND SCIENCE?

We are creatures, and so our intellect is limited. King Solomon, the author of much wisdom literature, explained that most of the complexity of the creation remains unknown to us.

> Just as you do not know the path of the wind and how bones are formed in the womb of the pregnant woman, so you do not know the activity of God who makes all things.
>
> Ecclesiastes 11:5

That is wisdom. Even today with all our supposed scientific knowledge, the path of a hurricane almost always fools us, and we certainly don't know how bones are formed. Only the arrogant boasts of his scientific knowledge. We must humbly accept our limitations and God's wisdom if we are to find fulfillment in any area of life.

THE TAKE AWAY: God is sovereign over all things; therefore, His Word is the final authority on all things. That truth remains inviolate whether we recognize it or not.

As an interesting side note, Matthew Maury (1806–1873), the father of oceanography, noticed the expressions "paths of the wind" and "path of the sea" in Scripture and believed they described true phenomena. He spent his life discovering and charting many of those paths. Navigators still use the information in his charts today. A monument to him in his home state of Virginia reads, "Matthew Fontaine Maury, Pathfinder of the Seas, the Genius Who First Snatched from the Ocean and Atmosphere the Secret of Their Laws. His Inspiration, Holy Writ, Psalm 8:8, Psalm 107:23, 24, and Ecclesiastes 1:6."[1] They could have included Ecclesiastes 11:5.

1. https://bible.ca/tracks/matthew-fontaine-maury-pathfinder-of-sea-ps8.htm.

God so wants us to rule His creation that He gives us many guideposts in His Word. As Maury reminded us, "The rule is, never to forget who is the Author of the great volume which nature spreads out before us."[2]

It makes scientific sense to consult the revelation of the Creator, as did Maury, rather than to plunge in blindly to try to understand the creation ourselves. As we've said many times, we are finite, and God is infinite. Science will never understand every aspect of the creation because, as Job learned, it remains beyond finite understanding. The Creator built the creation so we can know something about it but not exhaustively. We absolutely cannot understand even that small portion unless we first turn to God for illumination.

> God expects Christians to put His Word first in their lives, and to be cautious about any thoughts contrary to it. As Colossians 2:8 says, "See to it that no one takes you captive through philosophy and empty deception, according to the traditions of men, according to the elementary principles of the world, rather than according to Christ." If there is any group of professional people who should give glory to God, it should be scientists! Evolution has become a tradition of men. Mankind has assumed it to be true and it permeates virtually all disciplines of education and knowledge.
>
> **DR. JOHN G. LESLIE, BIOCHEMISTRY, MEDICINE, BIBLICAL ARCHEOLOGY**

Many biblical passages provide great insights on the natural realm. We have discussed a few in some detail, including Genesis 1, 3, 6-9, Job 38, and Psalms 104-105 in the Old Testament and Romans 1, 8, Philippians 2, 2 Peter 3, and Revelation 20-22 in the New Testament. The information in these passages and all others that God uses to reveal the creation should control all stages of every scientific investigation from observation to generalization to verification. In fact, the final experiment in any scientific study should be the test of Scripture. Do our conclusions agree with Scripture? Do they reflect the perfect character of the Creator who is the Author of Scripture? We should follow that pattern of comparing our conclusions with Scripture in every area of life.

Let's again look to Matthew Maury for insight. "I have been blamed by men of science, both in this country and in England, for quoting the Bible in confirmation of the doctrines of physical geography. The

2. https://creation.com/matthew-maury.

Bible, they say, was not written for scientific purposes, and is therefore no authority in matters of science. I beg pardon! The Bible IS authority for everything it touches."[3]

Much could be said about the ethics that should guide science. Ethical absolutes are at the heart of the biblical worldview but largely ignored by secularists. Only the infinitely righteous and holy God can reveal true ethics. He clearly spells out these moral principles in the pages of Scripture. Because the Bible is the ultimate source of absolutes, of all right and wrong, all goodness and sin, obviously, it presents the only inerrant framework for ethics for all areas of life, including science.

> Trying to "prove the Bible with science" is not the answer. The answer is to start with the assumption of the authority of the Bible, looking at all the evidence that it presents for trusting it. Then build your science from there, based upon the Bible's truth, and show how immensely consistent the weight of evidence in the real world is with the Bible's account of history.
>
> **DR. NEIL HUBER, PHYSICAL ANTHROPOLOGIST**

THE TAKE AWAY: For science to be constructive, not destructive, biblical standards must be at the heart of every scientific pursuit. Scientists must adhere to virtues like honesty in research, elimination of political or financial influences in conclusions, godly concern for fellow scientists, and self-control in limiting personal biases or prejudices. Scientific study without biblical ethics is like a rudderless ship, prone to accidents, even to the destruction of life.

Today, scientists are faced with many ethical dilemmas. Should humans and animals be used in scientific testing? If so, when does experimentation become cruelty? Does the genetic modification of plants improve or compromise their value? Do chemical compounds like plastics, pesticides, and herbicides help or harm? Should science try to control the weather? What about the development of chemicals for warfare? Do genetic engineering of the human genome and research in the area of transhumanism tamper with the image of God that the Creator established? Should limits be placed on artificial

3. Major, T. J. "Honor to Whom Honor...Matthew Fontaine Maury (1806-1873)." *Creation Research Society Quarterly*, 1995, 82-87, quote from page 83. https://www.thebereancall.org/content/matthew-maurys-paths-sea.

intelligence research that manipulates masses of people who are unaware of their manipulation? What about medical research in the area of transgenderism? Doesn't the Creator of the sexes distinguish only two, male and female, making transgenderism a mental, rather than physical illness? Should mental illnesses be treated with physical experiments, as was true only fifty years ago with the now discredited lobotomies promoted by too many medical scientists?

> I've learned to distinguish between the greatness of God and the inexcusable evil that has been done by those professing his name. And so I do not deduce...that God is not great, and that religion poisons everything. After all, if I failed to distinguish between the genius of Einstein and the abuse of his science to create weapons of mass destruction, I might be tempted to say science is not great, and technology poisons everything.
>
> **DR. JOHN LENNOX, MATHEMATICS**

More generally, should the evidence and standards used in scientific studies be examined and verified by scientists holding a different bias? Should those studies that contradict a popular view be suppressed? Should scientists who advocate unpopular views be silenced, or is debate the sign of a healthy scientific environment?

> There are problems to whose solution I would attach an infinitely greater importance than to those of mathematics, for example touching ethics, or our relation to God, or concerning our destiny and our future; but their solution lies wholly beyond us and completely outside the province of science.
>
> **CARL FRIEDRICH GAUSS, MATHEMATICS**

Ultimately, we must ask ourselves, should areas of scientific study be advocated and funded if they advance political or personal agendas but harm creation as God intended it? According to John K. Reed, a scientist himself with a PhD in geology:

> We all know that scientists are fallible, not always virtuous, and driven by paradigmatic commitments stemming from their worldview, their politics, greed, or passion.... Early scientists (thanks to Christianity) understood human nature better than moral relativists. They were realists,

> expecting man's fallen nature to produce "better," not "perfect," because that was for the next world. But as secularists like Hume displaced Christianity with mathematics and science, and later optimists preached science as a panacea for perfectible humans, the early ethical framework began to crumble. Events have shown its failure and illustrated the truth of Christian doctrine—men cannot attain perfect behaviour any more than they can attain perfect truth."[4]

> Science is the search for truth, that is the effort to understand the world: it involves the rejection of bias, of dogma, of revelation, but not the rejection of morality.
>
> **DR. LINUS PAULING, 1954 NOBEL PRIZE IN CHEMISTRY**

In our current technological culture, shoddy as well as inappropriate research has increased our desperate need for the ethics of the biblical worldview. Only the Creator knows what benefits His creation, including mankind, so only He can lay the ethical foundation of right and wrong that should guide its scientific study. Positive scientific advancement comes only through adherence to His ethical standards as found only in His Word. Scientists must be alert to not stray outside the honorable boundaries He has placed on their study.

> My life has been marked by two immense and fateful discoveries: the splitting of the atom, the recognition of the chemistry of heredity and its subsequent manipulation. It is the mistreatment of nucleus that, in both instances, lies at the basis: the nucleus of the atom, the nucleus of the cell. In both instances do I have the feeling that science has transgressed a barrier that should have remained inviolate. As happens often in science, the first discoveries were made by thoroughly admirable men, but the crowd that came right after had a more mephitic smell.
>
> **DR. ERWIN CHARGAFF, BIOCHEMISTRY**

4. https://creation.com/carol-cleland-historical-science.

WHAT DOES SCRIPTURE SAY ABOUT APPLYING DIVINE TRUTH TO EVERY AREA OF LIFE?

The Apostle Paul's exhortation applies to all of us:

> Test yourselves to see if you are in the faith! Examine yourselves! Or do you not recognize this about yourselves, that Jesus Christ is in you—unless indeed you fail the test?
>
> 2 Corinthians 13:5

Shouldn't we look to Christ in everything we do? Shouldn't we test to see if our scientific conclusions conform to God's revelations in His word? If our ideas fail the Scripture test, we need to do further study because our thinking is faulty, not the Bible, not perfect God.

Let me remind you that the emphasis of this book is the need for those working in science to depend on God and His Word to reach accurate conclusions. I remind you because the world has programmed us to think the Bible deals with theology while science deals with nature. I hope I have adequately shown that the Bible tells us all we must know about nature to do good science. While it doesn't tell us everything we will learn through scientific examination, Scripture does prevent us from making false turns into ideas that violate divine truths. It does lead us on right paths.

> Science is a game–but a game with reality, a game with sharpened knives. If a man cuts a picture carefully into 1000 pieces, you solve the puzzle when you reassemble the pieces into a picture; in the success or failure, both your intelligences compete. In the presentation of a scientific problem, the other player is the good Lord. He has not only set the problem but also has devised the rules of the game–but they are not completely known, half of them are left for you to discover or to deduce. The uncertainty is how many of the rules God himself has permanently ordained, and how many apparently are caused by your own mental inertia, while the solution generally becomes possible only through freedom from its limitations. This is perhaps the most exciting thing in the game.
>
> **DR. ERWIN SCHROEDINGER, 1933 NOBEL PRIZE IN PHYSICS**

Remember that the Bible is revelation from God Himself (Chapter 7). Remember also that He is the Creator of everything in heaven and on Earth and has revealed essential information that will affect our understanding of both. Remember that God directly orchestrated the writing of Scripture through inspiration and made sure His Word in the original languages was without mistake (Chapter 8).

Our Lord not only revealed truth, inspired writers to record that truth, prevented errors from entering that writing, but He also preserved His Bible so it continues to be available in its entirety for us today. It makes no logical sense to neglect that absolute truth when we study anything. If we neglect Scripture, we will come to wrong conclusions, especially concerning the major issues that affect our thinking about life.

THE TAKE AWAY: Not surprisingly, the naturalistic worldview is diametrically opposed to the biblical worldview. The conflicts are glaring, especially in historical science. By rejecting Scripture, secular scientists deny the major realities of life, including God as Creator, the Fall and subsequent sinful nature of man, the Genesis Flood and other divine judgments, His miracles, including Jesus' virgin conception and resurrection. They are blind to the future and, therefore, ignorant of future events that should influence their understanding of the present, including the future resurrection of all mankind, Christ's Second Coming, the Millennial Kingdom, and the Eternal State. Whew! That's a lot of truth to throw into the secular garbage heap!

When we deny any truth in Scripture, we have no recourse but to substitute our own fallible ideas and theories. When we hang onto our own explanations for the natural realm, we believe unscientific, even nonsensical ideas like naturalism and evolution. By now, I hope I have shown scientifically that man's ideas are inadequate to give us a picture of reality. We desperately need the Creator's perspective.

> Most people say that it is the intellect which makes a great scientist. They are wrong: it is character.
>
> **DR. ALBERT EINSTEIN, 1921 NOBEL PRIZE IN PHYSICS**

HOW HAVE THE FALLIBILITY OF SCIENCE AND INFALLIBILITY OF SCRIPTURE BEEN DISTORTED?

Sadly, separating science from God is a central tendency within the Christian world today and, thus, within the church. Many Christians make the mistake of attempting to harmonize science and Scripture, of putting scientific truth above biblical truth. Secular science dominates the thinking of most believers. The biblical worldview is seldom applied even by Bible-believing Christians.

> If the Bible is not reliable in its historical statements, how can it be true in other statements? If Jesus didn't speak the truth about Genesis, how could we trust what He was saying about sin, the cross, resurrection and everlasting life? Jesus Christ *is* the truth; and if He affirmed the literal creation of the world in six normal-length days, we Christians should do the same. If however we compromise and try to marry millions and billions of years with the creation account of the Bible, we may easily pull folk away from the truth of the Bible and the Gospel.
>
> **DR. MARKUS BLIETZ, ASTROPHYSICIST**

The worldly viewpoint attempts to use science to interpret biblical passages rather than letting Scripture stand on its own. To prove the lie to be true, secular scientists and those who believe them impose a faulty interpretive approach on the Scripture. They are forced to depart, sometimes drastically, from the Grammatical-Historical-Contextual hermeneutic to try to prove their faulty opinions (Chapter 11) and so they resort to false approaches that accommodate Scripture to secular science, resulting in distortions of the biblical text. Their false understanding of critical passages like Genesis 1-11 has led them to a false understanding of the creation and thus of science and also of earth history.

One Christian writer who believes science supersedes Scripture described scientific "facts" as the sixty-seventh book of the Bible! Imagine the danger of putting scientific knowledge on the same level if not above Scripture!

Hugh Ross, the main proponent of progressive creationism, wrote:

> God's revelation is not limited exclusively to the Bible's words. The facts of nature may be likened to a sixty-seventh book of the Bible...Some readers might fear I am implying that God's revelation through nature is somehow on an equal footing with His revelation through the words of the Bible. Let me simply state that truth, by definition, is information that is perfectly free of contradiction and error. Just as it is absurd to speak of some entity as more perfect than another, so also one revelation of God's truth cannot be held as inferior or superior to another."[5]

In spite of his conclusion that Scripture is inerrant, Ross clearly puts the authority of scientific knowledge above the Scriptures. In other words, he says Scripture is inerrant but scientific opinions are, too, maybe even more so. Both can't be inerrant because the authors are so different. God is perfect; scientists are like us, and we prove over and over every day our errancy.

In His commentary on Genesis 1-11,[6] Ross consistently twists the biblical text to fit the so-called truths of science. How is denying the accuracy and supremacy of Scripture any less blasphemous than denying God's existence as atheists do?

> There is abundant evidence that the Bible, though written by men, is not the product of the human mind. By countless multitudes it has always been revered as a communication to us from the Creator of the Universe.
>
> **DR. JOHN AMBROSE FLEMING, ELECTRICAL ENGINEERING**

If the Covid debacle taught us anything, it is that science is far from settled and scientists, many of whom think of themselves as gods, are, to the contrary, easily corrupted, deceived, and bought by the highest bidder. Anyone who worships science has forgotten scientists are fallible sinners like the rest of us.

---

5. Ross, Hugh, *Creation and Time: A Biblical and Scientific Perspective on the Creation-Date Controversy*, NavPress, 1994, 56-57.
6. Ross, Hugh, *The Genesis Question: Scientific Advances and the Accuracy of Genesis*, NavPress, 1998.

> The first gulp from the glass of natural sciences will turn you into an atheist, but at the bottom of the glass God is waiting for you.
> **WERNER HEISENBERG, 1932 NOBEL LAUREATE IN PHYSICS**

IS EQUATING SCIENTIFIC CONCLUSIONS WITH DIVINE TRUTH A FORM OF FALSE TEACHING?

Too often, creation science promotes supposedly biblical worldviews that degrade Scripture. A shocking number of creation-science books, organizations, and individuals take a faulty, God-belittling approach. They have rejected, never heard, or never understood the true biblical worldview as detailed in this book. Let me briefly describe some who try to pass their blasphemous opinions off as Christian. I will name names so you can beware.

First, the broad category of theistic evolution accepts Darwinian evolution as fact, asserting that God used evolution as His means of creation. No six-day creation for them! Physicist Howard van Till, a theistic evolutionist, wrote:

> I believe that God has so generously gifted the creation with the capabilities for self-organization and transformation that an unbroken line of evolutionary development from nonliving matter to the full array of existing life-forms is not only possible but has in fact taken place.[7]

God decided to allow creation to create itself? What? Obviously, creation could never have been perfect had perfect God not created it. You can see that such a view automatically rejects the perfection of the Garden of Eden.

Colin Humphreys, a theistic evolutionist and materials scientist, agreed:

> It seems that evolution is the general way in which God chooses to work, and it fills me with wonder that the whole of the universe and the whole of life were encapsulated

---

7. *Three Views on Creation and Evolution*. ed. by J. P. Moreland and John Mark Reynolds, Zondervan Publishing House, 1999), 171.

> in the very first concentration of matter and energy...I believe that God is in charge and that evolution is the way he chose to carry out his creation. If life emerged from a primeval soup then God was the master Chef.[8]

God is the chef who stepped back and allowed creation—whatever he means by that—to create itself? How does that even make sense?

> Theistic evolution is a significant threat to the Christian church. It undermines the very foundation of the Christian faith and causes people to doubt the truth of Scripture...I think people make that compromise because they do not know all of the scientific evidence that calls the theory of evolution into question and they also have not been made aware of the major theological/biblical problems that it generates... Evolution absolutely requires death... millions of years of it, struggle for existence, survival of the fittest millions of years before man comes on the scene. In this scenario, death is not the enemy but the very means by which God created everything. But the Bible is very clear about this: the wages of sin is death. Death came into the world through Adam' sin. Therefore there was no death prior to the fall of man and therefore there could be no evolution whatsoever before that time.
>
> **DR. DAVID A. DEWITT, BIOCHEMIST**

One problem with theistic evolution is that it jettisons the Bible in order to accommodate a failed scientific idea (Chapter 6). Another telling sign of its problems is that no secular evolutionist accepts it. They deny that God had anything to do with the creation at all. Most deny that God exists. If even evolutionists reject theistic evolution, which attempts to support their pet theory, shouldn't we?

Even though evolution today has been thoroughly and scientifically discredited (Chapter 6), books are still being published that hold to theistic evolution. In one such book, rocket scientist Michael O'Connell, who was involved in many NASA space projects, wrote:

> My intent here is not to recount the considerable scientific evidence for evolution, but to show that the physical theory is compatible with scripture.[9]

---

8. "Can Science and Christianity Both Be True?" in *Real Science, Real Faith*, ed. R. J. Berry (Eastbourne: Monarch, 1991), 124-125.
9. Michael O'Connell, *Finding God in Science*, Eigen Publishing, 2016, 40.

He makes many interesting points about his work in astronomy but seems unaware of the findings of cellular biology and genetics that debunk evolution. He promotes God as just another cog in the evolutionary wheel.

Other prominent proponents of theistic evolution include biologist Kenneth Miller, author of *Finding Darwin's God*; Eastern Nazarene University physicist Karl Giberson, author of *Saving Darwin*; former head of the Human Genome Project Francis Collins, author of *The Language of God*; and biologist/theologian Denis Lamoureux, author of *I Love Jesus and I Accept Evolution*.[10] All these men have decided that their interpretations of creation are superior to the Genesis account inspired by God. Sad, isn't it?

> God either exists or He doesn't. Either I believe in God or I don't. Of the four possibilities, only one is to my disadvantage. To avoid that possibility, I believe in God...Let us weigh the gain and the loss in wagering that God is. Let us consider the two possibilities. If you gain, you gain all; if you lose, you lose nothing. Hesitate not, then, to wager that He is.
>
> **BLAISE PASCAL, MATHEMATICIAN, SCIENTIST, INVENTOR**

Another very popular, compromising view is progressive creationism, the view of Hugh Ross, who, as mentioned, thinks science should be the final book in the Bible. Such arrogance! Progressive creationism in large measure also accepts evolution though with a more modern slant. Ross defines this view as "the hypothesis that God has increased the complexity of life on earth by successive creations of new life forms over billions of years while miraculously changing the earth to accommodate the new life."[11] Sounds erudite, doesn't it? Lies often do.

> For many years I have believed that God is the great designer behind all nature...All my studies in science since then have confirmed my faith. I regard the Bible as my principle source of authority.
>
> **SIR GHILLEAN T. PRANCE, BOTANIST, ECOLOGIST**

10. https://www.discovery.org/a/10091/.

11. From audiotape cited in Mark Van Bebber and Paul S. Taylor, *Creation and Time - A Report on the Progressive Creationist Book by Hugh Ross*, Eden Productions, 1994, 11.

Gleason Archer, a well-known Old Testament scholar, is also a proponent of this view.[12] Even the conservative hermeneutics expert Bernard Ramm (1916-1992) was a progressive creationist and a local flood and old earth advocate. He wrote:

> The writer is not a theistic evolutionist. He is a progressive creationist for he feels that in progressive creationism there is the best accounting for all the facts—biological, geological, and Biblical.[13]

Surprisingly, this is from the author of one of the most respected hermeneutics textbooks used in conservative Bible interpretation today.[14] How easily we can be swayed to believe the lie when we ignore the parts of truth that displease us or believe lies that contradict truth!

The major weakness of progressive creationism is that it imposes unprovable evolutionary theories, uniformitarian assumptions, and the deep time theory on the biblical text. Since all these have little scientific support, why would anyone, especially a Christian, want to promote them?

THE TAKE AWAY: Be sure to be a Berean (Acts 17:11) and study all literature, both theological and secular, from the biblical viewpoint. Try to keep even the tiniest lie out of your thinking. All lies, even that little white one, are dangerous to your spiritual life.

As we've seen, the most popular Flood accommodation theory imagines a local rather than worldwide flood. The main proponents of the local flood view are progressive creationists and old-earth advocates, like Hugh Ross and Davis A. Young.[15]

The most popular accommodation views relating to the age of the earth include the day age theory, the gap theory, and the days of revelation views. Most have been proven false. The rest are mere conjectures. Their basic flaw is the same as all accommodation theories,

12. Ross, Hugh, & Archer, Gleason, "The Day Age View" in *The Genesis Debate: Three Views on the Days of Creation*, ed. David G. Hagopian, Crux Press, 2001, 123-163.
13. Ramm, Bernard, *The Christian View of Science and Scripture*, William B. Eerdmans Publishing Company, 1954, 205.
14. I have taught the course on "Hermeneutics and Bible Study Methods" for seminaries for the past 35 years and have recommended Ramm's textbook, *Protestant Biblical Interpretation*. It is a classic.
15. Ross, Hugh, *The Genesis Question*, and Young, Davis A., *The Biblical Flood: A Case Study of the Church's Response to Extrabiblical Evidence*, William B. Eerdmans Publishing, 1995.

accepting the authority of scientific opinions over the authority of biblical revelation (Chapter 8).

Let me repeat: Man is finite and limited. No one is capable of observing all the data involved in any scientific inquiry. No one is capable of imagining all possible outcomes. No one can observe or evaluate the spiritual realm, which is more real than the physical realm (Hebrews 8:1-5). No one but God that is.

> When railways were invented, a popular scientific theory held that even though the train could go faster than 30 mph, it should not, because the passengers' lungs would collapse. Such absurd notions were grounded in the science of the day, which was found to be utterly flawed. I think this reverence for "science" rather than God's Word is a tremendous tool of Satan. As far as I'm concerned, the whole evolution/long ages thing requires a whole lot more faith than the creation perspective. There's a basic lack of honesty there–an unwillingness to say, "Look, this is my best guess." OK, friend, I don't have a problem with your guessing, but *tell* me it's a guess, and don't try to corrupt the authority of science by claiming your guess is science.
>
> **DR. SAAMI SHAIBANI, MEDICAL PHYSICIST**

HOW DOES SECULAR SCIENCE HANDLE THE SPIRITUAL REALM?

Secular science eliminates any consideration of the spiritual realm because it cannot be observed or evaluated. The religion of naturalism places its faith solely in naturalistic explanations. How can their science be complete when Scripture reveals many examples of the spirit world impacting the physical world?

No one can understand or evaluate unseen forces. God made that truth clear when Job attempted to understand his sufferings. Even with the advice of outside "objective" observers, his so-called friends, Job could not begin to imagine the spiritual forces involved in destroying his life. Eventually, God brought all Job's intellectual inadequacies to the surface by asking him a series of science questions (Job 38). Job was unable to provide any answers.

> Evolution destroys your faith in the Bible, because once you go down that slippery slope, you start questioning the Bible in other areas. Your human thoughts and decisions take precedence over the Bible, and we all know that humans are fallible, selfish, fallen creatures, so we'll take the easy way, and believe with "itching ears"? [2 Timothy 4:3] whatever sounds good to us. But we have to stick with what God says. And ultimately it makes the most sense.
>
> **DR. DUDLEY EIRICH, MOLECULAR BIOLOGIST**

Today, we have the benefit of God's revelation in Scripture, so we know what Job did not. We know Lucifer orchestrated Job's problems. We know Job was right about his close relationship with God. We know God's trust in Job was justified. Job should inspire us to be as certain as he was that our lives please God. We should not think suffering indicates God's displeasure. We live in Satan's world, so suffering is inevitable for everyone, even a great believer like Job.

THE TAKE AWAY: God's purpose in asking those questions was to help Job—and us by application—understand that much is simply beyond human comprehension. We are incapable of understanding much of life, both the physical and spiritual, because it comes from the hand of almighty God who is omniscient. We are so far from omniscient that we must rely on His revelations about both the physical and spiritual to gain any accurate understanding of either. We can understand Job's situation better than he did only because we have the benefit of God's revelation in Scripture. In the same way, we can understand creation better than we do now only if we turn to the words of the One who created it.

> Thousands of years ago Job saw the futility of finite man's attempting to define God when he cried, "Can man with searching find out God?." Similarly, wise men ever since have always looked in amazement at the wonderful orderliness of nature and then recognized their own ignorance and finiteness and have been content to stand in silence and in reverence before the Being who is immanent in Nature, repeating with the psalmist, "The fool hath said in his heart, there is no God."
>
> **ROBERT MILLIKAN, 1923 NOBEL LAUREATE IN PHYSICS**

God's questions to Job about the creation showed that naturalism could not give him the answers he needed. The philosophy and rationalism of his three friends could not give him adequate answers either. Job had to trust His Creator for all truth. Only divine revelation could give him any true understanding. Even so, God chose to not answer all his questions. Some things are just beyond our mental abilities or our need to know.

Like Job, we need to learn to be content with the limited extent of God's revelation to us and depend on that revelation alone to lead us to factual understanding in every area of life. Our Lord "has granted to us everything pertaining to life and godliness, through the true knowledge of Him who called us by His own glory and excellence" (2 Peter 1:3). That includes insights into science.

> Science is wonderfully equipped to answer the question "How?" but it gets terribly confused when you ask the question "Why?'"
> **DR. ERWIN CHARGAFF, BIOCHEMIST**

WHAT CAN WE CONCLUDE?

We are dangerously wrong to try to impose science on our interpretation of biblical passages. Scripture can and should stand alone. Why would we want to depend on ever-changing scientific truth when God puts absolute truth within easy reach? How can we not turn to Scripture to evaluate our scientific conclusions when Scripture is absolute truth? Depending on fallible scientists to discover absolute truth is like depending on a two-year-old to explain Einstein's theory of relativity. The mental capacity just isn't there.

Divine revelation sets an inspired and inerrant foundation for good science. Scripture's revelation of God's thoughts is the broad, overarching foundation stone for understanding the natural realm. All other foundation stones gain wisdom from this one. Secular science limits itself by eliminating any consideration of the spiritual. This omission makes science incomplete because Scripture reveals many examples of the spirit world impacting the physical world. Christians are equally blind when they try to impose secular science on their interpretation of biblical passages. Scripture can and should stand alone.

> People look at the sort of science that put men on the moon, and they put these "big bang" theories in the same basket. They're unaware of all the speculation and uncertainty (even rule-bending) there is in physical theories of origins. It's a tragedy that evangelicals are being urged to "re-interpret" the Bible because of the so-called "certain facts of science" in this area. It is even more tragic that there are professing Christians who are promoting evolutionary notions of the "big bang" and galactic and stellar evolution as supporting the Bible and belief in God, while at the same time denying literal six-day creation and the global nature of the Genesis Flood.
>
> **DR. KEITH WANSER, PHYSICIST**

Inquiries into historical science must have divine revelation as their starting point because no man was present to observe ancient events. No man observed the creation. Few observed the Genesis Flood, the change of languages at the tower of Babel, or God's miracles that freed Israel from Egyptian slavery. Even so, Scripture always claims at least one observer of ancient events, God Himself. Because we can trust His evaluations as found in Scripture, all the physical data and traces of the past should be interpreted through the lens of His biblical revelation. Man's rational abilities alone are not sufficient to give sound interpretations.

> Science is not, despite how it is often portrayed, about absolute truths. It is about developing an understanding of the world, making predictions, and then testing these predictions.
>
> **DR. BRIAN SCHMIDT, ASTROPHYSICIST**

THE TAKE AWAY: As we employ scriptural truths in our lives, we must acknowledge that in our finiteness we, like Job, cannot fully understand the inspired and inerrant revelation of God's Word. Despite our desire to understand every word, we are Adam's descendants and prone to misinterpret it. Though we will never understand it all, we must never stop reading, studying, and meditating on it.

We desperately need the illumination and guidance of the Holy Spirit because we are creatures, and He is our Creator. Our Lord wants us to understand His communication because He knows we are helpless without it. We need to be able to discern the truth from the lie, so we are not caught unaware at Christ's judgment. I hope you feel more prepared now to face that moment.

We can rejoice that the Lord Jesus Christ can and will destroy all false scientific claims and show by His glorious presence that He alone created every plant, rock, and monkey as well as every atom in them. Let's not relax our vigilance but keep studying the truth about all areas of life. Only then can we repel the sneaky attacks of every wave of worldly thinking.

> Mankind have been slow to believe that order reigns in the universe–that the world is a cosmos and a chaos....The divinities of heathen superstition still linger in one form or another in the faith of the ignorant, and even intelligent men shrink from the contemplation of one supreme will acting regularly, not fortuitously, through laws beautiful and simple rather than through a fitful and capricious system of intervention....The scientific spirit has cast out the demons, and presented us with nature clothed in her right mind and living under the reign of law. It has given us, for the sorceries of the alchemist, the beautiful laws of chemistry; for the dreams of the astrologer, the sublime truths of astronomy; for the wild visions of cosmogony, the monumental records of geology; for the anarchy of diabolism, the laws of God.
>
> **JAMES A. GARFIELD, 20TH AMERICAN PRESIDENT**

Let's keep our eyes focused on the horizon as we eagerly await the arrival of our Lord, who will deliver us from every false idea and absolutely crush the enemy once and for all. Do I hear a hallelujah or, better yet, a whole chorus of hallelujahs? Glory to God in the highest for giving us all we need to live wisely in service to Him!

> Oh, the depth of the riches both of the wisdom and knowledge of God! How unsearchable are His judgments and unfathomable His ways! For WHO HAS KNOWN THE MIND OF THE LORD, OR WHO BECAME HIS COUNSELOR? Or WHO HAS FIRST GIVEN TO HIM THAT IT MIGHT BE PAID BACK TO HIM AGAIN? For from Him and through Him and to Him are all things. To Him be the glory forever. Amen.
>
> Romans 11:33-36

But from the beginning of creation,
God made them male and female.

MARK 10:6

But pray that it may not happen in the winter.
For those days will be a time of tribulation
such as has not occurred since the beginning of the creation
which God created *until now,*
and never will.

MARK 13:18-19

JESUS, THE MESSIAH

CHAPTER 19

THE CONCLUSION

Okay, we made it through this together. Not being a writer, I am more surprised than you. What do you think? Have I convinced you God is shouting to all mankind through His creation?

The great believer Job proclaimed the importance of listening to God's shouts. God said of Job, "There is no one like him on the earth, a blameless and upright man, fearing God and turning away from evil." Shouldn't we pay attention to the words of such a wise man, especially since God preserved them for us?

> But now ask the beasts, and let them teach you; and the birds of the heavens, and let them tell you. Or speak to the earth, and let it teach you; and let the fish of the sea declare to you. Who among all these does not know that the hand of the Lord has done this, in whose hand is the life of every living thing, and the breath of all mankind?
>
> Job 12:7-10

This passage draws pictures of the natural realm that should awe and motivate us. Job wanted us to see as he did that the creation shouts about God. He encouraged us to "ask" questions of the beasts and the birds and the earth and the fish. Of course, he knew we couldn't sit down at the dinner table and converse with our pet rhinoceros or even our cat. He didn't want us to risk being taken away in a straitjacket for trying to speak to a patch of grass in front of our house. Instead, he used poetic language to vividly communicate that everyone can learn from creation ("let them teach you"). He encouraged us to use our senses to recognize God's hand in forming the beauty, symmetry, power, and complexities of the world around us.

> We are in the position of a little child entering a huge library filled with books in many different languages. The child knows someone must have written those books. It does not know how. It does not understand the languages in which they are written. The child dimly suspects a mysterious order in the arrangement of the books, but doesn't know what it is. That, it seems to me, is the attitude of even the most intelligent human being toward God. We see a Universe marvellously arranged and obeying certain laws, but only dimly understand these laws. Our limited minds cannot grasp the mysterious force that moves the constellations.
>
> DR. ALBERT EINSTEIN, 1921 NOBEL PRIZE IN PHYSICS

Was Job not encouraging us to study zoology, ornithology, ichthyology, geophysics, geology, anthropology, and every other area of science ("every living thing" and "all mankind") so we can learn about the Creator? Must we be full-time scientists to uncover the secrets of creation? Job suggested that God wants everyone to examine creation, to marvel at His creativity and kindness in revealing Himself to us in such differing, plentiful, and intriguing ways. He knew you and I would assuredly benefit from studying the general revelation of creation.

Wasn't Job encouraging us to make the creation a teaching tool in our lives? Shouldn't we listen to him and set aside a few minutes every day to look at a small portion of nature—really look at it? During those few minutes of rest as we try to understand God's purpose in putting thorns on roses and blackberries but not on daisies and blueberries and tails on cats and salamanders but cotton balls on bunnies, we will be transported away from Satan's dominion. The things of the world will truly grow strangely dim. When we slow down and take time to study the myriad treasures hidden in one small square foot of the earth or sky, God will, at the same time He astounds us, fill us with some of the comfort, peace, and joy He has embedded in all nature. Those few minutes of focusing on His love message to us in creation can become a daily oasis in our otherwise hectic lives, the least expensive but most uplifting vacations we will ever take.

Try it. Listen for both God's shouts and whispers. Be attuned to His voice so you benefit from the bountiful gifts in His creation.

> All creation is a mine, and every man a miner.
> **ABRAHAM LINCOLN, 16TH AMERICAN PRESIDENT**

WHAT DOES GOD WANT ME TO DO?

Whether we like it or not, we are soldiers fighting a spiritual culture war. For generations, Christians have sat back and allowed the unbelieving world to make our culture thoroughly secular, even down to the sciences. The absolute truth of Scripture has been slowly replaced by satanic lies. Few today, even Christians, realize they have been duped. Many have no idea they have allowed themselves to be blinded to the truth.

What can you do? How can you fight against this evil? How can you help others open their eyes to God and His magnificence? One way is to make them aware of the danger of believing secular lies. If you engage in any branch of science, use the scriptural investigative pattern presented in this book to develop a biblical foundation in your own specialty, whether it be biology, astrophysics, geophysics, physics, chemistry...The list of science areas that need the Christian perspective is long. The list of other areas of study that need the biblical perspective is even longer.

It is my hope that this book has given you greater confidence in God's Word. I hope I have shown that we do not need to fear the findings of science. Even though some scientific ideas are contrary to God's Word, they are not the final word concerning truth. Only God and His Word are absolute truth. Scientific truth, as wonderful as it may be, is not even close to being absolute truth. If we keep our eyes focused on God and His absolute truth, we won't be fooled or frightened by science's claims against it. Instead, we will rejoice in His control of all things in heaven and on Earth.

Let your excitement about truth overflow into every area of your life. Use every opportunity to tell others the source of your joy. Be bold in exposing the lie and presenting the truth. Be fearless because, as did David against Goliath, you know the battle is the Lord's (1 Sam. 17:47).

At the same time, remember that your battle must include an eternal perspective. What good is it if you educate someone about science, economics, or politics but neglect their eternal needs. Be a bold ambassador for Christ, giving the Gospel with grace, even as you use exciting creation truths to open the door to talk about God.

> The story of Daniel and his friends is a clarion call to our generation to be courageous; not to lose our nerve and allow the expression of our faith to be diluted and squeezed out of the public space and thus rendered spineless and ineffective. Their story will also tell us that this objective is not likely to be achieved without cost.
>
> **DR. JOHN LENNOX, MATHEMATICS**

Scripture speaks to every area of life, not just spiritual issues and our relationship to God. I have covered science (at least briefly, and now you know what that means). Whether your interest is music, art, politics, economics, government, education, law enforcement, medicine, plumbing, psychology, languages, cooking, or whatever, I hope this book has helped you focus your work on Christ because God holds you responsible to develop a biblical foundation for your area of expertise. If you do, you will have a better grasp of reality and the Creator of it. You will also help those of us who listen to you and—dare I suggest, read your books—appreciate the divine viewpoint related to your specialty. You may even influence an unbeliever to change his mind about Christ.

We all benefit from a fuller understanding of our awesome Creator. I benefited greatly from James Nickel's book on mathematics, which does for mathematics what I have tried to do for science generally. Now, maybe it's your turn to open our eyes to God's work in another area of creation.

> Nothing can be unworthy of being investigated by man, which was thought worthy of being created by God.
>
> **JOSEPH HENRY, EARLY AMERICAN PHYSICIST**

I enjoy reorganizing my thoughts around God's Word. I have done that many times in the past as I am sure you have, too. Every time I learn something new or have my mind opened to a new way of perceiving the creation, I benefit from reorganizing my thinking, from fitting the new puzzle piece in with the old. I hope this book has given you many new and intriguing pieces that make your own mental puzzle clearer and more complete. Maybe you have even thrown out some old, defective pieces! Getting rid of lies is as crucial as adding new truths. Thank you for letting me help add and delete pieces in your science-based mental puzzle.

> He that takes truth for his guide, and duty for his end, may safely trust to God's providence to lead him aright.
>
> **BLAISE PASCAL, 17TH CENTURY MATHEMATICIAN, PHYSICIST, INVENTOR**

Now, how do you answer the question I proposed in the title of the book? Is God shouting through His creation? Are you now better able to hear the heavens declare God's glory?

> The heavens are telling of the glory of God; and their expanse is declaring the work of His hands. Day to day pours forth speech, and night to night reveals knowledge. There is no speech, nor are there words; their voice is not heard. Their line has gone out through all the earth, and their utterances to the end of the world.
>
> Psalm 19:1-4

> For *by Him all things were created*, both in the heavens and on earth, visible and invisible, whether thrones or dominions or rulers or authorities—*all things have been created through Him and for Him*. He is before all things, and *in Him all things hold together.*
>
> Colossians 1:16-17

> Almighty God, who has created man in Thine own image, and made him a living soul that he might seek after Thee, and have dominion over Thy ceatures, teach us to study the works of Thy hands, that we may subdue the earth to our use, and strengthen the reason for Thy service; so, to receive Thy blessed Word, that we may believe on Him whom Thou has sent, to give us the knowledge of salvation and the remission of sins. All of which we ask in the name of the same Jesus Christ, our Lord.
>
> **JAMES CLERK MAXWELL, PHYSICIST**

APPENDIX A
CREATION STATEMENTS IN SCRIPTURE

The Bible not only teaches God as Creator in Genesis 1-2 but also throughout Scripture. This list is not exhaustive.

OLD TESTAMENT TESTIMONY WRITER OR SPEAKERS

ELOHIM
THE CREATOR OF ALL THINGS

GENESIS 1:1 In the beginning God created the heavens and the earth.

GENESIS 1:3 Then God said, "Let there be light;" and there was light.

YAHWEH
THE CREATOR OF ALL THINGS

JOB 38:4-7 "Where were you when I **laid** the **foundation** of the earth?Tell
Me, if you have understanding, **5** who **set** its measurements? Since you
know. Or who **stretched** the line on it? **6** On what were its bases sunk?
Or who **laid** its cornerstone, **7** when the morning stars sang together
and all the sons of God shouted for joy?"

ISAIAH 43:1 But now, thus says the Lord, your **Creator**, O Jacob, and He who formed you, O Israel, "Do not fear, for I have redeemed you; I have called you by name; you are Mine!"

44:2 Thus says the Lord who made you and **formed** you from the womb, who will help you, "Do not fear, O Jacob My servant; and you Jeshurun whom I have chosen."

44:24 Thus says the Lord, your Redeemer, and the one who **formed** you from the womb, "I, the Lord, am the **maker** of all things, stretching out the heavens by Myself and spreading out the earth all alone

45:7-12 "The One forming light and **creating** darkness, causing well-being
and **creating** calamity; I am the Lord who does all these. **8** Drip down,
O heavens, from above, and let the clouds pour down righteousness; let
the earth open up and salvation bear fruit, and righteousness spring up

with it. I, the Lord, have **created** it. 9 Woe to the one who quarrels with
his **Maker**—an earthenware vessel among the vessels of earth! Will the
clay say to the potter, 'What are you doing?' or the thing you are making
say, 'He has no hands?' 10 Woe to him who says to a father, 'What are you
begetting?' or to a woman, 'To what are you giving birth?'" 11 Thus says the
Lord, the Holy One of Israel, and his **Maker**: "Ask Me about the things
to come concerning My sons, and you shall commit to Me the work of
My hands. 12 "It is I who **made** the earth, and **created** man upon it. I
stretched out the heavens with My hands and I ordained all their host."

48:13 "Surely My hand **founded** the earth, and My right hand **spread** out the heavens; when I call to them, they stand together.

66:2 "For My hand **made** all these things, thus all these things came into being," declares the Lord. "But to this one I will look, to him who is humble and contrite of spirit, and who trembles at My word.

JEREMIAH 5:22 "Do you not fear Me?" declares the Lord. "Do you not tremble in My presence? For I have placed the sand as a boundary for the sea, an eternal decree, so it cannot cross over it. Though the waves toss, yet they cannot prevail; though they roar, yet they cannot cross over it."

27:5 "I have **made** the earth, the men and the beasts which are on the face of the earth by My great power and by My outstretched arm, and I will give it to the one who is pleasing in My sight."

JOB
A BLAMELESS, UPRIGHT, AND GOD-FEARING MAN

JOB 12:7-10 "But now ask the beasts and let them teach you; and the birds of
the heavens, and let them tell you. 8 Or speak to the earth, and let it teach
you; and let the fish of the sea declare to you. 9 Who among all these does
not know that the hand of the Lord has **done** this, 10 in whose hand is the
life of every living thing, and the breath of all mankind?"

26:7-10 "He **stretches** out the north over empty space and **hangs** the
earth on nothing. 8 He **wraps** up the waters in His clouds, and the
cloud does not burst under them. 9 He obscures the face of the full
moon and **spreads** His cloud over it. 10 He has **inscribed** a circle on the
surface of the waters at the boundary of light and darkness."

MOSES
PROPHET WHO SAW THE GLORY OF GOD

EXODUS 20:11 "For in six days the Lord **made** the heavens and the earth, the sea and all that is in them, and rested on the seventh day; therefore the Lord blessed the sabbath day and made it holy."

PSALM 90:2 Before the mountains were born or You gave birth to the earth and the world, even from everlasting to everlasting, You are God.

HANNAH
GODLY MOTHER OF A PROPHET

1 SAMUEL 2:8 "He raises the poor from the dust. He lifts the needy from the ash heap to make them sit with nobles, and inherit a seat of honor; for the pillars of the earth are the Lord's, and He **set** the world on them."

LEVITES
LEADERS OF WORSHIP IN ISRAEL

NEHEMIAH 9:5-6 "Arise, bless the Lord your God forever and ever! O may Your glorious name be blessed and exalted above all blessing and praise!
6 You alone are the Lord. You have **made** the heavens, the heaven of heavens with all their host, the earth and all that is on it, the seas and all that is in them. You **give** life to all of them and the heavenly host bows down before You."

PSALMISTS

ETHAN

PSALM 89:11-12 The heavens are Yours, the earth also is Yours; the world and all it contains, You have **founded** them.
12 The north and the south, You have **created** them; Tabor and Hermon shout for joy at Your name.

ASAPH

PSALM 74:16-17 Yours is the day, Yours also is the night; You have prepared the light and the sun.
17 You have **established** all the boundaries of the earth; You have **made** summer and winter.

OTHERS

PSALM 65:6 Who **establishes** the mountains by His strength, Being girded with might.

78:69 And He **built** His sanctuary like the heights, Like the earth which He has **founded** forever.

102:25 "Of old You **founded** the earth, And the heavens are the **work** of Your hands."

104:2-6 Covering Yourself with light as with a cloak, **Stretching** out heaven like a tent curtain. **3** He **lays** the beams of His upper chambers in the waters; He **makes** the clouds His chariot; He walks upon the wings of the wind. **4** He **makes** the winds His messengers, flaming fire His ministers. **5** He **established** the earth upon its foundations, so that it will not totter forever and ever. **6** You covered it with the deep as with a garment; The waters were standing above the mountains.

104:24 O Lord, how many are Your **works**! In wisdom You have **made** them all; the earth is full of Your possessions.

115:15 May you be blessed of the Lord, **Maker** of heaven and earth.

119:90 Your faithfulness continues throughout all generations; You **established** the earth, and it stands.

136:5-9 To Him who **made** the heavens with skill, for His lovingkindness is everlasting; **6** To Him who **spread** out the earth above the waters, for His lovingkindness is everlasting; **7** To Him who **made** the great lights, for His lovingkindness is everlasting. **8** The sun to rule by day, for His lovingkindness is everlasting, **9** The moon and stars to rule by night, for His lovingkindness is everlasting.

148:5 Let them praise the name of the Lord, For He commanded and they were **created**.

DAVID
THE KING AFTER GOD'S OWN HEART

PSALM 8:3-4 When I consider Your heavens, the **work** of Your **fingers**, the moon and the stars, which You have ordained, **4** What is man that You take thought of him.

19:1-2 The heavens are telling of the glory of God; and their expanse is declaring the **work** of His **hands**. 2 Day to day pours forth speech, and night to night reveals knowledge.

33:6-9 By the word of the Lord the heavens were **made**, and by the breath of His mouth all their host. 7 He **gathers** the waters of the sea together as a heap; He **lays** up the deeps in storehouses. 8 Let all the earth fear the Lord; let all the inhabitants of the world stand in awe of Him. 9 For He **spoke**, and it was done.

95:4-6 He **commanded**, and it stood fast. In whose hand are the depths of the earth, the peaks of the mountains are His also. 5 The sea is His, for it was He who **made** it, and His hands **formed** the dry land. 6 Come, let us worship and bow down. Let us kneel before the Lord our **Maker** for He is our God.

103:22 Bless the Lord, all you **works** of His, in all places of His dominion; bless the Lord, O my soul!

124:8 Our help is in the name of the Lord, Who **made** heaven and earth.

SOLOMON
ISRAEL'S THIRD KING AND THE WISEST MAN ON EARTH

PROVERBS 3:19 The Lord by wisdom **founded** the earth. By understanding He **established** the heavens.

8:22-31 The Lord possessed me at the beginning of His way, before His **works** of old. 23 From everlasting I was established, from the beginning, from the earliest times of the earth. 24 When there were no depths I was brought forth, when there were no springs abounding with water. 25 Before the mountains were **settled**, before the hills I was brought forth. 26 While He had not yet **made** the earth and the fields, nor the first dust of the world. 27 When He **established** the heavens, I was there, when He inscribed a circle on the face of the deep, 28 when He **made** firm the skies above, when the springs of the deep became **fixed**, 29 when He **set** for the sea its boundary so that the water would not transgress His command, when He **marked** out the foundations of the earth; 30 then I was beside Him, as a master **workman**; and I was daily His delight, rejoicing always before Him, 31 rejoicing in the world, His earth, and having my delight in the sons of men.

30:4 Who has ascended into heaven and descended? Who has **gathered** the wind in His fists? Who has **wrapped** the waters in His garment?

Who has **established** all the ends of the earth? What is His name or His son's name? Surely you know!

ECCLESIASTES 3:11 He has **made** everything appropriate in its time. He has also set eternity in their heart, yet so that man will not find out the work which God has done from the beginning even to the end.

11:5 Just as you do not know the path of the wind and how bones are formed in the womb of the pregnant woman, so you do not know the activity of God who **makes** all things.

HEZEKIAH
GODLY KING OF JUDAH

2 KINGS 19:15 Hezekiah prayed before the Lord and said, "O Lord, the God of Israel, who are enthroned above the cherubim, You are the God, You alone, of all the kingdoms of the earth. You have **made** heaven and earth."

JONAH
PROPHET TO NINEVEH

JONAH 1:9 He said to them, "I am a Hebrew, and I fear the Lord God of heaven who **made** the sea and the dry land."

AMOS
SHEPHERD AND PROPHET

AMOS 4:13 For behold, He who **forms** mountains and **creates** the wind and declares to man what are His thoughts, He who **makes** dawn into darkness and treads on the high places of the earth, the Lord God of hosts is His name.

5:8 He who **made** the Pleiades and Orion and changes deep darkness into morning, Who also darkens day into night, Who calls for the waters of the sea and **pours** them out on the surface of the earth, the Lord is His name.

9:6 The One who **builds** His upper chambers in the heavens and has **founded** His vaulted dome over the earth, He who calls for the waters of the sea and **pours** them out on the face of the earth, the Lord is His name.

ISAIAH
STATESMAN AND PROPHET

Also above when Yahweh spoke in Isaiah 43:1; 44:2, 24; 45:7-12; 48:13; 66:2.

ISAIAH 40:12 Who has **measured** the waters in the hollow of His hand, and **marked** off the heavens by the span, and **calculated** the dust of the earth by the measure, and **weighed** the mountains in a balance and the hills in a pair of scales?

40:26 Lift up your eyes on high and see who has **created** these stars, the One who leads forth their host by number, He calls them all by name; because of the greatness of His might and the strength of His power, not one of them is missing.

40:28 Do you not know? Have you not heard? The Everlasting God, the Lord, the **Creator** of the ends of the earth does not become weary or tired. His understanding is inscrutable.

42:5 Thus says God the Lord, who **created** the heavens and stretched them out, who spread out the earth and its offspring, who gives breath to the people on it and spirit to those who walk in it.

44:24 Thus says the Lord, your Redeemer, and the one who **formed** you from the womb, "I, the Lord, am the maker of all things, stretching out the heavens by Myself and spreading out the earth all alone.

45:18 For thus says the Lord, who **created** the heavens (He is the God who **formed** the earth and made it, He **established** it and did not **create** it a waste place, but **formed** it to be inhabited).

51:13 That you have forgotten the Lord your **Maker**, who **stretched** out the heavens and **laid** the foundations of the earth, that you fear continually all day long because of the fury of the oppressor, as he makes ready to destroy? But where is the fury of the oppressor?

54:5 For your husband is your **Maker**, whose name is the Lord of hosts; and your Redeemer is the Holy One of Israel, who is called the God of all the earth.

JEREMIAH
WEEPING PROPHET

JEREMIAH 10:12-13 It is He who **made** the earth by His power, Who **established** the world by His wisdom; and by His understanding He has **stretched** out the heavens. **13** When He utters His voice, there is a tumult of waters in the heavens, and He **causes** the clouds to ascend from the end of the earth; He **makes** lightning for the rain, and **brings** out the wind from His storehouses.

31:35 Thus says the Lord, Who **gives** the sun for light by day and the fixed order of the moon and the stars for light by night, Who **stirs** up the sea so that its waves roar; the Lord of hosts is His name.

32:17 Ah Lord God! Behold, You have **made** the heavens and the earth by Your great power and by Your woutstretched arm! Nothing is too difficult for You.

33:2 Thus says the Lord who **made** the earth, the Lord who **formed** it to **establish** it, the Lord is His name.

51:15-16 It is He who **made** the earth by His power, Who **established** the world by His wisdom, and by His understanding He **stretched** out the heavens. **16** When He utters His voice, there is a tumult of waters in the heavens, and He **causes** the clouds to ascend from the end of the earth; He **makes** lightning for the rain and brings forth the wind from His storehouses.

ZECHARIAH
POST-EXILIC PROPHET

ZECHARIAH 12:1 The burden of the word of the Lord concerning Israel. Thus declares the Lord who **stretches** out the heavens, **lays** the foundation of the earth, and **forms** the spirit of man within him.

MALACHI
POST-EXILIC PROPHET

MALACHI 2:10 "Do we not all have one father? Has not one God **created** us? Why do we deal treacherously each against his brother so as to profane the covenant of our fathers?"

NEW TESTAMENT TESTIMONY WRITER OR SPEAKERS

LUKE
HISTORIAN AND DOCTOR (PAUL SPEAKING)

ACTS 17:24-25 “The God who **made** the world and all things in it, since He is Lord of heaven and earth, does not dwell in temples made with hands;
25 nor is He served by human hands as though He needed anything, since He Himself **gives** to all people life and breath and all things.”

JOHN
APOSTLE AND PROPHET

JOHN 1:3 All things came into **being** through Him, and apart from Him nothing came into **being** that has come into being.

1:10 He was in the world, and the world was **made** through Him, and the world did not know Him.

REVELATION 4:11 “Worthy are You, our Lord and our God, to receive glory and honor and power; for You **created** all things, and because of Your will they existed, and were created.”

10:6 and swore by Him who lives forever and ever, who **created** heaven and the things in it, and the earth and the things in it, and the sea and the things in it, that there will be delay no longer.

14:7 and he said with a loud voice, “Fear God, and give Him glory, because the hour of His judgment has come; worship Him who **made** the heaven and the earth and sea and springs of waters.”

21-22 Creation of new heavens and new earth.

PAUL
SCHOLAR AND APOSTLE

ROMANS 1:19-20 Because that which is known about God is evident within
them; for God made it evident to them. 20 For since the **creation** of the
world His invisible attributes, His eternal power and divine nature, have been clearly seen, being understood through what has been **made**, so that they are without excuse.

EPHESIANS 3:8-9 To me, the very least of all saints, this grace was given, to preach to the Gentiles the unfathomable riches of Christ, 9 and to bring to light what is the administration of the mystery which for ages has been hidden in God who **created** all things.

COLOSSIANS 1:16-17 For by Him all things were **created**, both in the heavens and on earth, visible and invisible, whether thrones or dominions or rulers or authorities—all things have been **created** through Him and for Him. 17 He is before all things, and in Him all things hold together.

1 TIMOTHY 2:13-14 For it was Adam who was first **created**, and then Eve. 14 And it was not Adam who was deceived, but the woman being deceived, fell into transgression.

6:13 I charge you in the presence of God, who **gives life** to all things, and of Christ Jesus, who testified the good confession before Pontius Pilate.

PETER
APOSTLE

2 PETER 3:4-7 And saying, "Where is the promise of His coming? For ever since the fathers fell asleep, all continues just as it was from the beginning of **creation**." 5 For when they maintain this, it escapes their notice that **by the word** of God the heavens existed long ago and the earth was **formed** out of water and by water, 6 through which the world at that time was destroyed, being flooded with water. 7 But by His word the present heavens and earth are being reserved for fire, kept for the day of judgment and destruction of ungodly men.

ANONYMOUS AUTHOR OF HEBREWS
JEWISH SCHOLAR

HEBREWS 1:1-2 God, after He spoke long ago to the fathers in the prophets in many portions and in many ways, 2 in these last days has spoken to us in His Son, whom He appointed heir of all things, through whom also He **made** the world.

:10 "You, Lord, in the beginning laid the **foundation** of the earth, and the heavens are the **works** of Your hands."

2:10 For it was fitting for Him, for whom are all things, and **through whom** are all things, in bringing many sons to glory, to perfect the author of their salvation through sufferings.

3:4 For every house is **built** by someone, but the **builder of all** things is God.

4:3 For we who have believed enter that rest, just as He has said, "As I swore in My wrath, they shall not enter My rest," although **His works** were finished from the **foundation of the world**.

11:3 By faith we understand that the worlds were **prepared** by the word of God, so that what is seen was not made out of things which are visible.

JESUS
MESSIAH AND CREATOR

Also Matthew and Mark as writers and parallel passages.

MATTHEW 19:4 And He answered and said, "Have you not read that He who **created** them from the beginning made them male and female."

MARK 10:6 "But from the beginning of **creation**, God made them male and female."

13:18-19 But pray that it may not happen in the winter. **9** For those days will be a time of tribulation such as has not occurred since the beginning of the creation which God **created** until now.

APPENDIX B
TRUTH & KNOWLEDGE IN SCRIPTURE

Science is the search for truth and knowledge. The following word studies on the main biblical terms for truth and knowledge give the scriptural view of both. The data is presented in outline form for quick reference. The biblical examples are selective but representative of all uses of the terms.

TRUTH: OLD TESTAMENT TERMS

THE HEBREW TERMS

1. The primary term for "truth" occurs 127 times in the OT
2. Brown, Driver, Briggs lexicon: firmness, faithfulness, truth
    a. Reliability, sureness
    b. Stability, continuance
    c. Faithfulness, reliability
    d. Truth

THE USAGE OF HEBREW TERMS

1. **Truth in Reference to God (often translated faithful)**
    a. His character or as a perfection - Ex. 34:6; Ps. 30:9; 31:5; 57:10; 86:15; 91:4; 108:4; 115:1; Isa. 38:18-19; Jer. 10:10; Zech. 8:8
    b. Perfections of lovingkindness and truth together - Gen. 24:27; 32:10; Ex. 34:6; 2 Sam. 2:6; 1 Kings 3:6; Ps. 26:3, 10; 85:10; 115:1; 117:2; 138:2; Prov. 16:6; Isa. 16:5; many more
    c. Messiah - Isa. 42:3
    d. His works of truth - Ps. 25:10; 111:7-8; 146:6
    e. He sends truth - Ps. 57:3
    f. His judgment and truth - Ps. 19:9; 54:5; Zech. 7:9
    g. Truth expressed toward men

i. Abraham - Gen. 24:27

ii. Jacob - Gen. 32:10

iii. Those burying Saul - 2 Sam. 2:6

iv. Restoring Israel - Jer. 33:6

v. Israel - Jer. 32:41; Micah 7:20

2. **Truth in Reference to the Word of God, His Truth** - His truth - 2 Sam. 7:28; 1 Kings 17:24; Ps. 25:5; 40:10; 43:3; 69:13; 71:22; 86:11; 119:142, 151; Prov. 22:21; Jer. 23:28

3. **Truth in Reference to Man**

    a. Men of truth/faithfulness or as an attribute

        i. Sons of Jacob - Gen. 42:16

        ii. Israelites - Ex. 18:21

        iii. Joshua - Josh. 2:12

        iv. Hanani - Neh. 7:2

    b. Desired by God - Ps. 51:6

    c. Bestowed by God - 2 Sam. 2:6

    d. Walking in truth - Josh. 24:14; 1 Sam. 12:24; 1 Kings 2:4; 3:6; 22:16; Isa. 38:3; Zech. 8:19

    e. Speaking truth - Ps. 15:2; Jer. 9:5; Zech. 8:16

    f. Treating others in truth - Gen. 24:49; Josh. 2:14; Neh. 9:33; Ezek. 18:9

    g. Jerusalem a city of truth - Zech. 8:3

    h. Judging in faithfulness - Isa. 16:5

    i. Lack of truth - Hos. 4:1

4. **Truth as Reality, True Events, Things** - Deut. 13:14; 17:4; Prov. 11:18; 1 Kings 10:6

5. **Truth in General** - Ps. 45:4; 51:6; 85:11; Prov. 3:3; 8:7; 12:19; 14:22; 20:28; 23:23; Eccl. 12:10; Isa. 38:8

TRUTH: NEW TESTAMENT TERMS

THE GREEK TERMS

Bauer, Arndt and Gingrich Lexicon

1. ἀλήθεια - This primary noun occurs 110 times in the NT.
2. The Other Terms
    a. ἀληθεύω - The verb occurs 2 times in NT.
    b. ἀληθής - The adjective form occurs 27 times.
    c. ἀληθῶς - The adverb form occurs 21 times.
    d. ἀληθινός - This adjective form occurs 28 times.
    e. ἀμην - This particle form occurs 126 times.

THE USAGE OF TERMS

ἀλήθεια (unless noted)

1. **Truth in Reference to God**
    a. His Character
        i. Father - John 3:33 (ἀληθής); Rom. 3:7
        ii. Son - John 1:14, 17; 8:45-46; 14:6; 16:7, 18, 37-38; Eph. 4:21
        iii. Holy Spirit - John 14:17; 15:26; 16:13; 1 John 4:6, 5:7
    b. Jesus' Claim - Luke 4:24; John 8:40
    c. Noted by Others
        i. Pharisee's disciples - Matt. 22:16
        ii. Scribe - Mark 12:32
2. **Truth in Reference to the Word**
    a. The Scriptures described - John 17:17; Rom. 2:20; 15:8; 2 Co 4:2
    b. Scripture as truth - John 5:33; Gal. 5:7; Eph. 6:14; 1 Tim. 3:15; 4:3; 6:5; Titus 1:1; Jam. 1:18; 1 Pet. 1:22; 1 John 2:21; 3:19; 2 John 1-2, 4; 3 John 3-4, 8, 12

    c. The Gospel as truth - Eph. 1:13; Gal. 2:5, 2:14; Col. 1:5-6; 2 Thess. 2:12-13

3. **Truth in Reference to Man**
    a. Opposition of depraved men to truth - Rom. 1:18, 25; 2 Tim. 3:7-8; 4:4; Titus 1:14; 2 Pet. 2:2; 1 John 1:8; 2:4
    b. Man's tendency is to depart from truth - 2 Tim. 2:18, 25; Heb. 10:26; James 3:14; 5:19; 1 John 1:6
    c. God desires we know truth - John 8:32; 1 Tim. 2:4
    d. Judgment based on truth - 2 Thess. 2:10; Rom. 2:2 (rightly), 8
    e. Truth can be in us - 2 Cor. 11:10; 2 John 1-2
    f. God's will we walk in truth - John 3:21; 17:19; 2 John 4; 3 John 3-4
    g. Should be established in truth - 2 Pet. 1:12
    h. Communicating truth
        i. All of us - 2 Tim. 2:15
        ii. Woman healed in crowd - Mark 5:33
        iii. Peter - Luke 22:59; Acts 10:34 (certainly)
        iv. Paul - Acts 26:25; Rom. 9:1; 2 Cor. 7:14; 12:6
    i. Can support truth - 2 Cor. 13:8
4. **Truth in Reference to Actions**
    a. Worship - John 4:23-24
    b. Speech - 2 Cor. 6:7; Eph. 4:25
    c. Love - 1 John 3:18; 3 John 1
5. **Truth Absent in Satan** - John 8:44
6. **Speaking or Telling the Truth** (ἀληθεύω) - Gal. 4:16; Eph. 4:15

THE LEXICAL MEANINGS

Bauer, Arndt and Gingrich Lexicon

1. ἀλήθεια
    a. Truthfulness, dependability, uprightness
    b. Truth

            i. Generally, tell truth
            ii. Absolute truth, especially of content of Christianity
        c. Reality as opposed to mere appearance
2. The Other Terms
    a. ἀληθεύω - be truthful, tell the truth
    b. ἀληθής
        i. true: of persons, of things
        ii. real, genuine
    c. ἀληθῶς - truly, in truth, really, actually
    d. ἀληθινός - true
    e. ἀμην - so let it be, truly, amen

THE CONCLUSION

Truth Defined

1. Truth is objective.
2. Truth exists but so does falsehood.

Truth and God

1. Both the OT and NT terms are centered in God Himself.
2. Truth is not only a perfection of God, but He is truth.
3. Truth is personally embodied in the Father, Son, and Holy Spirit, making His truth absolute because He is absolute reality.
4. As truth, God works or expresses truth.
5. God is the ultimate source of truth and reveals truth in His Word.
6. God's truth is ultimate reality.

Truth and Mankind

1. God created man in His image and thus with the capability to obtain and understand truth.
2. Man can interact with and communicate truth.
3. God desires that man obtain and know truth, especially Himself as truth.

4. Sinful man tends to be lacking and departing from truth and so suppresses, distorts, and denies truth and often is involved in lies.

Truth and Science

1. God built truth into the created realm so that mankind can access it.
2. The purpose of science is to discover the truth.

KNOWLEDGE: OLD TESTAMENT TERMS

THE HEBREW TERMS

1. *da'at* - The primary term for "knowledge" occurs 89 times in the OT.
2. Brown, Driver, Briggs Lexicon
    a. Knowledge
    b. Discernment, understanding, wisdom (especially wisdom literature)

USAGE OF THE HEBREW TERM

1. **Source of Knowledge**
    a. God as source - Ps. 94:10; 119:66; 139:6; Prov. 2:6; 3:20; 22:12; Isa. 33:6; 40:14
    b. Messiah - Isa. 11:2
2. **Content of Knowledge**
    a. Good and evil - Gen. 2:9, 17
    b. God - Num. 24:16; Prov. 2:5; 9:10; Hos. 6:6; Isa. 58:2; Jer. 22:16
    c. Fear of Lord - Prov. 1:7
    d. Practical knowledge - Ex. 31:3; 35:31; 1 Kings 7:14
    e. Astrophysical knowledge - Ps. 19:2
3. **Godly Seeking, Receiving, Utilizing, and Knowing** - Prov. 1:4; 2:10; 8:9-10; 10:14; 11:9; 12:1; 13:16; 14:6-7; 15:7, 14; 18:15; 19:25; 21:11; 22:17; 24:4-5; Eccl. 1:16, 18; 2:21, 26; 9:10; Dan. 1:4; 12:4; Mal. 2:7
4. **Consequences of or Lack of Knowledge**

   a. Exile - Isa. 5:13
   b. Lack of knowledge - Isa. 44:19; Job 34:35; 35:16; 38:2
   c. Foolishness - Isa. 44:25; 47:10; Jer. 10:14; 51:17
   d. Destruction - Hos. 4:1, 6
5. **Unbelievers' Hate of Knowledge** - Prov. 1:22, 29
6. **In General** - Job 10:7; 15:2; 21:14, 22; 33:3

KNOWLEDGE: NEW TESTAMENT TERMS

THE GREEK TERMS

1. γινώσκω - This verb occurs 234 times in NT.
2. ἐπιγινώσκω - This verb with prefixed preposition occurs 44 times in the NT.
3. γνῶσις - This noun occurs 29 times in the NT.
4. ἐπίγνωσις - This noun with prefixed preposition occurs 20 times in the NT.
5. γνωστός - This adjective occurs 22 times in the NT.
6. οἶδα - This verb occurs 328 times in the NT.

THE USAGE OF TERMS

γινώσκω - This verb occurs 234 times in the NT.

1. **God Knowing**
   a. Father Knows
      i. Son - Luke 10:22
      ii. Hearts - Luke 16:15; John 4:1; Gal. 4:9; 1 Cor. 3:20; 8:2-3; 2 Tim. 2:19
   b. Son Knows
      i. Father - Luke 10:22; John 10:14-15; 17:25
      ii. All things/men, aware or knowing - Matt. 12:15; 16:8; 26:10; John 2:24-25; 5:6; 16:19; 21:17
      iii. His own believers - John 10:14, 27
      iv. Unbelievers or evil - Matt. 7:23; 22:18; Mark 15:10; John 6:15

v. Unbelieving Jews - John 5:42

vi. Nathaniel - John 1:48

vii. No Sin - 2 Cor. 5:21

2. **Man Knowing**

a. Common Knowledge

i. Natural realm, science - Matt. 16:3; 12:33; 24:32; Luke 6:44; 1 Cor. 14:7, 9

ii. Everyday things - Matt. 6:3; 25:24; Mark 5:29; 6:38; 15:45; Luke 2:43; 12:47-48; 16:4; 19:15; John 12:9; Acts 21:34; 1 Cor. 8:2

iii. People - Matt. 10:26; Luke 7:39; 18:34; John 4:53, 11:57, 13:28, 19:4; 2 Cor. 3:2

iv. Paul's life - Acts 20:34; 21:24; 22:30; 2 Cor. 2:4; Eph. 6:22; Phil. 1:12; Col. 4:8

v. Knowing sexually - Matt. 1:25; Luke 1:34

vi. Greek - Acts 21:37

vii. Status of Ephesus - Acts 19:35

viii. Timothy - Phil. 2:22; Heb. 13:23

ix. Onesiphorus - 2 Tim. 1:18

b. Relating to God

i. God - John 8:27, 55; 14:7, 9, 20, 31; 15:18; 17:3, 7, 23; Rom. 1:21; 11:34; 1 Cor. 2:16; Gal. 4:9; Heb. 3:10; 8:11; 1 John 2:3-5, 13

ii. Not knowing God - John 16:3; 1 Cor. 1:21; 2:11

iii. Knowing Messiah/Jesus - Luke 24:35; John 6:69; 8:28; 17:8; Acts 19:15; 2 Cor. 5:16; 8:9; Eph. 3:19; Phil. 3:10

iv. Not knowing Messiah/Jesus - John 1:10; 7:26-27; 8:52; 10:38

v. Holy Spirit - John 14:17

c. Unusual or Miraculous Events

i. His work - Matt. 9:30; 21:45; Mark 5:43; 7:24; 9:30; Luke 1:18; 9:11; 10:11; 24:18; Acts 9:24

ii. Evil conditions - Matt. 24:43, 33, 39; Luke 19:42, 44; 21:20, 31; Acts 23:6; Eph. 5:5; 2 Tim. 3:1

iii. Time - Matt. 24:50; Luke 12:46; Acts 1:7

iv. Israel

1. Messiah - Acts 2:36

2. Moses - Rom. 10:19

v. Foot washing - John 13:12

vi. Paul's teaching - Acts 17:19

d. Spiritual Things

i. Truth/doctrine/wisdom - Matt. 12:7; Mark 4:13; John 3:10; 7:17; 8:32, 43; 10:6, 38; 12:16; 13:7; Acts 8:30; 17:13; 1 Cor. 2:8, 14; Gal. 2:9; 3:7; Jam. 2:20; 2 Pet. 1:20

ii. Law - John 7:49, 51; Rom. 7:1

iii. Christian walk - Rom. 6:6; 7:7, 15; 1 Cor. 4:19; 2 Cor. 2:9; 13:6; Phil. 2:19; 4:5; 1 Thess. 3:5; James 1:3; 5:20; 2 Pet. 3:3

iv. God's will - Acts 22:14; Rom. 2:18

v. Mysteries of the Kingdom - Matt. 13:11; 1 Cor. 13:9, 12

vi. Satan - Rev. 2:24

vii. His disciples - John 13:35

viii. Peace - Rom. 3:17

ix. Better things - Heb. 10:34

ἐπιγινώσκω - This verb with prefixed preposition occurs 44 times.

1. **God Knowing**

a. Father and Son - Matt. 11:27

b. Jesus - Mark 2:8; 5:30; Luke 5:22

2. **Man Knowing**

a. Material things

i. Land - Acts 27:39; 28:1

ii. Facts - Acts 3:10; 22:24; 24:11

iii. Men - 1 Cor. 16:18

b. Spiritual things

i. Messiah - Matt. 14:35; Mark 6:54; Luke 7:37; 23:7; 24:16, 31
ii. People
iii. John the Baptizer - Matt. 17:12
iv. The 12 disciples - Mark 6:33; Acts 4:13; 12:14
v. Paul - Acts 9:30; 19:34; 22:29; 23:28; 24:8; 2 Cor. 6:9
vi. Spiritual things - Matt. 7:16, 20; Luke 1:4, 22; Acts 25:10; 1 Cor. 13:12; 14:37; 2 Cor. 13:5; Col. 1:6; 2 Pet. 2:21
vii. Truth - Rom. 1:32; 1 Tim. 4:3; 2 Cor. 1:13-14

γνῶσις - This noun occurs 29 times in NT.

1. **Knowledge of God**
   a. Its Nature - Rom. 11:33
   b. Christ - Col. 2:3
   c. Law - Rom. 2:20
2. **Knowledge Knowable by Man**
   a. Of God - 2 Cor. 2:14; 4:6; 10:5
   b. Of Christ - Eph. 3:19; Phil. 3:8
   c. Spiritual or biblical knowledge - Rom. 15:14; 1 Cor. 1:5; 2 Cor. 6:6; 8:7; 11:6; 1 Pet. 3:7; 2 Pet. 1:5-6; 3:18
   d. Of salvation - Luke 1:77; 11:52
   e. Spiritual gift of knowledge - 1 Cor. 12:8; 13:2, 8; 14:6
   f. False knowledge - 1 Tim. 6:20
   g. Of idols - 1 Col. 8:1, 7, 10-11

ἐπίγνωσις - This noun with prefixed preposition occurs 20 times.

1. **Source**
   a. God - Eph. 1:17
   b. Law - Rom. 3:20
2. **Content**
   a. God - Rom. 1:28; Col. 1:10; 2:2; 2 Pet. 1:2-3
   b. Christ - Eph. 4:13; 2 Pet. 1:8; 2:20

   c. Spiritual truth - Rom. 10:2; Phil. 1:9; Col. 3:10; 1 Tim. 2:4; 2 Tim. 2:25; 3:7; Titus 1:1; Heb. 10:26
   d. God's will - Col. 1:9
   e. Good things - Philem. 6

γνωστός - This adjective occurs 22 times in the NT.

1. **People Known**
   a. Disciple - John 18:15-16
   b. Friends of Mary and Joseph - Luke 2:44
   c. Friends of Jesus - Luke 23:49
2. **Information Known** - Acts 1:19
3. **Spiritual Truth**
   a. Sermon - Acts 2:14
   b. Miracles
      i. Lame man - Acts 4:10, 16
      ii. Tabitha - Acts 9:42
      iii. Demon possessed - Acts 19:17
   c. Forgiveness of sin - Acts 13:38
   d. Seeking the Lord - Acts 15:18
   e. Christianity - Acts 28:22
   f. Salvation for Gentiles - Acts 28:28
   g. God - Rom. 1:19

οἶδα - This verb occurs 328 times in the NT.

1. **God Knowing**
   a. Father knowing
      i. Our needs - Matt. 6:8, 32
      ii. Us - 2 Cor. 11:11, 31
   b. Christ knowing
      i. God the Father - John 7:27-29; 8:55; 11:42
      ii. About Himself
         1. His origin - John 8:14
         2. His purpose - John 13:1, 3; 18:4; 19:28

3. His own - John 13:17-18
4. His work - John 6:6
5. His name - Rev. 19:12

iii. About people

1. Jews - John 8:37
2. His betrayer - John 13:11
3. Their thoughts, hearts, and motives - Matt. 9:4; 12:25; Luke 6:8; 9:47; 11:17; 6:61, 64
4. Knowledge - John 3:11; 4:22; Matt. 15:12

iv. All about 7 churches - Rev. 2:2, 9, 13, 17; 3:1, 8, 15

c. Holy Spirit knowing - 1 Cor. 2:11-12

2. **Satan and Demons Knowing** - Mark 1:24, 34; Luke 4:41; Rev. 12:12
3. **Man Cognitively Knowing or Not Knowing**

a. General or unspecified (mainly not knowing) - John 1:26; 3:8; 14:4; 16:30; Luke 11:44; 12:39; 13:25, 27

b. Unbelievers knowing or not knowing

i. Mainly Jewish but other unbelievers - Matt. 22:29; Mark 12:14; John 6:42; 7:15; 11:49; 15:21; Acts 16:3

ii. Egyptian and Roman unbelieving leaders - Matt. 27:18, 65; John 19:10; Acts 7:18, 24:22

c. Believers knowing or not knowing

i. Specific individuals knowing - Mark 9:6; John 1:3, 33; 21:15-17; Acts 20:22, 25; Phil. 1:19, 25; 4:12

ii. Knowing certain things - Matt. 7:11; 20:25; 24:36, 42-43; 26:2; John 10:4

iii. Spiritual things believers know - Heb. 8:11; 1 Cor. 2:2; 2 Cor. 5:16; Rom. 3:19; 7:14; 8:22; Gal. 2:16

THE LEXICAL MEANINGS

Bauer, Arndt and Gingrich Lexicon

1. γινώσκω - to know, to learn, to understand, comprehend, perceive, notice, realize, have come to know, know, acknowledge, recognize, sex

2. γνῶσις - knowledge as attribute, Christian knowledge
3. ἐπιγινώσκω - to know, understand, recognize
4. ἐπίγνωσις - knowledge, recognition, a knowledge of the truth
5. γνωστός
    a. Known
        i. Of things
        ii. Of persons - acquaintance, friend, intimate
    b. Capable of being known, intelligible
6. οἶδα - to be cognizant or aware of a fact or a specific piece of information; possess knowledge or information about

THE CONCLUSION

1. Knowledge is similar to truth and could be considered synonymous with truth.
2. Both the OT and NT terms are centered in God Himself.
3. Knowledge includes both of the material and spiritual realms, especially of God.

APPENDIX C

GOD'S SOVEREIGNTY OVER THE PHYSICAL REALM

THE CREATION AND EXPANSION OF SCIENTIFIC STUDIES IN SCRIPTURE

The following passages show God's direct intervention in the creation (the natural realm) throughout the historical periods. They explicitly show supernatural interventions rather than simply God's work of providence.

PRIMEVAL HISTORY (GENESIS 1-11)

THE CREATION OF ALL THINGS

General references to the creation of the universe:

Old Testament
1 Chron. 16:26; 2 Chron. 2:12; Prov. 22:2; Neh. 9:6; Eccl. 11:5; Isa. 40:26, 28; Ps. 89:11-12; 94:9; 95:4-6; 96:5; 102:25; 119:73, 90.

New Testament
Acts 4:24; Rom. 1:20; Eph. 3:9; Col. 1:16; Heb. 1:2, 10; 3:4; 11:3.

Origin of the time frame	Absolute beginning: "In the beginning ..."	1:1
	Creation of all things out of nothing in six literal days	1:1-27 Rom. 4:17
	Establishment of all physical characteristics and constants (science)	1:1-27
Origin of cosmology and astrophysics	Creation of heaven and earth (universe)	1:1
Origin of physics, chemistry, hydrology, gravity, mechanics, thermodynamics, and energy	Earth created as water mass	1:1-2

Origin of communication and language	God speaking things into existence	1:3-27 Ps. 33:6-9
Origin of electromagnetic spectrum and optics	Creation of light	1:3
	Creation of dawn and sunset	Ps. 65:8
Origin of outer space	Creation of the expanse (outer space)	1:6-8
	Stretching of the heavens	Ps. 104:2
Origin of oceanography, geophysics, geology, tectonics, climatology, and orogeny	Separation of waters from land	1:6-7, 9-10
	Establishment of mountains	Ps. 65:7
	Control of seas	Ps. 65:8
	Sea turned to dry land	Ps. 66:6
	Establishment of earth	Ps. 104:5
Origin of biology, botany, and genetics	Creation of all plant life	1:11-13
Origin of astronomy, astrophysics, galactic astronomy, planetary science, and stellar astronomy	Creation of sun, moon, and stars	1:14-19 Ps. 104:19-20
Origin of zoology	Creation of all animal life	1:20-25
Origin of anthropology	Male and female	1:26-27
	Creation of Adam and Eve	Matt. 19:4-5 Mark 10:6-8
Origin of hydrological system	A mist and rivers	2:5-6, 10-14
	Productivity of streams and rivers	Ps. 65:9-13
Origin of metallurgy	Gold, bdellium, onyx	2:11-12
Origin of geography	Eden, Pishon, Gihon, Tigris, Euphrates	2:10-14
Origin of agriculture	Garden cultivation	2:15
Origin of morality and ethics		2:16-17

Origin of medical science	Deep sleep, removal of rib, fashioning of Eve	2:21-22
1st Law of Thermodynamics established	Completion of creation	1:31
Physics and audio	Speaking to Adam	2:15-17

THE FALL OF MANKIND

Judgments of God that directly changed the physical creation.

Herpetology	Cursing of serpent	3:14
Biology, Zoology	Cursing of animal realm	3:14
Anthropology	Pain in childbearing	3:16
Geophysics	Cursing of ground	3:17
Botany	Thorns and thistles	3:18
Physics and the 2nd Law of Thermodynamics	Sweat and toil	3:17, 19
Anthropology	Death imposed	3:19
Sociology	Provision of first clothing	3:21
Resurrection	Rapture of Enoch	5:24 Heb. 11:5

THE UNIVERSAL FLOOD

A judgment of God with direct intervention in many details.

Geophysics and volcanism	Bringing of the Flood	6:7, 13, 17 7:11 Luke 17:26-27
Climatology	40 days of rain	7:11
Zoology	Animals coming to ark	7:9, 15-16
Physics	Closing door of ark	7:16
Biology	Blotting out of all living creatures	7:23
Physics and hydraulics	Rising of tsunami flood waters for 150 days	7:24
Climatology, oceanography, hydraulics, and geophysics	Bringing of wind and subsiding of waters	8:1

	Closing of fountains and floodgates and restraining of rain	8:2
Oceanography and hydraulics	Receding of waters for 150 days	8:3
Zoology	Instilling fear in animals	9:2
Physics and geophysics	Establishing of stable earth with no Flood again	9:9-11, 15-16
Physics and optics	Creation of the rainbow as the sign of the Noahic Covenant	9:12-14, 17
Physics	Deeds of deliverance: Flood, leviathan crushed, light of sun, boundaries of earth, seasons	Ps. 74:12-17

CONFUSION OF LANGUAGES

Another divine judgment through direct intervention.

Origin of multi languages and linguistics	Confusion of language and creation of new ones	11:7-9
Anthropology	Scattering of the peoples on earth	11:7-9

THE PATRIARCHAL HISTORY (GENESIS 12-50, JOB)

THE PATRIARCHS

Medicine	Disease plagues brought on Pharaoh for Abraham	Gen. 12:17
Military	Deliverance of northern armies into Abraham's hand	14:20
Medicine	Sodomites struck with blindness	19:11

Astrophysics, geophysics, and physics	Destruction of Sodom and cities	18:17, 23-33 19:24-25, 28-29 Deut. 29:23 Luke 17:28-29
Medicine	Turning Lot's wife into a pillar of salt	19:26
Conception	of Isaac as promised to Sarah	21:1-2, 19
Prosperity	Wealth given to Abraham	24:35
Guidance	Divine guidance in seeking a wife for Isaac	24:48
Anthropology	Fertility after barrenness for Rebekah	25:21-22
Prosperity	Wealth given Isaac	26:12
Conception	Opening of Leah's womb	29:31-35
Conception	Fruit withheld from womb	30:2
Conception	Giving of sons to Bilhah and Leah	30:6, 17-18, 20
Conception	Rachel remembered to conceive	30:22-24
Medicine	Dislocation of Jacob's hip and blessing after wrestling	32:24-29
Conception	Gift of sons to Joseph in Egypt	41:51-52
Various physical areas	Summary in the formation of the nation of Israel	Ps. 105:1-44

THE BOOK OF JOB

Job probably lived in Patriarchal time.

Physical and medical	Permission for Satan to afflict Job	Job 1:8-12
	Loss of health	2:7-10
Hydrology	Giving of rain/water	5:8
	Bringing waters and clouds	26:8
Orogeny, tectonics, astrophysics, and oceanography	Removing mountains, shaking earth, commanding sun to not shine, trampling waves, making constellations	9:5
Political science	Making nations great, then destroying them	12:23
Astrophysics	Stretching the north over empty space	26:7 Ps. 104:2

	Obscuring moon and earth horizon	26:9-10
	Pillars of heaven trembling	26:11
	Clearing heavens	26:13
Geophysics	Quieting sea	26:12
Creation	Creating Elihu	33:4, 6; 36:3
Climatology		38:9-13
	Storehouse of snow and hail	38:21-23, 25-30
	Precipitation, rain, snow, weather	38:31-37
	Answering Job from a storm	40:6-14; 42:5
Creation	Answering Job with questions from whirlwind	38:1-3
Oceanography	Boundaries of seas	38:4-7
Optics	Controlling dawn	38:12-15
	Deep darkness	38:16-18
	Light	38:19-21, 24
Zoology	Caring for animals	38:39-41
	Births of animals	39:1-4
	Freedom of wild animals, donkey, oxen	39:5-12
	Strength of horses	39:19-25
Ornithology	Feeding ravens	38:41
	Carefree ostrich	39:26-30
	Hawks and eagles up high	39:26-30
Paleontology	Created with no fear	40:15-24
	Leviathan's independence and power	41:1-34
Restoration	of wealth and family for Job	42:10-17

THE FORMATION OF ISRAEL (EXODUS-JUDGES)

THE EXODUS GENERATION

Conception	Blessing of midwives	Ex. 1:20-21
Physics	Combustion of bush without consumption	3:2-4 Mark 12:26
Physics, herpetology	Staff turned into a snake	4:2-4
Medicine	Moses' hand turned leprous and restored to normal	4:6-7
Physics	Physical miraculous signs and judgments for Moses	4:1-8
	for Jews	4:8-9, 17, 30-31
	for Pharaoh	7:3-5, 9-18, 20; 8:1-2, 8-10, 19-21, 24, 29; 9:1-3, 7-8, 13-18, 27-28; 10:1, 3-4; 11:1, 9-10; 12:29-30
	for Egypt	7:21, 24; 8:24; 9:11; 10:2, 6-7, 11:9; 12:29-30
	in Egypt	Ps. 78:11-12, 42-52; 105:25-36
Chemistry and physics	Plague #1: Instruction for Nile turned to blood	Ex. 7:14-21
Herpetology	Plague #2: Instruction for and removal of frogs	8:2-6, 10-14
Entomology	Plague #3: Gnats	8:16-19
Entomology	Plague #4: Infestation of insects	8:20-24, 31
	Removal of insects in Egypt but not in Goshen	8:22-23
Zoology	Plague #5: Destruction of Egyptian livestock	9:1-6
	No livestock in Israel destroyed	9:4, 6-7

Anthropology	Pharaoh's heart spiritually hardened	9:7, 12, 34-35; 10:1-2, 20, 27; 11:9-10; 14:4-9
Chemistry and medicine	Plague #6: Soot causing boils	9:8-11
Climatology	Plague #7: Hail with fire falling from heaven and ceasing	9:18-26, 29-33
	No hail in Goshen	9:26
Entomology	Plague #8: Infestation of locusts and cessation	10:4-6, 12-15, 19
Physics and optics	Plague #9: Darkness in daytime	10:21-23
Anthropology	Plague #10: Death of only firstborn	11:4-8; 12:29-30
	Distinction of plagues effects between Egypt and Israel	11:7; 12:23, 27
Hydrodynamics, hydraulics, physics, climatology	Parting of sea	14:21-22, 29-31; Lev. 26:13 1 Cor. 10:1-2
	Closing of waters	Ex. 14:26-28
	Praise song for God's deliverance	15:1, 4-5, 8, 10, 12, 19, 21
	Retelling of the Exodus story to Jethro	18:8-11
	Retelling of Exodus story to Israel	Num. 20:16; 23:22; 24:8
	Reminders of the Exodus: Old Testament Deut. 4:34, 37; 5:6, 15; 6:12, 21-23; 7:18-19; 8:14; 11:2-4; 26:5-9; 29:2-3, 16, 25; 34:11-12; Josh. 2:9-11; 4:23; 9:9; 24:5-7, 17; Judg. 2:1, 12; 6:8-9; 1 Sam. 4:8; 6:6; 8:8; 10:18; 12:6-8; 2 Sam. 7:23; 1 Kings 9:9; 2 Kings 17:7, 36; 21:15; 2 Chron. 7:22; Neh. 9:11; Ps. 78:13, 53; 82:10; 105:37-38; 106:7-12. New Testament Acts 7:7, 25; 13:17.	

THE SINAI AND WILDERNESS EXPERIENCE

Chemistry	Changing of bitter water to sweet	Ex. 15:23-25
Medicine	Preventing Egyptian diseases	15:26
Nutrition	Provision of manna for 40 years	16:12-30, 35; John 6:31-33; 1 Cor. 10:3
	Provision of manna and quail	Num. 11:6-32
Sustenance	Provision of water at Meribah	Ex. 17:5-7; Deut. 6:16; 1 Cor. 10:4-5
Climatology, chemistry, tectonics	Thunder, lightning, clouds, trumpet sound, smoke, quaking of mountain	Ex. 19:16-19; 20:18, 22
Physics	Writing on two tablets with His finger	31:18; 32:16; 34:1-4, 28
Optics, physics	Glowing of Moses' skin	34:29-34; 2 Cor. 3:7, 13
Medicine	Miriam's leprosy	Num. 12:10-15
Geophysics, physics	Earth splitting and closing for judgment	16:31-33
Physics	Fire consuming 250 men	16:35
Medicine	Plague killing 14,700	16:43-50
Botany	Rapid (next day) sprouting of rod with almonds	17:8
Physics	Water from rock	20:9-11
Zoology	Angel blocking Balaam's donkey, donkey speaking	22:21-30
Optics	Balaam's eyes opened to see angel	22:31-35
Anthropology	God's burial of Moses	34:5-6
Reminders of wilderness		Deut. 29:4-8
	Geophysics: earth quaked in wilderness and Sinai	Ps. 68:7
	Physics: Miracles in wilderness	Ps. 78:14-16, 23-27; 105:39-43

THE CONQUEST AND JUDGES PERIOD

Hydrodynamics	Stopping waters of Jordan in one heap	Josh. 3:13-17; 4:23-24
	Waters returned	4:18
Hydrology	Drying ground immediately for crossing	5:1
Nutrition	Last of manna	5:12
Manufacturing	Stone cutting	8:31
Climatology	Large hail-stones falling from heaven	10:10-11, 19
Astrophysics	Sun standing still at noon for whole day	10:12-15
Physics	Sign of fleece	Judg. 6:36-40
Super human strength	Holy Spirit freeing Samson, killing thousands	15:14, 18
	Lord leaving Samuel and thus his strength leaving him	16:20
	Toppling house, killing thousands	16:28-31
Social/political	Hannah's prayer: shattering mighty, killing, and making alive, making poor and rich, preserving godly, and silencing the wicked	1 Sam. 2:4, 6-10
Conception	Three sons for Hannah	2:21
Hydrology	Provision of thunder and rain	12:16-18
Conquest	of Canaanites	Ps. 105:44; Acts 13:19

THE KINGDOM PERIOD (1 SAMUEL-2 KINGS, PSALMS)

THE UNITED KINGDOM

Political science	Raising David	Acts 13:22
Physics	Fire from heaven consuming sacrifices and glory filling temple	2 Chron. 7:1-3

THE PSALMS

Geophysics	Deliverance with earthquake, smoke, fire, and hail	Ps. 18:7-15
	God sitting as king at Flood	29:10
	Worldwide flood, creation of mountains and valleys	104:3-9
	Earth trembling and mountains smoking	104:32
Astrophysics	Heavens declare the glory of God	19:1-3
Physics, botany	Voice breaking cedars of Lebanon	29:5
Physics	Flames of fire shaking wilderness	29:7-8
Zoology	Making deer calve	29:9
	Caring for animals	38:39-41
	Quenching thirst of beasts and wild donkeys	104:10-11
	Grass and vegetation for food and food for lions	104:14-15, 21
Anthropology	Giving strength to His people	29:11
Climatology	Precipitation, rain, and snow	104:3-4, 13
Botany	Grass and begetation as food for animals and man	104:14-15
Ichthyology	Feeding of sea creatures	104:25-28
Paleontology	Leviathan's independence and power	104:26

THE DIVIDED AND SURVIVING KINGDOMS

Medicine	Jeroboam's hand dried up and restored	1 Kings 13:4, 6
Physics	Altar split apart	13:5

Miracles and Prophecies of Elijah

Nutrition	Provision by ravens	1 Kings 17:2-6
	Expanding of food for widow	17:15-16
Physics	Raising of son of Gentile widow	17:17-24

	Exposing of 450 prophets with fire from heaven	18:21-40
Hydrology	Bringing rain in famine	18:43-45; Jam. 5:17
	Famine in Elijah's day	Luke 4:25-26
Physics	Calling fire down from heaven to consume two groups	2 Kings 1:10-14
Hydraulics	Striking water of Jordan and splitting for crossing	2:8
Physics, climatology	Taken up to Heaven in whirlwind	2:11
Predictions	Calamity on Jehoram for his evil	2 Chron. 21:12-15
	Philistines and Ahab raised up against Judah	21:16-17

Miracles and Prophecies of Elisha

Hydraulics	Striking water of Jordan and splitting for crossing	2 Kings 2:14
Chemistry	Purifying water	2:19-22
Zoology, death	Cursing 42 lads with bears tearing them up	2:23-24
Nutrition	Increasing of oil for widow	4:1-6
Medicine, anthropology	Bearing of son for Shunammite widow	4:13-17
Medicine, physics	Raising of son of Shunammite widow	4:31-37
Nutrition	Deadly stew not dangerous in famine	4:38-41
Medicine	Curing of leprosy of Naaman, king of Aram	5:8-14; Luke 4:27
	Judging Gehazi and descendants with leprosy	5:26-29
Physics, hydraulics	Making iron axe head float	6:5-6
Optics, physics	Opening of Elisha's servant's eyes to see angelic army	6:17
Judgment	Striking Aramean army with blindness	6:18

Optics, physics	Opening eyes to see in midst of Samaria	6:20
Prophecies	Victory over Moabites	2 Kings 3:14-20
	Cost of goods, not famine cost	7:1-2, 16-19
	Famine from Lord	8:1
	Ben-hadad king of Aram recovering from sickness but dying	8:10
	Hazael as king of Aram doing evil to Israel	8:12-13, 15
	Destruction of house of Ahab	9:5-10
	Fulfillment	10:10-11, 17
	Fulfillment of Jezebel's death	9:36-37
	Victory over Arameans	13:17
Medicine	Revival of man thrown in Elisha's grave	13:21
Physics	Causing sound of great army to scare Arameans to flee	7:6-7
Events	Limiting of Israel and Jehu	10:30-33
Judgment	Giving Israel over to king Hazael, king of Aram	13:3
	Raising up deliverer	13:5
Events	Seeing affliction of Israel under Jeroboam II and saving them	14:25-27
Medicine	Hezekiah's recovery from sickness	20:7
Astrophysics	Shadow on steps going backward ten steps	20:8-11; Isa. 38:7-8
Events	Kings scattered	Ps. 68:12, 14
	People who love war scattered	68:30

THE FIRST CENTURY (NEW TESTAMENT)

THE LIFE OF CHRIST (GOSPELS)

Creation of all		John 1:3
Ultimate intervention	Incarnation	1:1-14
Medicine	Virgin conception	Matt. 1:18-23
	Baby leaping in Elizabeth's womb	Luke 1:41, 44
	Zacharias becoming mute	1:20
	Zacharias' healing from muteness	1:64
Astrophysics	Star appearing to magi	Matt. 2:2, 7, 9-10
	Heavens opening and Holy Spirit descending from heaven	Mark 1:9-11

Various Miracles of Christ

General Miracles	Mark 6:2, 14, 55-56; Luke 9:11; 13:32; 19:37; John 2:23; 6:2; Acts 22; Heb. 2:3-4	
	Son = Father = Father's work	John 14:9-11
	Many others unmentioned	20:30
Physics, climatology	Stilling of storm	Matt. 8:23-27; Mark 4:37-41; Luke 8:23-25
Chemistry	Turning water to wine	John 2:8-11
Ichthyology, physics	Catch of fish	Luke 5:4-7, 9
Medicine, physics	Raising from dead of widow's son from Nain	7:12-16
	of synagogue official's daughter	Matt. 9:18, 24-25; Mark 5:22-24, 35-42; Luke 8:41-42, 49-56
	Lazarus	John 11:40-44; 12:9, 17-18

Nutrition	Feeding of 5,000	Matt. 14:19-21; Mark 6:41-44; 8:19; Luke 9:16-17; John 6:10-14
	Feeding of 4,000	Matt. 15:35-38; Mark 8:6-9, 20; Luke 15
Physics	Jesus and Peter walking on water, storm stilled	Matt. 14:25-26, 29; Mark 6:48-51; John 6:19-21
Climatology	Wind stopped	Matt. 14:32
Ichthyology, events	Coin in fish's mouth	17:27
Botany	Withering of fig tree	Matt. 21:19-20; Mark 11:14, 20-21; Luke 21
Physics, Ichthyology	Catch of fish on right side of boat	John 21:6-9

Medical Miracles and Healings

General sickness and disease		Matt. 4:23-24; 9:35; 11:5; 14:14, 35-36; 15:30; 19:2; Mark 1:32-34; 3:10; 6:5; Luke 4:40-41; 5:15; 6:18-19; Acts 10:38-39
Report to John	Blind, lame, lepers, deaf, dead raised	Matt. 11:5; Luke 7:21-22
Leprosy		Matt. 8:3; 17:12-14; Mark 1:40-42; Luke 5:12-13; 17:12-14
Paralysis	of Centurion's son	Matt. 8:6-7, 13; Luke 7:2-10
	of man and sins forgiven	Matt. 9:2-7; Mark 2:2-11; Luke 5:18-26
	Paralyzed	John 5:5-9, 15
	Lame and crippled	Matt. 15:30-31; 21:14; Luke 13:11

Healing of royal official's son		John 4:50-54
Fever of Peter's mother-in-law		Matt. 8:14-15; Mark 1:30-31; Luke 4:38-39
Demon possession	Legion and casting into swine at Gadarenes	Matt. 8:28-33; Mark 5:1-17; Luke 8:26-36
	Others	Matt. 8:16; 9:32-33; 12:22; 15:22-28 (Mark 7:25-30; Luke 15); 17:14-18 (Mark 9:17-27; Matt. 9:42-43; 11:14, 13:32); Mark 1:23-26, 32-34, 39 (Luke 4:33-36); 3:11; 5:2-16, 19-20
	Accused of casting out demons by Satan	Matt. 12:24-30; Mark 3:22-30
Hemorrhage of woman for twelve years		Matt. 9:20-22 (Mark 5:25-34; Luke 8:43-48)
Blindness		Matt. 9:27-30; 12:22; 15:30-31; 20:30-34 (Mark 10:46-59; Luke 18:35-43); 21:14; Mark 8:22-25; John 9:1-7, 39
Muteness		Matt. 9:32-33; 12:22; 15:30-31 (Mark 7:32-37)
Deafness		Mark 7:32-37
Deformed hand of man		Matt. 12:10-13, 15 (Mark 3:1-5; Luke 6:8-10)
Lunatic		Matt. 17:14-18
Dropsy		Luke 14:1-5
Restored ear of slave of high priest		22:51

Medicine	Delegating powers of healing and exorcisms to disciples	Matt. 10:1, 8; Mark 6:7, 13; Luke 9:1-2; 10:9, 12, 19
	Delegating of signs including demons tongues, serpents, drinking poison, healing	*Mark 16:17-18
Omniscience	Aware of men's reasoning	Mark 2:8
Physics, optics, audiology	Transfiguration with Moses and Elijah, voice from heaven	Matt. 17:2-3, 5; Mark 9:1-7; Luke 9:29-32
Biology, zoology	Provision for bird, lilies, ravens, and believers	Matt. 6:26, 28-30; Luke 12:24-30
Physics, optics	Darkness from noon to 3:00 in all the land	Matt. 27:45; Mark 15:33; Luke 23:44-45
Physics, geophysics	Veil torn from top to bottom and earthquake	Matt. 27:51, 54; Mark 15:38; Luke 23:44-45
Physics	Tombs opened and saints raised and appearances	Matt. 27:52-53
Geophysics	Earthquake and angel rolling stone away	28:2-4
Sovereign power	Authority over all men	John 17:2
Omnipotence	Ability to supply all needs	2 Cor. 9:8-10
Physics: Resurrections		
	of Christ	Matt. 28:5-6; 1 Cor. 15:4; John 10:17-18; Heb. 11:19; 12:2; 13:20; 1 Pet. 1:21
	of believers	Luke 14:14; John 5:21, 25-29; 6:40, 44; 11:23-26
Christ's resurrection appearances	for 40 days	Acts 1:3

	Mary touching His feet	Matt. 28:9-10; Mark 16:9; Luke 24:1-10 (angel); John 20:11, 14-18
	Two Emmaeus travelers	*Mark 16:12; Luke 24:13-31
	10 disciples	John 20:19-23
	11 disciples while eating with Thomas	*Mark 16:14; John 20:26-29
	to Cephas and 11 disciples	1 Cor. 15:5; Luke 24:34
	to disciples at Sea of Galilee	Matt. 28:16-20; Luke 24:36-43; John 21:1-14
	to 500	Matt. 15:6
	to James and all apostles	Matt. 15:7
	to Paul	Matt. 15:8
Restoration of all: Elijah		Mark 9:12-13
All things possible		Mark 10:27
Physics	Ascension to right hand	*Mark 16:19; Luke 24:51; Acts 1:2, 9-11, 22
	Exaltation	2:33-35

*Textual problems

THE EARLY CHURCH

Miracles in general		Acts 2:43; 4:30; 5:12; Gal. 3:5
Anthroplogy, linguistics	Tongues, outpouring of Holy Spirit, and supernatural signs	Acts 2:1-4, 17-21; 1:4
Resurrection	Major theme of preaching in early church	2:24, 31-32; 3:15; 4:10; 5:30-31; 10:40-41; 13:30-31, 33-37; 17:3, 31-32

Miracles of Peter	Medicine: Healing of lame man	3:2-10, 12, 16; 4:10, 14, 16
	Medicine: Healings, not specified	5:15-16
	Medicine: Healing of Aeneas	9:32-34
	Physics: Raising of Dorcas	9:36-42
Physics	Angel opening prison gate to release apostles	5:19-20
Miracles of Stephen		6:8
Miracles of Philip	Casting out demons and healings	8:7, 13
Dynamics, physics	Holy Spirit transporting Philip to Azostos	8:39-40
Medicine	Blinding of Saul and sight restoration	9:3-9, 17-18
Miraculous power	of Holy Spirit with Cornelius family	10:44-47
Miracles of Paul	in general with Barnabas	15:12; 19:11-12
	Medicine: Healing of lame man at Lystra	14:8-10
	Medicine: Casting out of demon from slave girl at Philippi	16:16-18
	Medicine: Restoring back to life of Eutychus	20:9-12
Creation	Christ as creator	Col. 1:16
	Christ as upholder of the universe	1:17
Biology, physics	Giving life to dead	Rom. 4:17
	Jesus' resurrection	4:24; 8:11
	Our resurrection	6:4-5, 9; Phil. 3:11, 21; Col. 3:4; 1 Thess. 1:10
Physics	Second coming and rapture	1 Thess. 1:10; 2:19; 3:13; 4:13-18; 5:2-3
	Second law of thermodynamics, future freedom from it	Rom. 8:20-23

Power	All authority and consequences from God	13:1-4
Medicine	Gifts of healing and miracles	1 Cor. 12:9-10, 28-30
Resurrection	of believers and defense of historicity	1 Cor. 15:12-34; 2 Cor. 1:9
	Description of resurrection	1 Cor. 15:35-58
Events	Physical deliverance	2 Cor. 1:10
Revelation	All wisdom and knowledge hidden in Him	Col. 2:3
Sovereignty	As head over all rule and authority	2:10
Zoology	Deliverance of Paul out of lion's mouth	2 Tim. 4:17
Physics	Christ upholding all things	Heb. 1:3
	Second law of thermodynamics described	1:11

APPENDIX D

GOD'S DIRECT INTERVENTIONS IN EVENTS OR HISTORY

The following passages illustrate God's direct, supernatural intervention in historical events or history itself, the invisible spiritual realm affecting the material realm.

THE VISIBLE APPEARANCES AND VISIONS OF GOD

THE APPEARANCES OF GOD HIMSELF

(Based on John 1:18, these may be manifestations of the pre-incarnate Son.)

Audible and visible	God speaking and hearing in the Garden of Eden	Gen. 3:8-19, 22-24
	Discussion with Cain	4:6-7, 9-15
	Explanation for Noah	6:13-7:4
Appearances in glory	to Abram	15:12-18; Acts 7:2
Appearances to Patriarchs	Isaac	26:2-5, 24
	Rebekah	25:23
	Jacob	28:12-16; 31:3, 11-13; 32:22-30; 35:1, 9-13
Physics, optics	Speaking to Moses from the burning bush	Ex. 3:2-4
	Revelation of glory of the Lord on mountain, fire	24:15-18; Deut. 4:12, 15; Acts 7:38

	Revelation of the Lord's glory filling tabernacle, cloud, fire	40:34-38
	Appearing in the Holy of Holies	16:2; Num. 14:10-12, 14
	Glory appeared as sacrifice consumed	Lev. 9:23-24
	Appearance in wilderness	Num. 16:19; 20:6
Audible and visible	Cloud and fire in wilderness	9:15-23; 10:11-12; 12:4-9; 16:42; Deut. 4:33, 36; 5:23-26
	Appearance and instruction to Balaam	Num. 22:9-12, 20
	Speaking face to face from fire	Deut. 5:4-5, 22
	Calling of Samuel and prediction about Eli's sons	1 Sam. 3:4-18
	Appearance at Shiloh	3:21
Audible	Appearance to Solomon with promises at Gibeon	1 Kings 3:5-9; 2 Chron. 1:7-12
	Assuring of answered prayer and dynasty for obedience	2 Chron. 7:12
	Request of wisdom granted	1 Kings 3:6-14, 28; 4:29-30; 5:12
	Promise and warning to Solomon	9:1-9; 11:9
Visible	Vision for Isaiah of Lord on His throne	Is. 6:1-5
	Paul caught up to 3rd heaven, receiving revelation	2 Cor. 12:3-4

THE APPEARANCES OF SHEKINAH GLORY

(These may be manifestations of the pre-incarnate Son.)

Garden of Eden	Presence in Garden and flaming sword	Gen. 3:8, 23-24
Abrahamic Covenant	Flaming torch that passed between sacrificial pieces	15:12-18

Burning Bush	Did not consume the bush	Ex. 3:1-5; 13:21-22; 14:19-20, 24; 16:6-12
Exodus	Pillar of cloud by day and the pillar of fire by night	Book of Exodus
Mount Sinai	10 Commandments written by finger of God, thunder, lightning, thick cloud	Ex. 19:16-20; 24:15-18; Deut. 5:22-27
Meeting with Moses	Afterglow of Moses' face	Ex. 33:17-23; 34:5-9, 29-35; 29:42-46; 40:34-38
Tabernacle and Ark of Covenant	Glory-cloud presence often	Lev. 9; 1 Sam. 4:21-22
Judgment for sin and disobedience		Num. 13:30-14:45; 16:1-50; 20:5-13
Temple		1 Kings 8:10-11; 2 Chron. 5:2-7:3
Departure		Ez. 1:28; 3:12, 23; 8:3-4; 9:3; 10:4, 18-19; 11:22-23
2nd Temple	Not present but promised in future	Hag. 2:3, 9

THE APPEARANCES OF THE ANGEL OF THE LORD

(These may be manifestations of the pre-incarnate Son.)

Audible and visible	to Gideon	Judg. 6:11-27
	to Manoah's wife and revelation of Samson's birth	13:3-7
	to both Manoah and wife, including wonders	13:8-23
	to Elijah in wilderness, provides food	1 Kings 19:4-8
Audible	Promise at Bochim that Canaanites would be a snare	Josh. 2:1-4

	On Pestilence upon Israel with 700 killed	2 Sam. 24:15-17
	Instruction to Elijah about death of messengers of king	2 Kings 1:3-4, 6
	Instruction to Elijah to predict death of king Ahaziah	1:15-16
Military	Struck 185,000 Assyrians	19:35; Is. 37:36-38
Birth of Jesus	to the Shepherds	Luke 2:8-9

THE APPEARANCES OF JESUS

Anthropology	Incarnation	4 Gospels; John 1:1-14
Audible and visible	Descent of Holy Spirit on Jesus and Father's voice	Matt. 3:16-17; Mark 1:10-11; Luke 3:21-22; 9:35; John 12:28
Physics, optics	Transfiguration to 3 disciples	Matt. 17:1-8; Mark 9:2-8; Luke 9:28-36; Heb. 1:1-3; 2 Pet. 1:16-18; Rev. 1:12-16
Optics	Stephen observed glory and Son at right hand	Acts 7:55-56
Physics	Cloven tongues of fire and blinding light on Paul	Acts 2:1-3; 9:3-8; 22:6-11; 26:13-18
Audible and visible	Conversion of Paul on road to Damascus, blinded	Acts 9:3-9, 27; Gal. 1:15
Visible	Vision for Ananias	Acts 9:10-16
Audible and visible	Vision for Paul to go to Macedonia	16:9-10
	Vision and instruction for Paul to stay in Corinth	18:9-10
Vision of Jesus Christ	to John at Patmos	Rev. 1:12-16
Tribulation bowl judgments		Rev. 15:8

Second Coming		Matt. 16:27; 24:30; Mark 13:26; Luke 21:27
Millennium		Ez. 43:1-7; 44:1-2; Is. 4:5-6; 11:10; 35:1-2; 40:5; 58:8-9; 60:1-3; Zech. 2:4-5; 11:10
Eternal State		Rev. 21:1-3, 10-11, 23-24
Resurrection appearances (40 days)		Acts 1:3
	Mary touching His feet	Matt. 28:9-10; Mark 16:9; Luke 24:1-10 (angel); John 20:11, 14-18
	Two Emmaeus travelers	Mark 16:12; Luke 24:24:13-31
	Ten disciples	John 20:19-23
	Eleven disciples while eating with Thomas	Mark 16:14; John 20:26-29
	to Cephas and twelve disciples	1 Cor. 15:5; Luke 24:34
	to disciples at Sea of Galilee	Matt. 28:16-20; Luke 24:36-43; John 21:1-14
	to 500	Matt. 15:6
	to James and all apostles	Matt. 15:7
	to Paul	Matt. 15:8

THE APPEARANCES OF ANGELS

Appearance	to Moses	Acts 7:2, 30-34
Revelation	Law ordained by angels	Gal. 3:9
Audible and Visible	to Joshue near Jericho	Josh. 5:13-15

	to David between heaven and earth, directing him not to build temple but Solomon would	1 Chron. 21:14
	to Zachariah, prediction of John's birth and ministry	Luke 1:11-17
	Gabriel sent to Mary about birth and Jesus' ministry	Luke 1:26-28
	to shepherds with announcement	Luke 2:9
	Explaining ascension to disciples	Acts 1:11
	Vision for Cornelius	Acts 10:30-33; 11:13-14
	Vision of sheets with animals to Peter	Acts 10:9-17, 19; 11:5-12
Optics, Physics	Elisha's servant's eyes opened to see angelic army	2 Kings 6:17
	Angel like lightning, white garments	Matt. 28:2-4; Mark 16:5-7
Military	Destruction of Sennacherib of Assyria by angel	2 Chron. 32:20-22
Audible	Instruction of angel, messages of illumination	Isa. 6:6
	Angel speaking to Philip	Acts 8:26, 29
Medical	Angel strengthening Jesus	Luke 22:43
	Angel stirring waters and healing	John 5:3-4
Visible	Angel releasing Peter from jail	Acts 12:6-12, 17

MILITARY VICTORIES IN BATTLE WITH GOD RECEIVING CREDIT

IN THE PENTATEUCH

Victory over four kings of east	Gen. 14:20
Defeat of Amelek	Ex. 17:8-14
Defeat of Canaanites at Hormak	Num. 21:1-3; Deut. 4:38

IN THE CONQUEST PERIOD

Defeat of Sihon from Arnon to Jabbok		Num. 21:21-26
	Defeat remembered	Deut. 2:24-37; Josh 2:10
Defeat of Og king of Bashan from Arnon to Mt. Hermon		Num. 21:34-35
	Defeat remembered	Deut. 3:1-11; Josh 2:10; 9:10
Moses anticipating Lord's conquest		Num. 32:20-23
Giving of land		Josh. 1:2-4, 11, 13, 15; 2:14, 24; 3:10
Conquest of Jericho		6:16, 20, 27
Conquest of Ai		8:18-29
Victory by swords and miracles over kings of Jerusalem, Hebron, Jarmuth, Lachish, Eglon		10:10-11, 19
Example of God's victory		10:24-25
Victory over cities	Makkedah	10:28
	Libnah	10:29-30
	Lachish	10:31-32
	Gezer	10:33
	Hebron	10:36-37
	Debir	10:38-39
Victory over south regions	hill country, Negev, Shephelah	10:40-42
Victory over: Hazor, Madon, Shimron, Achshaph, north hill country, Arabah		11:7-9, 10-15
Summary of conquest of Moses east of Jordan		12:1-6
Summary of conquest of Joshua west of Jordan		12:7-24
Fulfillment of promises of land		21:43-45
Rest given from enemies		23:1, 3, 10, 14
Conquest remembered		24:8-13, 18

IN THE JUDGES PERIOD

Victory over Canaanites and Perizzites at Bezek		Judg. 1:2-7
Victory over Hill country, Negev, and lowlands		1:9-10, 19
Victory over cities	Jerusalem	1:8
	Hebron	1:10, 20
	Debir	1:11
	Zephath	1:17
	Gaza and Ashkelon and Ekron	1:18
General raising of judges as deliverers		2:16-18
Deliverance of Othniel (40 years)		3:9-11
Deliverance of Ehud (8 years)		3:15, 30
Deliverance of Deborah and Barak (40 years)		4:4, 6, 14-16
Victory over Jabin		4:23-24
Deliverance of Gideon (40 years)		6:11-27; 8:28
Victory over Midian		7:22-25
Victory over sons of Ammon by Jephthah		11:29-33; 12:3
Victory over Benjamin		20:35; 21:15
Victory over Philistines by confusion		Sam. 7:10-11, 13
Victory over Ammonites		Judg. 11:11-13
Deliverance in general in time of Judges		12:11

IN THE KINGDOM PERIOD

Victory and deliverance of people of Keilah	1 Sam. 23:5
Victory over Amalekites: two wives recover what they captured	30:17-21, 23
Victory over Philistines at Baal-Perazim	2 Sam. 5:19-20
Victories from Geba to Gezer	5:22-25
David's rest from enemies	7:1
David's help in all victories	8:6, 14

Victory of Judah and Abijah over Israel due to trust	2 Chron. 13:8-20
Success due to seeking of Lord	14:7
Victory over Ethiopians by Abijah	14:9-15
Victory over Aramean multitude	1 Kings 20:20-21
Victory over Ammon, Moab, and Mt. Seir without fighting, destroying slaves	2 Chron. 20:22-23
Victory over Moabites as Elisha predicted	2 Kings 3:24-25
Angel of Lord striking 185,000 Assyrians	19:35
Reminder of nations defeated in past	21:9
Victory and destruction of Sennacherib of Assyria by angel	2 Chron. 32:20-22
Reminder of Lord destroying nations around Israel	33:9

GOD'S INTERVENTION IN OUTCOME OF EVENTS

GOD'S DIRECT INVOLVEMENT IN EVENTS

Creation of heaven and earth in six days	Gen. 1
Worldwide Flood	6:7, 13, 17; Luke 17:26-27
Confusing of languages and scattering of peoples	Gen. 11:7-9
Sodom and cities destroyed	Gen. 18:17, 23-33; 19:24-25, 28-29; Deut. 29:23; Luke 17:28-29
Brought Abraham and made covenant	Neh. 9:7-8
Sovereign display to Pharaoh	Ex. 9:13-17; Rom. 9:17, 23
Hardening of Egyptians	Ex. 14:17-18
Confusing of Egyptians	14:23-25
Miraculous signs in Egypt	Neh. 9:9-10
Leading and providing for Israel in wilderness	9:12, 14-21
Law given to children of Israel	9:13

Conquest of land	9:22-25
Providing deliverers (judges) for nation	9:26-29
Admonishing of nation through prophets	9:30-31
Captivity and slavery of nation	9:32-38
No direct mention of intervention but implied providence	Book of Esther
Death of Naomi's husband and two sons	Ruth 1:20-21
Closing of Hannah's womb and then Hannah remembered	1 Sam 1:5-6, 19-20, 27
Raising up of faithful priests	2:35-36
Knocking Dagon down two times and cutting hands off	5:3-5, 7
Ashdodits and Gathites ravaged with tumors	5:6-7, 9
Deadly confusion in Ekron over ark of covenant	5:11-12
Saul's transformation by Holy Spirit	10:6-7, 9-11
Holy Spirit on Saul and dread of Lord on all Israel	11:6-7
King over Israel	12:13
Saul's rejection as king for partial obedience	15:10-23, 28-29
Selection and anointing of David	16:1-13
Sending of evil spirit on Saul	19:9
Holy Spirit on Saul and his messengers, causing prophesying	19:19-24
David escaping Saul by answered prayer	23:10-12
David escaping by sound sleep of Saul and men	26:12
David growing greater since the Lord was with him	2 Sam. 5:10, 12
David's praises for promise	7:18-29
David's establishment as king over Israel and kingdom's exaltation	1 Chron. 14:2
Nations' fear of David	14:17
Sheba's acknowledgement of God's establishment of Solomon	2 Chron. 9:8, 23
Success due to hand of Lord	Ezra 7:6, 9, 28; Neh. 2:8, 20; 6:16

Raising up of cyrus of Persia, acknowledgement of God, God's commission of temple building		2 Chron. 36:22-23; Ezra 1:1-4
Lord's turning of king's heart to complete temple		Ezra 6:22
Setting eternity in hearts of all men		Eccl. 3:11
Sovereign over all nations		Ps. 33:10-17
	Power belongs to God	62:11
Rescue from famine		33:19-20; Prov. 10:3
Driving out nations and planting Israel in land		Ps. 44:2-3
Raising His voice to melt earth, causing desolations, making and ending wars, subduing nations, reigning over nations		46:6, 8-9; 47:3, 7-9
Mary praising His bringing down rulers and exalting the humble		Luke 1:52
	Reversing fortunes of poor and rich, helping Israel	1:53-54
Orchestrating outcomes in general		Rom. 10:18-20; 11:8-10

DIVINE JUDGMENT

GENERAL REFERENCES

1 Chron. 16:33

IN GENESIS

Fall of man	for 1st sin (see above)	Gen. 3:14-19
Curse of Cain	no fertility, wandering, sign for protection	4:11-15
Worldwide Flood	for depravity and evil (see above)	6:7, 13, 17; 7:11; Luke 17:26-27
Confusion of language and scattering	for organized rebellion at Babel	11:7-9
Sodom and cities destroyed		18:17, 23-33; 19:24-25, 28-29; Deut. 29:23; Luke 17:28-29

IN THE PENTATEUCH

Death of people	Er (1st born of Judah)	Gen. 38:8, 10; Num. 26:19
	Israel for idolatry	Ex. 32:30-35
Death of priests	Nadab and Abihu with fire for offering strange fire	Lev. 10:1-8
Physics	Fire judfment for complaining at Taberah	Num. 11:1-3
Medical	Plague for greed	11:33-34
	Leprosy in Miriam	12:10-15; Deut. 24:9
Death of people	Death of ten spies	Num. 14:36-38; 1 Cor. 10:3-5
Geophysics	Earth splitting and closing over Korah and family	Num. 16:20-34
Physics	Fire consuming 250 men of Israel	16:35
Death of people	Moses disobeying, denied going into land	20:7-13; Deut. 3:23-28
	Execution of judges who joined to Baal of Peor	Num. 25:3-5
	24,000 dying	25:8-9; 26:65
	Korah	26:9-10
Physics and deaths	Reminder of 40-year wilderness judgment	Num. 32:9-13; Deut. 11:5-7
	Judgment of Israel and Canaanites	Ps. 106:13-33; Acts 7:42-43, 45

IN THE CONQUEST AND JUDGES PERIOD

Death of people	for violating ban, stoning of Achan/ family	Josh. 7:1, 4-5, 11-15, 24-26
Military	General plunder of enemies for disobedience	Judg. 2:14-15, 20-23

	for failure to eliminate Canaanites, oppressors in Judges	Ps. 106:34-46
	8 year domination by Cush-rishathaim of Mesopotamia	3:8
	Israel domination by Eglon of Moab (18 years)	3:12-14
	Israel domination by Jabin of Hazor (20 years)	4:2
	Israel domination by Midian (7 years)	6:1-2
	Evil spirit between Abimelech and men of Shechem	9:23
Death	Abimelech for wickedness	9:56-57
Military	Israel domination by Philistines and Annon (8 years)	10:7-8
	50,070 casualties at Beth-shemesh, looking into ark	1 Sam. 6:19
Social	Israel judgment for rejecting God as king for Saul	10:17-19
Individual	Holy Spirit departing Saul and evil spirit terrorizing him	1 Sam. 16:14-15; 18:10, 12
Military	Judgment of nations	Acts 7:8

IN THE KINGDOM PERIOD

Death	Saul	1 Chron. 10:14
Death and misery	in wilderness	Ps. 78:21-22, 32-33
	of captivity	78:59-64
Death	Nabal	1 Sam. 25:38

	Uzzah for taking hold of ark	2 Sam. 6:6-7
	David's son with Bathsheba	12:15
Plague	Pestilence upon Israel with 700 killed	24:15-17
	Plague on Israel limited	24:16, 25
	David's census results in 70,000 deaths	1 Chron. 21:7, 14
Military	Hadad (14) and Rezon (23) against Solomon	1 Kings 11:14, 23
	Shishak of Egypt invades Israel to Jerusalem	2 Chron. 12:2-7
	Deliverance (Rehoboam) after repentance but discipline	12:7-12
Medical	Jeroboam's hand dried up and restored	1 Kings 13:4, 6
Death	Prophet by lion for disobedience	13:20-25
	Ahab as predicted	22:34-38
Calamity	Azariah's interpretation	2 Chron. 15:1-6
	The blessing of Asa's reforms	15:14-15
Death	Death of king Ahaziah as predicted by Elijah	2 Kings 1:15-16; 2 Chron. 22:7-9
Medical	Uzziah's leprosy after unfaithfulness	2 Chron. 26:16-21

IN THE KINGDOM DESTRUCTION PERIOD

Political	Removal of Jeroboam's kingdom because of evil	1 Kings 14:6-16
Military	Philistines and Ahab against Jehoram and Judah	2 Chron. 21:16-17
	Death of Jehoram as predicted	21:18-19

Medical	Elisha praying for Aramean army with blindness	2 Kings 6:18
Political	Prediciton of destruction of house of Ahab	9:5-10
	Wrath for idolatry before Joash	2 Chron. 24:18
Death	Prophet predicts judgment on Amaziah for idolatry	25:14-16
Military	Ahaz's captivity to king of Aram for idolatry	28:5-10
	Ahaz causing Judah's humiliation by Philistines, Assyria	28:16-21
	Wrath because of Judah's evil	29:8-10
	Rezin king of Aram and Pekah of Israel against Judah	2 Kings 15:37; 16:5-6
	Captivity of Israel by Shalmaneser of Assyria	17:7-23; 1 Chron. 6:15
Death	Lions and death from Babylon on Samaria	17:25
Military	Invading enemies' judgment	24:1-3
	Manasseh and Judah captivity to Babylon	2 Chron. 33:10-11
	Babylonian captivity	2 Kings 24:10-16, 20; 1 Chron. 6:15
	due to evil	Ezra 9:7, 13
	due to evil and the words of Jeremiah	2 Chron. 36:13-21
Wrath	Josiah and Huldah attribute wrath to disobedience to Word	24:21, 23-25
Military	Babylonian destruction by God's wrath	Ezra 5:12
Climatology	Withholding of rain	Is. 5:6

FINAL JUDGMENT

Of nations	Ps. 2:4-9
Babylon	Is. 13:1-14:23; 21:1-6
Assyria	14:24-27
Philistia	14:28-32
Moab	14:15-16
Damascus	14:17
Ethiopia	14:18
Egypt	14:19
Edom	21:11-17
Jerusalem	21:22
Tyre	21:23

IN THE TIME OF THE LIFE OF CHRIST

Discipline	Zacharias made mute by angel	Luke 1:18-20, 22

DURING THE EARLY CHURCH

Death	Ananias and Sapphira	Acts 5:5, 9-11
	Herod by angel	12:23
Blindness	Bar-Jesus blinded by Paul	13:11
Medical	Giving over to sinful destruction of individuals	Rom. 1:24, 26, 28
Physics and deaths	Judgment of Ex. 32	1 Cor. 10:7
Death	Judgment of 33,000 in Num. 25	10:8
	Judgment of serpents in Num. 21	10:9
Medical	Sickness, death for violation of communion	11:30-32, 34

GOD GIVING DIVINE GUIDANCE, LEADING, OR INSTRUCTING

PATRIARCHS

Circumstances	Driving Adam and Eve out of Garden	Gen. 3:23-24
	Driving Cain out of land	4:11-15
	Keeping Abimelech away from Sarah with dream	20:6
	Events working to lead servant to find Rebekah	24:7, 12, 40, 48
	Isaac trusting God to bless Jacob	27:27-29
	Counseling in dream concerning Jacob	31:24, 29
Events	Command to return to Canaan	31:3, 11-13
	Command to Jacob to live in Bethel	35:1
	Interpretation of dreams of cupbearer and baker	40:8
	Interpretation of Pharaoh's dreams on famine	41:16, 25, 28, 32, 39

EXODUS

Events	Command to Moses to warn Pharaoh to release Israel	Ex. 6:11-13, 29; 7:2
	Instructions for exodus and Passover	11:1-3; 12:1-20, 36
	Leading people out of Egypt	12:42, 51: 13:3, 14, 16

SINAI AND WILDERNESS

Events	Leading people in wilderness with pillar of fire/cloud	Ex. 13:17-22; 14:1-2, 19-20
	Guidance with silver trumpets	Num. 10:1-10
	Command to send out spies	13:1-2
Craftsmanship	Giving of wisdom of craftsmanship	Ex. 31:1-11
Circumstances	Warning of death for seeing the Lord	19:21-24
	Israel warned	Lev. 18:28
	Instructions for Aaron's death at My Hor	Num. 20:23-25
	Instruction to Balaam not to curse Israel	22:9-12, 20
	Word of blessing put in Balaam at Baal high places	23:5-12
	2nd word of blessing at top of Pisgah	23:16-26
	3rd word of blessing at top of Peor	24:10-13
	4th oracle predicting Israel's victory	24:15-24
	Instruction to take census	26:1-2
Circumstances	Instruction on dividing land	Num. 26:52-56
	Instruction for families with no sons	27:6-11
	Instruction for Moses' death	27:12-14
	Commission of Joshua	27:15-21
	Command to do vengeance on Midianites	31:1-2, 7

	Instruction on distribution of spoils	31:25-31
	Instruction for officials who divide land	34:16-29
	Instruction for cities for Levites and cities of refuge	35:1-34
	Review of God's leading, instructions in wilderness	Deut. 1:6-46
	Review of God's leading from Kadesh to Moab	2:1-23
	Instruction for success in land	8:1-10
	Review of humbling during wilderness and miracles	8:2-4, 15-16
	Reminder of complacency of success	8:11-20
	Reminder of rebelliousness	9:7-8, 12-29
	Reminder of Sinai law	9:9-11; 10:1-5, 8-11
Military	Instruction to be hostile to Midianites	25:16-18
	Instruction for conquest and warning	33:50-56
	Instruction on borders	34:1-12

CONQUEST AND JUDGES

Events	Instructions for Joshua to take land	Josh. 1:1-9
	Promise of victory	1:5
	Instructions for Joshua for memorial stone	4:1-7
	Promise and instructions for taking Jericho	6:2-5

	Instructions for taking Ai	8:1-2, 18
	Instructions for conquest of areas not yet taken	13:1-7
	Instructions for cities of refuge	20:1-6
	Prophet's reminder and rebuke	Judg. 6:7-10
	Instructions to Gideon to reduce army	7:2-11
Dream	of victory over Midian from man	7:13-15
Inward	Spirit stirring Samson	13:24
	Spirit empowering Samson to kill lion	14:6
	Spirit empowering Samson to kill 30 men from Ashkelon	14:19
Events	Instructions for Judah to go 1st in battle against Benjamin	20:18
	Instructions for Israel's request for king	1 Sam. 8:6-18
	Instruction to give Israel a king	8:21-22
	Instructions to anoint Saul	8:15-17

KINGDOM

Events	Warning of consequences for obedience or rebellion	1 Sam. 12:14-25
	Instructions to David to attack Philistines to deliver Keilah	23:2-4
	Instructions to David to take Amalekites	30:8

	Instructions to David to take Philistines	2 Sam. 5:19-20
	David reminding Solomon of God's promise	1 Kings 2:4
	Instructions with promise to Solomon concerning temple	6:11-13; 8:12-21
	Instructions to Elijah to go to Ahab	17:8-9
	Instructions to Elijah to go to wilderness of Damascus to anoint king Hazael of Aram and Jehu and Elisha	19:9-18
	Prophet's command to strike him as sign for Ahab	20:35-43
	Shemaiah's warning against battle with Jeroboam	2 Chron. 11:2-4
	Asa's rebuke by Hanani for trusting king of Aram, not the Lord	16:7-10

LIFE OF CHRIST

Events	Warning in dreams to Egypt and Galilee	Matt. 2:12, 13, 19-20, 22
Chemistry	God able to turn stones into sons	3:9

ECONOMIC OR MATERIAL BLESSING

Economics	Wealth from herds and money	Gen. 24:25
	Blessing of Isaac	25:11; 26:12-14
	Abimelech observing God's blessing on Isaac	26:28

	Laban's blessing thru Jacob	30:30
	Rachael and Leah crediting God for wealth	31:16
	Jacob crediting God for blessing	33:11; 48:3-4
	Joseph crediting God for enriching brothers	43:23
Success	Joseph's success with Potiphar and in jail	39:2-5, 21-23
	Jacob crediting God for seeing Joseph	48:11
	David's success in all	1 Sam. 18:14
Sustenance	Provision of quail	Num. 11:16-32
	Water supply from hollow place for Samson	Judg. 15:18-19
	Food in famine	Ruth 1:6
Blessing	Obed-edom's blessing for keeping ark	2 Sam. 6:10-12
	Queen of Sheba praising God for blessing Solomon	1 Kings 10:9
	Reforms of Asa bringing blessing	2 Chron. 15:14-15
	Jehoshaphat's blessings for obedience	17:3-6
	God prospering king Uzziah as he sought Him	26:5, 7
Restoration	Manasseh's restoration after prayer and repentance	33:12-13
Medical, prosperity	from wisdom of the Lord	Prov. 3:8-10, 33-35

DIVINE PROTECTION

Circumstances	Jacob's preservation in Mesopotamia	Gen. 31:5, 7, 9
	Jacob's protection	31:42
Events	Jacob's trust in God's compassion	42:14
	Joseph telling brothers God used evil to save them	45:5-9; 50:19-20
Military	to defeat Amelek	Ex. 17:15-16
Medical	to heal in Canaan	23:25-26
Physical	Deliverance from lion and bear	1 Sam. 17:37
Military	Protection from enemies	Ezra 8:31; Neh. 4:14, 20
Events	Prayer and answered prayer	Many Psalms
	David's prayer for physical deliverance	Ps. 31
Physics	Temptations: stones to bread, cast from pinnacle	Matt. 4:3-7; Luke 4:1-12
Medical	Jonah in sea monster	12:40
Events	Works all things for good of believers	Rom. 8:28
	Deliverance from serpents by looking at bronze serpent	Num. 21:8-9

DIVINE LAWS REGULATING LIFE

Legal	10 Commandments with consequences	Ex. 20:1-17; Deut. 5:6-21
	Laws, ordinances, etc. with consequences	20:22-23:33
	Instruction for tabernacle	25:1; 31:18; 39:43; 40:16, 19-33

Instruction for observing feasts	34:18-28
Laws of offerings	Lev. 1:1-8:3
Executing of anointing	8:4-9:24
Laws on purity	11-15
Day of Atonement instructions	16
Laws of sacrifices	17
Laws for sanctification	18-27; Num. 5-6
Instruction on Levites	Num. 3:5-4:33; 8:1f
Instruction on Passover	9
Instruction for living in land	15
Instruction after judgment	16:36-40
Instruction for rods	17:1-5
Instruction for Aaron for priesthood	18-19
Regulations for offerings and vows	28-30
Exposition on entire law that God gave	Deut. 1:5-6
Exposition of laws	12-26

GOD'S COVENANTS WITH ISRAEL

Noahic	Gen. 9:8-17
Abrahamic	15:1-21; Luke 1:55, 73
Mosaic: review of covenant for land, including 10 Commandments	Deut. 4:1-31
Palestinian	29-30
Davidic: through Nathan	2 Sam. 7:4-17
New	Jer. 31:31-34; Ezek. 36:25-32

PROMISES OR PROPHECIES OF FUTURE INTERVENTIONS IN HISTORY

FALL

Messiah	Protevangelium	Gen. 3:15
Agriculture	No fertility of land for Cain	4:12

FLOOD

Promise of Flood and instructions for Noah	Gen. 6:13-7:4
Promise to not curse ground or destroy all again with a flood	8:21
Noahic Covenant promising control of natural realm	9:9-17

ABRAHAM

Anthropology: fetility of child bearing		
	to make Abraham a great nation	Gen. 12:2-3; 18:18; Rom. 4:16, 18
	to produce descendants like dust of earth	13:16; 22:17
	to produce son and descendants like stars of heaven	15:4-5; 22:17; Deut. 10:22
	to give many descendants to Hagar	16:10-11; 21:13, 18
	to give Abraham many nations and kings	17:2-6
	to give Sarah at age 90 many nations and kings	17:16, 19, 21
	to give Ishmael nations	17:20
	to give Abimelech children after closing wombs	21:17

	to produce two nations through Rebekah, covenant renewed	25:23
To judge Egypt and deliverance of Israel		15:14
Promise to give Israel land of Canaan (anticipates conquest)		17:8; 15:18; Deut. 3:22

ISAAC THROUGH JOSEPH

Covenant renewed with Isaac		Gen. 26:24
Covenant renewed with Jacob		28:13-15; 35:9-12
	Jacob's reminder of God's protection and descendants of covenant	32:12
Events	Jacob's promise of a great nation in Egypt	46:2-4
Blessing	God's blessing on Ephraim and Manasseh	48:15-21
Events	Joseph's assurance of God's fulfillment of covenant	50:24-25

EXODUS GENERATION

Events	to intervene to deliver and give land	Ex. 3:8-10, 12, 16-18
	to bring plagues on Egypt	3:19-22; 4:8-9, 21-26
	to guide Moses' speech	4:12, 15
	to bring Jews out of Egypt to Canaan	6:1, 6-8
	to harden Pharaoh's heart	7:3-5, 22; 8:15, 19, 32
	command to miracle	7:9
	to bring people to Canaan	13:5, 11
	to save Israel and destroy Egypt	14:13-14
	Instruction for parting of sea	14:15-16
Nutrition	to provide manna and meat	16:4-7, 12

SINAI

Legal	Offer of Covenant at Sinai	19:3-6
Events	Revelation of Himself	19:9-13
	Protection in conquest, judgment	23:20-24, 27-33
	Revealing His visible presence	29:45-46
Judgment	Threat of judgment	32:7-14
Military	to give victory over Canaanites with miracles	34:10-17, 24
	to spew Canaanites out of land	Lev. 18:24-28; 20:22-24
Promises for obedience		26:1-13; Deut. 28:1-15; 30:8-10, 15-16, 19-20
Climatology, agriculture:	Rain and crop fertility	26:3-5; Deut. 11:13-15; 28:12
Anthropology:	Fertility of childbearing	26:9; Deut. 7:13-14; 28:4, 8, 30; 30:9
Medical:	Medical: good health	7:15
Zoology:	Fertility of herds	Deut. 7:13; 28:4, 8; 30:9
	Removal of beasts	Lev. 26:6
Military:	Victory and security in land	26:5-8; Deut. 7:16-24; 11:8, 22-25; 12:10, 29; 28:7; 31:3-8
Glory:	to manifest His presence	26:11-12
Agriculture:	Fertility of crops	Deut. 7:13; 11:9-12; 28:4, 8; 30:9
Economic:	Success	28:5-6, 11
Promises of adversity for disobedience		Lev. 26:14-45; Deut. 28:16-67; 30:15-19
Economic:	Cursed in city and country	Deut. 28:15-17, 19
	No success or prosperity but oppression from enemies	28:29-33, 43-47
Agriculture:	Infertility of crops	28:18
	Locust and worms consume crops	28:38-40, 42

Anthropology:	Curse in childbearing	28:18
Zoology:	Infertility of herds	28:18
	Beasts killing children and animals	26:22
Medical:	Sickness, disease, and plagues	Lev. 26:14-16, 21; Deut. 28:22, 58-61
	Incurable diseases	Deut. 28:27-28, 35
	Death	28:26; 30:15, 19
	Cannibalism	28:53, 57
	Few in number	28:62-63
Mental health:	Confusion, madness, and darkness	28:20, 28-29, 34
Military:	Enemies to eat crops	Lev. 26:16-18
	Defeat, sword, pestilence, famine	26:23-33; Deut. 28:21, 25
	Siege, captivity, rotting away of those left	26:33-39, 41, 43; Deut. 28:36-37, 41, 48-57, 62-63, 68
	Scattering, despair, and dread	28:64-67
Climatology, agriculture:	No rain and crop failure	26:19-20; Deut. 11:16-17, 23-24
Promise of restoration		26:40-45
Promise to set Israel above nations		Deut. 28:9-10

WILDERNESS

Promise to reveal glory in all earth	Num. 14:21-23
Promise that 1st generation will not enter land	14:21-35
Promise for success with obedience	Deut. 4:30-31, 40; 5:31-33; 6:1-3, 17-19, 24-25; 10:12-13; 11:26-32; 12:25-28; 28:1, 3, 12-13; 29:9; 30:8-10, 15-16, 19-20
Promise of land	6:10-11

Promise of judgment if rebellious	6:15; 7:4
Promise of victory over Canaanites (7 nations)	7:1-5; 9:3-6; 20:1-4, 13-16; 25:19
Promise of judgment with curses and scattering	29:18-29
Promise of regathering	30:8-10, 15-16, 19-20
Prediction of idolatry	31:16-21

CONQUEST AND JUDGES

Promise for success with obedience	Josh. 1:8
Promise to exalt Joshua	3:7
Promise of victory over kings of Jerusalem, Hebron, Jarmuth, Lachish, Eglon	10:8
Promise of victory over Hazor, Madon, Shimron, Achshaph, north hill country, Arabah	11:7-9
Promise of curses if unfaithful	23:15-16
Promise of death of Hophni and Phinehas to Eli by man of God	1 Sam. 2:27-34

KINGDOM

Promise of David of victory over Goliath	1 Sam. 17:46-49
Samuel's prophecy from grave defeat of Philistines	28:16-19
Nathan's prediction that sword never to depart from David after sin	2 Sam. 12:10-12
Predict Solomon to have riches/honor greater than any king before or after	2 Chron. 1:12
Shemaiah's word of instruction against Rehoboam	1 Kings 12:21-24
Man of God's prophecies at Bethel and alter split	13:1-3
Abijah's prophecy of removal of Jeroboam's kingdom because of evil	14:6-16
Jehu's prophecy against Baasha's dynasty	16:1-4

Elijah's prophecy against Ahab	17:1
Prophet's prediction of victory over Aramean multitude	20:13-15
Elijah's prediction of horrible deaths of Ahab and Jezebel	21:17-24, 29
Prophets of Israel predict victory for Jehoshaphat	22:5-6, 12
Micaiah's prediction of disaster for Ahab	22:13-23
Prediction of victory over Ammon, Moab, and Mt. Seir without any fighting	2 Chron. 20:14-19
Victory with these destroying themselves	20:22-23
Elisha's predictions of 8 different events	2 Kings 3:14-13:17
Isaiah's prediction of Assyria	19:6-7; Isa. 37:6-7
Assyria to be turned back, remnant preserved	19:20-34; Isa. 37:36-38
Prophet's prediction of judgment on Amaziah for idolatry	2 Chron. 25:14-16
Isaiah to Hezekiah on healing, 15 years added, God to defend city against Assyria	2 Kings 20:4-6; Isa. 28:5-7
Hezekiah's recovery	20:7
Isaiah predicts Babylonian captivity	20:16-19
Prophets' predictions of destruction of Jerusalem and nations	21:10-15
Huldah's prediction of evil on Jerusalem and nation but peace on Judah	22:14-20; 2 Chron. 34:26-28
Lord's predictions of destruction of nation	23:27
Predictions of Isaiah	
All nations will stream to house of Lord	Isa. 2:2-4
Sign of virgin	7:14
Attack by Assyria	8:5-8
Birth and rule of Messiah	9:1-7
Messiah's ministry	11:1-5
Millennium	11:6-10
Restoration of Israel	12:1-6

	Judgment on all nations	13-23
	Tribulation of earth	24
	Israel in Kingdom: fulfillment of Abrahamic Covenant	27:12
	Babylonian captivity	39:5-8

LIFE OF CHRIST

Revelation	Holy Spirit on Zacharias, seeing Messiah before dying and prophecy	Luke 2:25-32, 34-35
Judgment	Future judgment in general	11:22-24; 12:41-45; Luke 11:29-32; John 12:47-48
	In kingdom by disciples	19:28-29
	Removal of kingdom from 1st century Israel and 70 AD	21:43-44; 23:36
	70 AD destruction	Luke 19:43-44; 21:6, 20-24
	Prediction of some taken in judgment	24:40-42
Circumstances	Future church	Luke 16:18-19
	Promise to Mary to be remembered	Matt. 26:13; Mark 14:9
	Predicting Passover circumstances	26:18-19; Mark 14:13-16; Luke 19:19-31; 22:10-13
Astrophysics	2nd coming	Luke 12:40; 21:25-28
	Sun and moon darkened, stars falling	Matt. 24; Mark 8; 13
	Climatology: sign in sky, coming in clouds	Matt. 24:29-30; Mark 8:38; 13:24-27
Events	John the Baptizer's prophecies on Jesus' ministry	3:2, 7-14, 16-17
	Predicting future outcomes for disciples	Matt. 10:15-23
	Predicting angel ascending and descending on Son of Man	John 1:51

Future in parables	Luke 13:3-52
His death and resurrection	Matt. 16:21; 17:9-12, 22-23; 19:18-19; 26:31-32; Mark 8:31; 9:9-10, 31; 10:33-34; 14:28; Luke 9:44; 18:31-33; John 2:19-22; 12:32-33; 2 Cor. 5:15
Coming in glory	16:27-28; Luke 9:26; 12:40
Final days eschatology	24-25; Mark 13:2-33; Luke 21
Judas' betrayal	26:21; Mark 14:18-21
Departure prediction	John 14:1-3, 28
Rapture predicted	14:1-3
Entering kingdom	Luke 13:24-30; 22:30
Communion in kingdom	Matt. 26:29; Mark 14:25
Scattering of disciples	Matt. 26:31-32; Mark 14:27
Peter's denial	Matt. 26:34; Mark 14:30; Luke 22:34, 61
Prediction of arrest	Matt. 26:45; Mark 14:41-42; Luke 22
Jesus at right hand and return	Matt. 26:64; Mark 14:62
Predicting session at right hand	Luke 22:69
Promise of Holy Spirit	John 14:16-17, 26
Sending of Holy Spirit	15:26-27; 16:7-15
Kind of death for Peter	21:18-19
Angel explaining return of Lord after ascension to disciples	Acts 1:11
All to be subjected to Son	Heb. 2:7-8

CHURCH

Transformation	Refreshing = kingdom promised	Acts 3:19-21
Judgment	Future judgment	10:42; 17:31; Heb. 12:23, 26-27
Restoration	Israel's future salvation predicted	Rom. 11:25-27
Physics	Promise to abolish all authority and power	1 Cor. 15:24
	Promise to put all in subjection, even death	15:25-28
Resurrection	Bodily of believers	1 Cor. 15; 2 Cor. 4:14; 5:1-3
Glory	Eternal glory	2 Cor. 4:17-18
Rewards	Promise of rewards	Col. 3:24-25
Judgment	Future judgment of persecutors at 2nd coming	2 Thess. 1:5-10; 2 Tim. 4:1
Apostasy	Prediction of apostasy	2 Thess. 2:1-3
Physics	Prediction of coming and gathering to Him	2 Thess. 2:1; 2 Tim. 4:1
Events	Prediction of antichrist	2 Thess. 2:3-10
	Prediction of slaying antichrist	2:8
	Prediction of restrainer removed	2:7

APPENDIX E

BIBLIOGRAPHY

Arndt, William F. and Gingrich, F. Wilbur, *A Greek-English Lexicon of the New Testament*, The University of Chicago Press, 1957.

Beckwith, Francis J., William Lane Craig, and J. P. Moreland, ed., *To Everyone an Answer: A Case for the Christian Worldview*, InterVarsity Press, 2004.

Berry, R. J., ed. *Real Science, Real Faith*, Monarch, 1991.

Bethell, Tom, *The Politically Incorrect Guide to Science*, Regnery Publishing, 2005.

Brown, Francis, Driver, S.R., and Briggs, Charles A., *A Hebrew and English Lexicon of the Old Testament*, Clarendon Press, 1974.

Bruce, F.F., *New Testament Documents: Are They Reliable*?, Eerdmans, 1960.

Chaffey, Tim, *Fallen: The Sons of God and the Nephilim*, Rise Books, 2019.

Collins, Steven, *Championing the Faith: A Layman's Guide to Proving Christianity's Claims*, Virgil W. Hensley Inc Publishing, 1991.

Davidheiser, Bolton, *Evolution and Christian Faith*, The Presbyterian and Reformed Publishing Company, 1969.

Davis, John J., *Conquest and Crisis: Studies in Joshua, Judges, and Ruth*, Baker Book House, 1969.

De Beer, Gavin. *Homology: An Unsolved Problem*. London: Oxford University Press, 1971.

Denton, Michael, *Evolution: A Theory in Crisis*, Adler and Adler Publishers, 1985.

de Vaux, Roland, *Ancient Israel, Volume 1: Social Institutions*, McGraw-Hill Book Company, 1965.

Dewar, Douglas, *Difficulties of the Evolution Theory*, E. Arnold & Co, 1931.

DeYoung, Don, *Thousands … Not Billions*, Master Books, 2005.

Dye, David L., *Faith and the Physical World: A Comprehensive View*, Eerdmans, 1966.

Eiseley, Loren, *Darwin's Century*, Doubleday Anchor Books, 1958.
Geisler, Norman L., *Christian Apologetics*, Baker Book House, 1976.
Geisler, Norman L., & Watkins, William D., *Worlds Apart*, Baker Book House, 1989.
Gish, Duane T., *Evolution: The Fossils Say NO!*, Creation-Life Publishers, 1973.
Godde, Sandra, *Reaching for Immortality: Can Science Cheat Death?* Wipf and Stock, 2022.
Greenlee, J. Harold, *Introduction to New Testament Textual Criticism*, William B. Eerdmans Publishing Co., 1964.
Guillen, Michael, *Amazing Truths—How Science and the Bible Agree*, Zondervan, 2016.
Halley, H.H., *Halley's Bible Handbook,* 24th ed., Zondervan, 1965.
Ham, Ken, *The Lie: Evolution*, Master Books, 2012.
Hagopian, David G., *The Genesis Debate: Three Views on the Days of Creation*, Crux Press, 2001.
Jaki, Stanley L., *Cosmos and Creator*, Scottish Academic Press, 1980.
Johnson, Phillip E., *Objections Sustained: Subversive Essays on Evolution, Law, & Culture*, InterVarsity Press, 1998.
Jones, James B., and Hawkins, George A., *Engineering Thermodynamics: An Introductory Textbook*, John Wiley & Sons, Inc, 1960.
Kitchen, Kenneth A., *The Bible in Its World*, Inter-Varsity, 1978.
Lamont, Ann, *21 Great Scientists Who Believed the Bible,* Creation Science Foundation, 1995.
Latourette, Kenneth Scott, *A History of Christianity,* Harper Row Publishers, 1953.
Leupold, H. C., *Exposition of Genesis*, Vol I, Baker Book House, 1942.
Lewis, C.S., *God in the Dock*, Wm Eerdmans Publishing, 1970.
Mayr, Ernst, *Populations, Species and Evolution*, Harvard University Press, 1970.
Medawar, Peter B., *The Limits of Science*, Harper Row, 1984.
Moreland, J.P., & Reynolds, John Mark ed, *Three Views on Creation and Evolution*, Zondervan, 1999.
Morris, Henry M., *The Biblical Basis for Modern Science*, Baker Book House, 1984.
_____ *The Genesis Record: A scientific & devotional commentary on the book of beginnings*, Baker Book House, 1976.
_____*Impact #107*, Institute for Creation Research, 1982.
_____*The Long War Against God*, Master Books, 2000.

_____ *Men of Science, Men of God*, Master Books, 1988.
_____ *Scientific Creationism*, Master Books, 1974.
Newman, Horatio Hockett, *Evolution, Genetics, and Eugenics*, University of Chicago Press, 1932.
Nickel, James, *Mathematics: Is God Silent*, Ross House Books, 2001.
Noebel, David A., *Understanding the Times: The Collision of Today's Competing Worldviews*, Summit Press, 2006.
Oard, Michael J., *An Ice Age Caused by the Genesis Flood*, Institute for Creation Research, 1990.
O'Connell, Michael, *Finding God in Science*, Eigen Publishing, 2016.
Parker, Gary, *Creation Facts of Life*, Master Books, 1994.
_____ *Impact #89*, Institute for Creation Research, 1980.
Pearcey, Nancy R., and Charles B. Thaxton, *The Soul of Science–Christian Faith and Natural Philosophy*, Crossway Books, 1994.
Quagliano, James V., *Chemistry*, Prentice-Hall, 1964.
Ramm, Bernard, *The Christian View of Science and Scripture*, William B. Eerdmans, 1954.
Rensberger, Boyce, *How the World Works*, William Morrow, 1986.
Robertson, A. T., *Introduction to the Textual Criticism of the New Testament*, Broadman, 1925.
Ross, Allen P., *Creation & Blessing: A Guide to the Study and Exposition of Genesis*, Baker Books, 1988.
Ross, Hugh, *Creation and Time: A Biblical and Scientific Perspective on the Creation-Date Controversy*, NavPress, 1994.
_____ *The Genesis Question: Scientific Advances and the Accuracy of Genesis*, NavPress, 1998.
Ryrie, Charles C., *Basic Theology*, Victor Books, 1986.
Sarfati, Jonathan, *By Design: Evidence for Nature's Intelligent Designer–The God of the Bible,* Creation Book Publishers, 2008.
_____ *The Genesis Account: A theological, historical, and scientific commentary on Genesis 1-11*, Creation Book Publishers, 2015.
_____ *Refuting Compromise: A Biblical and Scientific Refutation of "Progressive Creationism" (Billions of Years) as Popularized by Astronomer Hugh Ross,* Master Books, 2011.
_____ *Refuting Evolution: A Responding to the National Academy of Sciences' Teaching about Evolution and the Nature of Science*, Master Books, 1999.
Sears, F. W., and Zemansky, M. W., *University Physics*, Addison-Wesley Publishing Co, 1964.

Shedd, Willian G.T., *Dogmatic Theology*, Vol 1, Thomas Nelson Publishers, 1980.

Simpson, George Gaylord, *Notes on the Nature of Science*, Harcourt, Brace, and World, 1962.

_____ *Tempo and Mode in Evolution*, Columbia University Press, 1949.

Sire, James W., *The Universe Next Door*, InterVarsity Press, 1988.

Snelling, Andrew, *Earth's Catastrophic Past – Geology, Creation, & The Flood*, Institute for Creation Research, 2009.

Thiessen, Henry C., *Lectures in Systematic Theology*, William B. Eerdmans, 1990.

Tozar, A. W., *The Knowledge of the Holy*, Harper & Row Publishers, 1961.

Van Bebber, Mark, and Taylor, Paul S., *Creation and Time - A Report on the Progressive Creationist Book by Hugh Ross,* Eden Productions, 1994.

Vardiman, Larry, Snelling, Andrew, and Chaffin, Eugene, editors, *Radioisotopes and the Age of The Earth: A Young Earth Creationist Research Initiative*, Institute for Creation Research, 2000.

Waltke, Bruce K., *Genesis: A Commentary*, Zondervan, 2001.

Webster's New World Dictionary of the American Language, The World Publishing Company, 1957.

Whitcomb, John C. and Henry M. Morris, *The Genesis Flood: The Biblical Record and Its Scientific Implications*, Baker Book House, 1961.

Whitcomb, Jr., John C., *The World That Perished*, Baker Book House Company, 1973.

Williams, Alex, and Hartnett, John, *Dismantling the Big Bang: God's Universe Rediscovered*, Master Books, 2005.

Williams, Stephen, *What Your Atheist Professor Doesn't Know (But Should),* RFH, 2013.

Young, Davis A., *The Biblical Flood: A Case Study of the Church's Response to Extrabiblical Evidence*, William B. Eerdmans Publishing, 1995.

Zuck, Roy B., *Creation: Evidence from Scripture and Science*, Scripture Press Publications, 1975.

NOTES

BOXED QUOTES & FIGURES

PREFACE

1. **Dr. Ernest Walton quote**: Scott Youngren, "Quotes about God to consider…if you think science leads to atheism," God Evidence, August 12, 2010, https://godevidence.com/2010/08/quotes-about-god-atheism/ as quoted in V.J. McBrierty (2003): Ernest Thomas Sinton Walton, The Irish Scientist, 1903-1995, Trinity College Dublin Press.

CHAPTER 1

1. **Figure 1 Venus**: Venus - Computer Simulated Global View of the Northern Hemisphere NASA ID: PIA00271, photograph, NASA, August 15, 2024, https://images.nasa.gov/details/PIA00271.
2. **Dr. Max Planck quote**: Tihomir Dimitrov, 50 Nobel Laureates and Other Great Scientists Who Believe in God, 1995-2008, accessed July 2024, www.academia.edu/33919294/50_nobel_laureates_and_other_great_scientists_who_believe_in_god.
3. **Isaac Newton quote**: "Isaac Newton," History.com editors, History, October 16, 2023, https://www.history.com/topics/inventions/isaac-newton.
4. **Dr. Markus Blietz quote**: "Galaxies, Black Holes, and Creation," Creation.com, Creation Ministries International, January 11, 2023, creation.com/markus-blietz-astrophysicist-interview. This article first appeared in Creation 35(2):34–36, April 2013.
5. **Figure 2 Andromeda Galaxy**: Our Neighbor Andromeda NASA ID: PIA12832, photograph, NASA, August 15, 2024, https://images.nasa.gov/details/PIA12832.
6. **Niels Bohr quote**: Garson O'Toole, "Quote Origin: An Expert Is a Person Who Has Made All the Mistakes Which Can Be Made in a Very Narrow Field," quoteinvestigator.com, January 15, 2023, https://quoteinvestigator.com/2023/01/15/expert-mistakes/ as quoted from "LIFE" magazine, 1954, in a profile of Edward Teller who ascribed it to Bohr.
7. **Enrico Medi quote**: Volubrjotr, "Enrico Medi: Look There Is A New Star," Political Vel Craft, October 10, 2010, politicalvelcraft.org/2010/10/24/enrico-medi-look-there-is-a-new-star/.
8. **Albert Einstein quote**: "Albert Einstein: In Brief," Institute for Advanced Study by The Center for History of Physics, on the occasion of the Einstein Centennial in 1979, accessed July 2024, www.ias.edu/albert-einstein-brief.

CHAPTER 2

1. **Dr. Isidor Isaac Rabi quote**: Tihomir Dimitrov, 50 Nobel Laureates and Other Great Scientists Who Believe in God, 1995-2008, accessed July 2024, www.academia.edu/33919294/50_nobel_laureates_and_other_great_scientists_who_believe_in_god.
2. **Erwin Schrodinger quote**: Professor H. Pleijel, "Award ceremony speech," The Nobel Prize Foundation, December 10, 1933, accessed July 2024, www.nobelprize.org/prizes/physics/1933/ceremony-speech/.
3. **Dr. Geoff Downes quote**: Dr. John Ashton, editor, "Geoff Downes, forestry research," Creation.com, Creation Ministries Inc., accessed July 2024, creation.com/geoff-downes-forestry-research-in-six-days, as quoted from John F Ashton, In Six Days, Why 50 scientists choose to believe in creation, April 1, 2000.
4. **Nikola Tesla quote**: Otis Chandler, "Nikola Tesla > Quotes," Goodreads, accessed July 2024, www.goodreads.com/author/quotes/278.Nikola_Tesla.
5. **Sir Ronald Ross quote**: "This Day Relenting God," Encyclopedia.com, accessed July 2024, www.encyclopedia.com/science/medical-magazines/day-relenting-god as quoted from Ronald Ross, Memoirs with a Full Account of the Great Malaria Problem and Its Solution, [London]: John Murray, 1923.

CHAPTER 3

1. **Dr. David Russell quote**: "David Russell Quotes," Brainy Media, Inc, BrainyQuote, accessed July 2024, www.brainyquote.com/quotes/david_russell_381231.
2. **Isaac Newton quote**: Otis Chandler, "Isaac Newton > Quotes," GoodReads, accessed July 2024, www.goodreads.com/quotes/7952880-opposite-to-godliness-is-atheism-in-profession-and-idolatry-in.
3. **Dr. Robert Millikan quote**: Tihomir Dimitrov, 50 Nobel Laureates and Other Great Scientists Who Believe in God, 1995-2008, accessed July 2024, www.academia.edu/33919294/50_nobel_laureates_and_other_great_scientists_who_believe_in_god.
4. **Charles Townes quote**: Tihomir Dimitrov, 50 Nobel Laureates and Other Great Scientists Who Believe in God, 1995-2008, accessed July 2024, www.academia.edu/33919294/50_nobel_laureates_and_other_great_scientists_who_believe_in_god.
5. **Sir Fred Hoyle quote:** John G. Leslie, Ph.D. and Charles K. Pallaghy, Ph.D, "The Religious Nature of Evolution Theory and its Attack on Christianity," Creation.com, Creation Ministries International, 6/3/13, ia601000.us.archive.org/view_archive.php?archive=/33/items/sermonaudio-dot-com-sermon-pdf/Sermonaudio.com%20Sermon%20PDF.zip&file=52813162492.pdf.

6. **Dr. Stanley Jaki quote**: Ellen Myers, "Creation and Science: The Work of Stanley L. Jaki," Creation Social Science & Humanities Org., Vol IX, 1987, accessed July 2024, www.creationism.org/csshs/v09n2p17.htm.
7. **Max Planck quote**: Biblium, "10 Quotes about God by Famous Scientists," 2021, bibilium.com/quotes-about-god-scientists-quotes/.
8. **R.H. Nash quote:** Ronald H. Nash, *Worldviews in Conflict*, Grand Rapids: Zondervan, 1992, 16.
9. **Blaise Pascal quote**: Christian Classics Ethereal Library, "Blaise Pascal," Pensees, accessed July 2024, ccel.org/ccel/pascal/pensees/pensees.viii.html.
10. **Dr. Arthur Eddington quote**: Javier Ordovas, "25 Famous Scientists on God," June 26, 2016, www.catholiceducation.org/en/science/faith-and-science/25-famous-scientists-on-god.html.
11. **Isaac Newton quote**: Isaac Newton, "Untitled Treatise on Revelation" (section 1.1), Newton Project, accessed July 2024, www.newtonproject.ox.ac.uk/view/texts/normalized/them00135.
12. **Dr. Werner Von Braun quote**: David F. Coppedge, "Werner von Braun," Creation/Evolution Headlines, accessed July 2024, crev.info/scientists/wernher-von-braun/.
13. **Dr. Werner Heisenberg quote**: Daniel C. Peterson, "Werner Heisenberg on Religion," Sic et Non, Patheos, Nov 10, 2019, /www.patheos.com/blogs/danpeterson/2019/11/werner-heisenberg-on-religion.html.
14. **Sir James Young Simpson quote**: Paul James-Griffiths, "The Nineteenth Century Awakening," Christian Heritage Edinburgh, Aug 23, 2016, www.christianheritageedinburgh.org.uk/2016/08/23/sir-james-young-simpson-1811-1870/ as quoted from "his own personal testimony, Free Church."
15. **Richard Feynman quote**: Otis Chandler, "Richard Feynman > Quotes," GoodReads, accessed July 2024, www.goodreads.com/author/quotes/1429989.Richard_P_Feynman.
16. **Galileo Galilei quote**: Otis Chandler, "Galileo Galilei > Quotes," GoodReads, 2024, www.goodreads.com/quotes/32934-the-sun-with-all-those-planets-revolving-around-it-and.
17. **Max Planck quote**: "Max Planck Quotes," AZ Quotes, accessed July 2024, www.azquotes.com/author/11714-Max_Planck.

CHAPTER 4

1. **Dr. John Eccles quote**: Tihomir Dimitrov, 50 Nobel Laureates and Other Great Scientists Who Believe in God, 1995-2008, accessed July 2024, www.academia.edu/33919294/50_nobel_laureates_and_other_great_scientists_who_believe_in_god.
2. **James Clerk Maxwell quote**: Curt Blattman, "Quotes from Great Christian Scientists on God," Bible Apologetics – A Daily Devotional, October 17, 2021, bibleapologetics.org/quotes-from-great-christian-scientists-on-god/.

3. **James Joule quote**: "James Prescott Joule Quotes," Brainy Media, Inc, BrainyQuote, accessed July 2024, www.brainyquote.com/quotes/james_prescott_joule_317814.
4. **Dr. George Ellis quote**: "George F. R. Ellis Quotes," AZ Quotes, accessed July 2024, www.azquotes.com/quote/587236.
5. **Dr. Nevill Mott quote**: Daniel C. Peterson, "The religious faith of two Nobel laureate physicists," Sic et Non, Patheos, Nov 18, 2019, www.patheos.com/blogs/danpeterson/2019/11/the-religious-faith-of-two-nobel-laureate-physicists.html.
6. **Dr. Arthur Schawow quote**: Tihomir Dimitrov, "Nobel Laureate Arthur L. Schawlow on Science & Religion," 2012 Community of Scientific God Inc, Jan 3, 2012, 2012daily.com/community/blogs/126.
7. **Dr. Guglielmo Marconi quote**: Tihomir Dimitrov, 50 Nobel Laureates and Other Great Scientists Who Believe in God, 1994-2008, accessed July 2024, issuu.com/tihomirdimitrov1/docs/50_nobelists_on_god.
8. **Dr. Arthur Compton quote**: Tihomir Dimitrov, 50 Nobel Laureates and Other Great Scientists Who Believe in God, 1994-2008, accessed July 2024, issuu.com/tihomirdimitrov1/docs/50_nobelists_on_god.
9. **Albert Einstein quote**: J.M. Rankin, University of Vermont, accessed July 2024, www.uvm.edu/~jmrankin/Einstein_quote.htm as quoted from Albert Einstein's address to the student body, California Institute of Technology.
10. **Henrich Rohrer quote**: "Henrich Rohrer Quotes," Brainy Media, Inc, "BrainyQuote," accessed July 2024, www.brainyquote.com/quotes/heinrich_rohrer_736250.
11. **Dr. John Polkinghorne quote**: John Polkinghorne, "Reason and Reality: The Relationship Between Science and Theology," Society for Promoting Christian Knowledge, Reissued 2011, biblia.com/api/plugins/embeddedpreview?resourceName=lls:rsnrltyrltnshps.

CHAPTER 5

1. **C.V. Raman quote**: "C. V. Raman," Famous Scientists, famousscientists.org, June 30, 2016, Web 8/6/2024, www.famousscientists.org/c-v-raman/.
2. **Albert Einstein quote**: Garson O'Toole, "Anyone Who Doesn't Take Truth Seriously in Small Matters Cannot Be Trusted in Large Ones Either," quoteinvestigator.com, April 15, 2014, quoteinvestigator.com/2014/04/15/large-truth/.
3. **Dr. John Lennox quote**: "John Lennox Quotes: 100+ Powerful Quotes to Shift Your Perspective," StillFaith.com, accessed July 2024, stillfaith.com/christian-quotes/john-lennox-quotes/.
4. **Figure 3 Hurricane Matthew**: Hurricane Matthew Hits Haiti NASA ID: GSFC_20171208_Archive_e000205, photograph, NASA, August 15, 2024, https://images.nasa.gov/details/GSFC_20171208_Archive_e000205
5. **Figure 4 Cartoon**: Michael Ramirez, 2014, used with permission.

6. **Jaroslav Joseph Havel quote**: "Seeing a way through the forest, Gary Bates and Don Batten talk to forest and ecology authority Dr Jaroslav Joseph (Joe) Havel," Creation.com, Creation Ministries International, creation.com/seeing-a-way-through-the-forest. This article is from Creation 28(2):52–55, March 2006.
7. **Richard Feynman quote**: "Richard Feynman," wikiquote, July 17, 2024, en.wikiquote.org/wiki/Richard_Feynman.
8. **Dr. Lewis Thomas quote**: The Doc, "What is Science? Quotes from Great Scientists," Famous Scientists: The Art of Genius, accessed July 2024, www.famousscientists.org/what-is-science-quotes/.
9. **Dr. Alexis Carrel quote**: Tihomir Dimitrov, 50 Nobel Laureates and Other Great Scientists Who Believe in God, 1995-2008, accessed July 2024, www.academia.edu/33919294/50_nobel_laureates_and_other_great_scientists_who_believe_in_god.
10. **Albert Einstein quote**: "Quotation Details," The Quotation Page, accessed July 2024, www.quotationspage.com/quote/29604.html as quoted from New York Times, March 19, 1940.
11. **Dr. John G. Hartnett quote**: Gary Bates, "Exploding the big bang!" Creation.com, Creation Ministries International, accessed July 2024, creation.com/exploding-the-big-bang#:~:text=by%20Gary%20Bates%20How%20did, as originally in Creation 25(4):37–39, September 2003.
12. **Dr. E.H. Andrews quote**: John G. Leslie, Ph.D. and Charles K. Pallaghy, Ph.D., "The Religious Nature of Evolution Theory and its Attack on Christianity," Creation.com, Creation Ministries International, creation.com/evolution-religious. This article is from Creation 7(4):42–48, June 1986.
13. **Marie Curie quote**: National Geographic, media.nationalgeographic.org/assets/file/breakthrough_quote_5.pdf.
14. **Dr. Charles Townes quote**: Charles H. Townes, "Logic and Uncertainties in Science and Religion," Science and the Future of Mankind, Pontifical Academy of Sciences, Scripta Varia 99, Vatican City, 2001, www.pas.va/content/dam/accademia/pdf/sv99/sv99-townes.pdf.
15. **Paul Kalanithi quote**: Paul Kalanithi, "When Breath Becomes Air," accessed July 2024, www.endinmindproject.org/wp-content/uploads/2016/05/EiM_Discussion_Guide_Kalanithi.pdf.
16. **Richard Feynman quote**: "Richard Feynman Quotes," Libquotes, accessed July 2024, libquotes.com/richard-feynman/quote/lbc4m0h quoting from The Quotable Feynman (ed. Princeton University Press, 2015).
17. **James Clerk Maxwell quote**: James Clerk Maxwell, "The Molecule," The Victorian Web: literature, history, and culture in the age of Victoria, accessed July 2024, victorianweb.org/science/maxwell/molecules.html. "In 2001 John van Wyhe, Ph.D., Cambridge University, digitized the following text from Maxwell's 'Molecules', which appeared in the September 1873 issue of Nature and, according to a note, was a "Lecture

delivered before the British Association at Bradford, by Prof. Clerk-Maxwell, F.R.S." In April 2008 Yaakov Eisenberg of New York corrected the transcription and George P. Landow reformatted the text."

CHAPTER 6

1. **Figure 5 Comparative Anatomy**: Skeletons of Humans and Gorillas Compared, diagram/illustration, Encyclopedia Britannica, August 6, 2024, https://www.britannica.com/science/comparative-anatomy.
2. **Michael Denton quote:** Denton, Michael, *Evolution: A Theory in Crisis*, Burnett Books, 1985, 358.
3. **Dr. Dudley Eirich quote**: Carl Wieland and Jonathan Sarfati, "Manipulating life? Genetic engineering researcher backs Genesis," accessed July 2024, Creation.com, Creation Ministries International, creation.com/manipulating-life-dr-eirich-interview. This article is from Creation 27(1):46–49, December 2004.
4. **Dr. John K.G. Kramer quote**: John K.G. Kramer, "John K.G. Kramer, biochemistry," Creation.com, Creation Ministries International, accessed July 2024, creation.com/john-k-g-kramer-biochemistry-in-six-days.
5. **Dr. Jerry Bergman quote**: Jerry Bergman, "The palmar grasp reflex is neither useless nor evidence of an ape past," Creation.com, Creation Ministries International, accessed July 2024, creation.com/palmar-grasp-reflex. This article is from Journal of Creation 35(2):9–10, August 2021.
6. **Dr. Jerry Bergman quote**: Jerry Bergman, "Do any vestigial organs exist in humans?" Creation.com, Creation Ministries International, accessed July 2024, creation.com/do-any-vestigial-organs-exist-in-humans. This article is from Journal of Creation 14(2):95–98, August 2000.
7. **Dr. André Eggen quote**: Ken Ham, "French Creation: Interview with French Scientist Dr André Eggen," Answers in Genesis, accessed July 2024, answersingenesis.org/ministry-news/ministry/french-creation/ Originally published in Creation 20, no 4 (September 1998): 17-19.
8. **Werner Arber quote**: Graeme, "Evidence for and against God (quotes)," Wise Trove, accessed July 2024, wisdomtrove.com/evidence-for-and-against-god-quotes/.
9. **Dr. Samuel Gan quote:** "Taking a stand, Don Batten interviews high-profile Singapore researcher Dr Samuel Gan," Creation.com, Creation Ministries International, accessed July 2024, creation.com/sam-gan-interview. This article is from Creation 43(2):22–24, April 2021.
10. **Dr. Jim Allan quote**: Dr Don Batten and Carl Wieland, "Jumping ship: Dr Jim Allan, a geneticist, tells of his 'double conversion,'" Creation.com, Creation Ministries International, accessed July 2024, creation.com/jumping-ship. This article is from Creation 20(3):26–27, June 1998.

11. **Figure 6 Geologic Column**: Major Evolutionary Events, Encyclopedia Britannica, August 6, 2024, https://www.britannica.com/science/evolution-scientific-theory/images-videos#/media/1/197367/1650.
12. **Dr. Alan Galbraith quote**: Carl Wieland, "Recovery from evolution: A chat with hydro-scientist Alan Galbraith," Creation.com, Creation Ministries International, accessed July 2024, creation.com/recovery-from-evolution. This article is from Creation 22(2):16–17, March 2000.
13. **Dr. Ralph Matthews quote**: Ralph Matthews, "Reflections on the emperor's new clothes," Creation.com, Creation Ministries International, accessed July 2024, creation.com/reflections-on-the-emperors-new-clothes. This article is from Creation 17(4):35–37, September 1995.
14. **Dr. Jerry Bergman quote**: Jerry Bergman, "Coelacanth: The transitional fossil that wasn't," Creation.com, Creation Ministries International, accessed July 2024, creation.com/coelacanth-not-transitional-fossil. This article is from Creation 43(4):39–41, October 2021.
15. **Dr. Niles Eldredge quote**: David Plaisted, "General Quotes about Evolution…Horse evolution," March 2006, accessed July 2024, www.cs.unc.edu/~plaisted/ce/challenge9.html.
16. **Werner Von Braun quote**: Ray & Gale Lawson, "Werner von Braun: The Christian Rocket Scientist," Dec 11, 2020, Crosswalk, www.crosswalk.com/family/homeschool/wernher-von-braun-the-christian-rocket-scientist-1351753.html.
17. **Dr. Erwin Chargaff quote**: "Erwin Chargaff," accessed July 2024, en.wikipedia.org/wiki/Erwin_Chargaff as quoted from Chargaff, Erwin (1978). Heraclitean Fire: Sketches from a Life Before Nature. Rockefeller University Press.
18. **Richard Feynman quote**: Otis Chandler, "Richard Feynman > Quotes," GoodReads, accessed July 2024, www.goodreads.com/quotes/29894-no-government-has-the-right-to-decide-on-the-truth.
19. **Ana Maria Mihalcea quote**: Ana Maria Mihalcea, MD, PhD, "Nanotechnology, Cybernetic Hive Minds, Artificial Intelligence and Mind Control - DARPA and CIA Insider Dr. Robert Duncan's Interviews Confirms Hijacking Of Human Soul Possible.," Humanity United Now, Apr 24, 2023, anamihalceamdphd.substack.com/p/nanotechnology-cybernetic-hive-minds.
20. **Edsger W. Dijkstra quote**: "Edsger Wybe Dijkstra," A.M. Turing Award, 1972, mturing.acm.org/award_winners/dijkstra_1053701.cfm.
21. **Ana Maria Mihalcea quote**: Ana Maria Mihalcea, "Plasmons and Demons in Graphene - Using Artificial Intelligence, Graphene, Metals and Plasmonic Demonic Fields for Spiritual Warfare," Humanity United Now, Feb 18, 2023, anamihalceamdphd.substack.com/p/plasmons-and-demons-in-graphene-using.

22. **Albert Einstein quote**: Otis Chandler, "Albert Einstein > Quotes," GoodReads, accessed July 2024, www.goodreads.com/quotes/528444-the-fanatical-atheists-are-like-slaves-who-are-still-feeling.

CHAPTER 7

1. **Sir Ambrose Fleming quote**: Curt Blattman, "Quotes from Great Christian Scientists on God," Bible Apologetics – A Daily Devotional, October 17, 2021, bibleapologetics.org/quotes-from-great-christian-scientists-on-god/
2. **Figure 7 Yod in Hebrew**: originally created for this publication.
3. **Figure 8 Hebrew Serif**: originally created for this publication.
4. **Arno Penzias quote**: John Bally, "Arno A. Penzias (1933–2024), co-discoverer of the cosmic microwave background," February 22, 2024, www.nature.com/articles/d41586-024-00555-1. The article first appeared in Nature 627, 30 (2024).
5. **Figure 9 Vaticanus Codex**: Codex Vaticanus, Public Domain via Wikimedia Commons, https://commons.wikimedia.org/wiki/File:Codex_Vaticanus_B,_2Thess._3,11-18,_Hebr._1,1-2,2.jpg
6. **Galileo Galilei quote**: "Galileo Galilei - Letter to the Grand Duchess Christina (1636)," Libquotes, accessed July 2024, libquotes.com/galileo-galilei/works/letter-to-the-grand-duchess-christina. Originally in a Letter to "the Grand Duchess Christina (1615)."

CHAPTER 8

1. **Galileo Galilei quote**: Huping Hu, "Galileo Galilei: Human Intelligence Is a Work of GOD," Scientific God, Inc., December 10, 2011, 11prompt.com/?q=node/89.
2. **Erwin Schrödinger quote**: Tihomir Dimitrov, "4. Erwin Schrödinger, Nobel Laureate in Physics," accessed July 2024, atheismexposed.tripod.com/nobelistsgod.htm.
3. **Sir Derek Barton quote**: Kuruvilla Pandikattu SJ, "God as the Ultimate Truth: Derek Barton," Light of Truth, March 30, 2021, lightoftruth.in/valaranvalarthan/god-ultimate-truth-derek-barton/.
4. **Sir Joseph J. Thomson quote**: Kevin M. Downard, "J. J. Thomson Goes to America," Science Direct, July 18, 2009, www.sciencedirect.com/science/article/pii/S1044030509005534 as quoted from Journal of the American Society for Mass Spectrometry, Volume 20, Issue 11, November 2009, Pages 1964-1973.

WHAT HAVE WE LEARNED

1. **Johannes Kepler quote**: "Johannes Kepler Quotes," AZ Quotes, accessed July 2024, www.azquotes.com/author/7921-Johannes_Kepler.
2. **Leonhard Euler quote**: "Quotations, Leonhard Euler," MacTutor, School of Mathematics and Statistics, University of St Andrews, Scotland, accessed July 2024, mathshistory.st-andrews.ac.uk/Biographies/Euler/quotations/.

3. **George Washington Carver quote**: "George Washington Carver Quotes," AZ Quotes, accessed July 2024. www.azquotes.com/author/2580-George_Washington_Carver.
4. **Dr. Joseph Henry quote**: "Joseph Henry Quotations," Smithsonian Institute Archives, accessed July 2024, siarchives.si.edu/sites/default/files/pdfs/jhpp/JHP_Joseph_Henry_Quotations.pdf as quoted first in the Smithsonian Annual Report for 1852, p. 15.
5. **Blaise Pascal quote**: Otis Chandler, "Blaise Pascal > Quotes > Quotable Quote," GoodReads, accessed July 2024, www.goodreads.com/quotes/250235-no-religion-except-ours-has-taught-that-man-is-born as quoted from Pensées by Blaise Pascal.
6. **James Clerk Maxwell quote**: "Discovering the Unknown: 50 Famous Scientists' Quotes to Expand Your Mind," Empuls by Giift, October 4, 2023, blog.empuls.io/famous-scientists-quotes/.
7. **Michael Faraday quote**: Ian H. Hutchinson, "The Genius and Faith of Faraday and Maxwell," The New Atlantis, No. 41, Winter 2014, pp. 81-99, published by The Center for the Study of Technology and Society, www.thenewatlantis.com/publications/the-genius-and-faith-of-faraday-and-maxwell.
8. **Dr. Allen Rex Sandage quote**: "Allen Sandage Quotes," AZ Quotes, accessed July 2024, www.azquotes.com/author/28127-Allan_Sandage as quoted from John Noble Wilford, "Sizing up the Cosmos: An Astronomer's Quest" www.nytimes.com. March 12, 1991.
9. **Dwight L. Moody quote**: "Dwight L. Moody Quotes - Page 3," AZ Quotes, accessed July 2024, www.azquotes.com/author/10304-Dwight_L_Moody?p=3.
10. **Isaac Newton quote**: David Ng, "Isaac Newton, Standing on the Shoulder of Giants. Excerpts from His Diary," The Science Creative Quarterly, 3/11/2014, www.scq.ubc.ca/isaac-newton-standing-on-the-shoulder-of-giants-excerpts-from-his-diary/ as quoted from Newton's "letter to his rival Robert Hooke, 1676."
11. **James Irwin quote**: Jeremy Kocal, "NASA Quotes: The Apollo Missions to the Moon," Pilgrimage: Exploring good, evil, science, religion, accessed July 2024, pilgrimage.space/astronaut-quotes/ as quoted from Kevin V. Kelley, The Home Planet.
12. **Carl Linnaeus quote**: Curt Blattman, "Quotes from Great Christian Scientists on God," Bible Apologetics – A Daily Devotional, October 17, 2021, bibleapologetics.org/quotes-from-great-christian-scientists-on-god/.

CHAPTER 9

1. **Dr. Martin H. Fischer quote**: "Home > Dictionary of Science Quotations > Scientist Names Index F > Martin H. Fischer Quotes," Today in Science History, accessed July 2024, todayinsci.com/F/Fischer_Martin/FischerMartin-Quotations.htm.

2. **Dr. Max Planck quote**: Curt Blattman, "Quotes from Great Christian Scientists on God," Bible Apologetics – A Daily Devotional, October 17, 2021, bibleapologetics.org/quotes-from-great-christian-scientists-on-god/.
3. **Robert Boyle quote**: Curt Blattman, "Quotes from Great Christian Scientists on God," Bible Apologetics – A Daily Devotional, October 17, 2021, bibleapologetics.org/quotes-from-great-christian-scientists-on-god/.
4. **Sir Karl Popper quote**: "Karl Popper Quotes," AZ Quotes, accessed July 2024, www.azquotes.com/quote/234474 attributed to "Lecture to British Academy, 20 Jan. 1960, in Proceedings of the British Academy (1960) vol. 46, p. 69."
5. **Dr. Ilya Prigogine quote**: "Ilya Prigogine," Wikiquote, September 29, 2023, en.wikiquote.org/wiki/Ilya_Prigogine#:~:text=The%20probability%20that%20at%20ordinary%20temperatures%20a%20macroscopic,coordinated%20functions%20characterizing%20living%20organisms%20is%20vanishingly%20small as quoted from "Part 1, p. 23; As cited in: Meyer, Stephen C. "DNA by design: An inference to the best explanation for the origin of biological information." Rhetoric & Public Affairs 1.4 (1998): 551."
6. **Dr. Robert Jastrow quote**: Sarah Salviander, "In their own words—Robert Jastrow," SixDay Science, accessed July 2024, sixdayscience.com/2015/09/04/in-their-own-words-robert-jastrow/.
7. **Dr. John G. Leslie quote**: John Leslie, "Religion, Science, Philosophy And How Each Interfaces In The Noah Flood Account," May 28, 2013, creation.com/religion-science-philosophy-interface-noah-flood.
8. **James Clerk Maxwell quote**: "James Clerk Maxwell Quotes," AZ Quotes, accessed July 2024, www.azquotes.com/author/21477-James_Clerk_Maxwell as quoted from "James Clerk Maxwell, W. D. Niven (2003). "The Scientific Papers of James Clerk Maxwell", p.376, Courier Corporation."
9. **Dr. John Archibald Wheeler quote**: Dennis Overbye, "John A. Wheeler, Physicist Who Coined the Term 'Black Hole,' Is Dead at 96," New York Times, April 14, 2008, www.nytimes.com/2008/04/14/science/14wheeler.html.
10. **Johannes Kepler quote**: John Stonestreet and Glenn Sunshine, "Johannes Kepler, Thinking God's Thoughts After Him," Breakpoint: Colson Center, November 27, 2023, breakpoint.org/johannes-kepler-thinking-gods-thoughts-after-him/.
11. **Dr. Arthur L. Schawlow quote**: "Arthur Leonard Schawlow Quotes," AZ Quotes, accessed July 2024, www.azquotes.com/author/21540-Arthur_Leonard_Schawlow.
12. **Johannes Kepler quote:** Pettinger, Tejvan, "Johannes Kepler Biography," Biography Online, June 27, 2019, www.biographyonline.net/scientists/johannes-kepler.html.
13. **Joseph Henry quote**: "Joseph Henry Quotations-- On Science," Smithsonian Institution Archives, accessed July 2024, siarchives.si.edu/sites/default/files/pdfs/jhpp/JHP_Joseph_Henry_Quotations.pdf. Attributed

to "Presidential address to the American Association for the Advancement of Science, [August 22, 1850], in Henry Papers, vol. 8, p. 99."

14. **Michael Faraday quote**: Jürgen Martschukat , "'The Art of Killing by Electricity': The Sublime and the Electric Chair," American History, accessed July 2024, academic.oup.com/jah/article-abstract/89/3/900/832569 as quoted from Journal of American History, Volume 89, Issue 3, December 2002, Pages 900–921.
15. **Dr. Nikola Tesla quote**: Arjun Walia, "Why Nikola Tesla Was A Spiritual Mystic & Told The World To Study Non-Physical Phenomena," Stillness in the Storm, June 19, 2018, stillnessinthestorm.com/2018/06/19/why-nikola-tesla-was-a-spiritual-mystic-told-the-world-to-study-non-physical-phenomena/09/01/27/63204/consciousness/stillness-in-the-storm/.
16. **Dr. Thomas Gold quote**: The Doc, "30 Brilliant Scientist Quotes," Famous Scientists: The Art of Genius, accessed July 2024, www.famousscientists.org/30-brilliant-scientist-quotes/.
17. **Dr. Stephen Taylor quote**: Dr. John Ashton and Prof. Steve Taylor, "Chapter 35, Stephen Taylor, Electrical Engineering," Answers in Genesis, accessed July 2024, answersingenesis.org/answers/books/in-six-days/stephen-taylor-electrical-engineering/. "The book In Six Days has been graciously provided at no charge to Answers in Genesis by Master Books, a division of New Leaf Press (Green Forest, Arkansas)."
18. **Dr. D.B. Gower quote**: Dr John Ashton, editor, "D.B. Gower, biochemistry," accessed July 2024, Creation.com, Creation Ministries International, creation.com/d-b-gower-biochemistry-in-six-days. "The book In Six Days has been graciously provided at no charge to Answers in Genesis by Master Books, a division of New Leaf Press (Green Forest, Arkansas)."
19. **Leonardo Da Vinci quote**: "Home > Dictionary of Science Quotations > Scientist Names Index D > Leonardo da Vinci Quotes," Today in Science History, accessed July 2024, todayinsci.com/D/DaVinci_Leonardo/DaVinciLeonardo-Quotations.htm. Quoted from Introductory College Physics by Oswald Blackwood (1939).
20. **Dr. Charles Townes quote**: Tihomir Dimitrov, 50 Nobel Laureates and Other Great Scientists Who Believe in God, 1995-2008, accessed July 2024, www.academia.edu/33919294/50_nobel_laureates_and_other_great_scientists_who_believe_in_god.
21. **Thomas Hill quote**: "Home > Dictionary of Science Quotations > Scientist Names Index H > Thomas Hill Quotes," Today in Science History, accessed July 2024, todayinsci.com/H/Hill_Thomas/HillThomas-Quotations.htm. As quoted from 'The Uses of Mathesis', Bibliotheca Sacra, 31, 532.
22. **Tony Hewish quote**: Scott Youngren, "Quotes about God to consider… if you think science leads to atheism," God Evidence, October 25, 2014, https:// godevidence.com/2014/10/reality-miracles/ as quoted in the

foreword to John Polkinghorne and Nicholas Beale's book Questions of Truth: Fifty-one Responses to Questions about God, Science, and Belief.

23. **Robert Boyle quote**: "Robert Boyle Quote," Libquotes, accessed July 2024, libquotes.com/robert-boyle/quote/lbh9s1q. Quoted in Peter Gay, The Enlightenment (1977).
24. **Dr. James Tour quote**: "Top World Chemist, James Tour, To Speak at COSM 2021," Discovery Institute, September 27, 2021, mindmatters.ai/2021/09/top-world-chemist-james-tour-to-speak-at-cosm-2021/.
25. **Dr. John P. Marcus quote**: Dr John Ashton, editor, "John P. Marcus, biochemistry," Creation.com, Creation Ministries International, accessed July 2024, creation.com/john-p-marcus-biochemistry-in-six-days.

CHAPTER 10

1. **Thomas Paine quote: (1)** "Home > Dictionary of Science Quotations > Scientist Names Index P > Thomas Paine Quotes," Today in Science History, accessed July 2024, todayinsci.com/P/Paine_Thomas/PaineThomas-Quotations.htm as quoted from 'Address to the Addressers,' The Political Writings of Thomas Paine (1824), Vol. 2, 284. **(2)** "Home > Dictionary of Science Quotations > Scientist Names Index P > Thomas Paine Quotes," Today in Science History, accessed July 2024, todayinsci.com/P/Paine_Thomas/PaineThomas-Quotations.htm as quoted from The Age of Reason: Being an Investigation of True and Fabulous Theology (27 Jan O.S. 1794), 44. **(3)** "Home > Dictionary of Science Quotations > Scientist Names Index P > Thomas Paine Quotes," Today in Science History, accessed July 2024, todayinsci.com/P/Paine_Thomas/PaineThomas-Quotations.htm as quoted from Age of Reason (1794, 1818), 35.
2. **Dr. David Eagleman quote**: Advaita Suresh, "Inspirational Quotes From Famous Scientists," CareerIndia, February 22, 2024, https://www.careerindia.com/general-knowledge/inspirational-quotes-from-famous-scientists-039711.html.
3. **Dr. Edward A. Boudreaux quote**: John Ashton, editor, "Edward A. Boudreaux, theoretical chemistry," Creation.com, Creation Ministries International, accessed July 2024, creation.com/edward-a-boudreaux-theoretical-chemistry-in-six-days.
4. **Dr. Ken Funk quote**: "Gary Bates interviews peptide chemist Ken Funk Ph.D.," Creation.com, Creation Ministries International, accessed July 2024, creation.com/god-left-handed. This article is from Creation 36(1):32–34, January 2014.
5. **Dr. Richard P. Feynman quote**: "Top 25 Quotes by Richard P. Feynman (of 377)," AZ Quotes, accessed July 2024, www.azquotes.com/author/4774-Richard_P_Feynman.

6. **Guglielmo Marconi quote**: "Guglielmo Marconi Quotes," AZ Quotes, accessed July 2024, www.azquotes.com/author/23859-Guglielmo_Marconi.
7. **Dr. John Arbuthnot quote**: "Science Quotes by John Arbuthnot (12 quotes)," Today in Science History, accessed July 2024, todayinsci.com/A/Arbuthnot_John/ArbuthnotJohn-Quotations.htm as quoted from An Essay on the Usefulness of Mathematical Learning (1701), 3-4.
8. **Dr. John K.G. Kramer quote**: John K.G. Kramer, "Chapter 3, John K.G. Kramer, Biochemistry," Answers in Genesis, accessed July 2024, answersingenesis.org/answers/books/in-six-days/john-kg-kramer-biochemistry/. "The book In Six Days has been graciously provided at no charge to Answers in Genesis by Master Books, a division of New Leaf Press (Green Forest, Arkansas)."
9. **Ken Ham quote: Otis Chandler**, "The Lie Quotes," GoodReads, accessed July 2024, www.goodreads.com/work/quotes/625208-the-lie-evolution as quoted from Ken Ham, The Lie: Evolution.
10. **Sir Oliver Joseph Lodge quote**: "Root Quotes (121 quotes)," Today in Science History, accessed July 2024, todayinsci.com/QuotationsCategories/R_Cat/Root-Quotations.htm as quoted from Continuity: The Presidential Address to the British Association (1913), 92-93.
11. **Dr. George David Birkhoff quote**: "George David Birkhoff Quote," LibQuotes, accessed July 2024, libquotes.com/george-david-birkhoff/quote/lbi4x6h.
12. **G.K. Chesterton quote**: Otis Chandler, "G.K. Chesterton > Quotes > Quotable Quote," GoodReads, accessed July 2024, www.goodreads.com/quotes/39954-it-is-absurd-for-the-evolutionist-to-complain-that-it.
13. **Sir James Jeans quote**: "Home > Dictionary of Science Quotations > Scientist Names Index J > Sir James Jeans Quotes," Today in Science History, accessed July 2024, todayinsci.com/J/Jeans_James/JeansJames-Quotations.htm as quoted from The Mysterious Universe (1930, 1932), 97-98.
14. **Sir Arthur Eddington quote**: Scott Youngren, "Quotes about God to consider…if you think science leads to atheism," God Evidence, August 12, 2010, godevidence.com/2010/08/quotes-about-god-atheism/ as quoted in Arthur Eddington, The Nature of the Physical World.
15. **Alexander Graham Bell quote:** "Quotes by Alexander Graham Bell," AZ Quotes, accessed July 2024, www.azquotes.com/author/1156-Alexander_Graham_Bell.
16. **Arthur Schawlow quote**: "Arthur Leonard Schawlow Quotes," AZ Quotes, accessed July 2024, www.azquotes.com/author/21540-Arthur_Leonard_Schawlow.
17. **Dr. Arthur Jones quote**: John Ashton, editor, "Arthur Jones, biology," Creation.com, Creation Ministries International, accessed July 2024, creation.com/arthur-jones-biology-in-six-days.

18. **Dr. Timothy G. Standish quote**: Timothy G. Standish, "Chapter 9, Timothy G. Standish, Biology," Answers in Genesis, accessed July 2024, answersingenesis.org/answers/books/in-six-days/timothy-g-standish-biology/. "The book In Six Days has been graciously provided at no charge to Answers in Genesis by Master Books, a division of New Leaf Press (Green Forest, Arkansas)."
19. **Tycho Brahe quote**: "Tycho Brahe Quotes," AZ Quotes, accessed July 2024, www.azquotes.com/quote/586495.
20. **George Washington Carver quote**: "Sermons and Biblical Studies," Biblia.work, accessed July 2024, www.biblia.work/sermons/carver-george-washington/ as quoted from George Washington Carver in His Own Words, ed. Gary R. Kremer (Columbia: University of Missouri Press, 1987), specifically the chapter titled "The Scientist as Mystic: Reading God out of Nature's Book" (pp. 138–43).
21. **Dr. Lane P. Lester quote**: Lane Lester, "Genetics: No Friend of Evolution," Answers in Genesis, accessed July 2024, answersingenesis.org/genetics/genetics-no-friend-of-evolution/. Originally published in Creation 20, no 2 (March 1998): 20-22.
22. **Dr. Andrew Hodge quote**: Gary Bates, "Dealing with the heart of the problem, Speaking with leading heart and lung surgeon Andrew Hodge," Creation.com, Creation Ministries International, accessed July 2024, creation.com/dealing-with-the-heart-of-the-problem. This article is from Creation 25(4):16–18, September 2003.
23. **John Ray quote**: "Science Quotes by John Ray (7 quotes)," Today in Science History, accessed July 2024, todayinsci.com/R/Ray_John/Ray-John-Quotations.htm. This is from The Wisdom of God Manifested in the Works of the Creation (1691), 169-70.
24. **Dr. Patrick H. Young quote**: Don Batten and Jonathan Sarfati, "Bullet-proofing belief, A talk with research chemist Patrick Young," Creation.com, Creation Ministries International, accessed July 2024, creation.com/bullet-proofing-belief. This article is from Creation 23(2):19–21, March 2001.
25. **Dr. Lawrence O. Richards quote**: Otis Chandler, "Lawrence O. Richards," GoodReads, accessed July 2024, www.goodreads.com/author/show/72217.Lawrence_O_Richards.
26. **Dr. William D. Phillips quote**: William D. Phillips, "Does Science Make Belief in God Obsolete?," Fair Observer, November 10, 2013, www.fairobserver.com/culture/does-science-make-belief-god-obsolete/.

CHAPTER 11

1. **Dr. Robert Breault quote**: "Robert Breault Quotes," AZ Quotes, accessed July 2024, www.azquotes.com/quote/554790.
2. **Dr. Saami Shaibani quote**: Carl Wieland and Don Batten, "Physics, faith and the FBI, A talk with physicist Saami Shaibani," Creation.com,

Creation Ministries International, accessed July 2024, creation.com/physics-faith-and-the-fbi. This article is from Creation 23(1):18–21, December 2000.

3. **William McDonough quote**: "Science Quotes by William McDonough (15 quotes)," Today in Science, accessed July 2024, todayinsci.com/M/McDonough_William/McDonoughWilliam-Quotations.htm as quoted from Anne Underwood, 'Designing For The Future', Newsweek (15 May 2005).
4. **Joseph H. Taylor, Jr. quote**: "Joseph Hooton Taylor, Jr. Quotes," AZ Quotes, accessed July 2024, www.azquotes.com/quote/587231.

CHAPTER 12

1. **Dr. Werner Heisenberg quote**: Otis Chandler, "Werner Heisenberg > Quotes," GoodReads, accessed July 2024, www.goodreads.com/author/quotes/64309. Werner_Heisenberg, from Werner Heisenberg, Across the Frontier.
2. **Thomas A. Edison quote**: Erin Allen, "Imagination and Invention," Library of Congress Blogs, Timeless Stories from the Library of Congress, August 30, 2013, blogs.loc.gov/loc/2013/08/imagination-and-invention/.
3. **Samuel Morse quote**: "Samuel Morse Quotes," AZ Quotes, accessed July 2024, www.azquotes.com/author/37705-Samuel_Morse.

CHAPTER 13

1. **Dr. A.J. Monty White quote**: Dr. Monty White, "Chapter 27, A. J. Monty White, Physical Chemistry," Answers in Genesis, accessed July 2024, answersingenesis.org/answers/books/in-six-days/a-j-monty-white-physical-chemistry/. "The book In Six Days has been graciously provided at no charge to Answers in Genesis by Master Books, a division of New Leaf Press (Green Forest, Arkansas)."
2. **A.W. Tozer quote**: Otis Chandler, "A.W. Tozer > Quotes," GoodReads, accessed July 2024, www.goodreads.com/author/quotes/1082290.A_W_Tozer.
3. **Dr. Stephen Taylor quote**: Dr John Ashton, editor, "Stephen Taylor, electrical engineering," Creation.com, Creation Ministries International, accessed July 2024, creation.com/stephen-taylor-electrical-engineering-in-six-days.
4. **Dr. Ian Macreadie quote**: Carl Wieland and Don Batten, "Creation in the research lab, An interview with leading Australian molecular biologist and microbiologist Ian Macreadie," Creation.com, Creation Ministries International, accessed July 2024, creation.com/creation-in-the-research-lab. This article is from Creation 21(2):16–17, March 1999.
5. **Blaise Pascal quote**: "Blaise Pascal Quotes," Brainy Media, Inc, BrainyQuote, accessed July 2024, www.brainyquote.com/quotes/blaise_pascal_159868.
6. **Dr. John Lee Comstock quote**: "Quotes by others about Law (242)," Today in Science History, accessed July 2024, todayinsci.com/L/Law/Law-Quotations.htm. Quoted from Elements of Chemistry (1845), 84.

7. **Dr. Vera Kistiakowsky quote**: "Teleological Argument and Scientists," Discover All about Philosophy, accessed July 2024, www.allaboutphilosophy.org/teleological-argument-and-scientists-faq.htm. Quoted from H Margenau and R.A. Varghese, ed. 1992. Cosmos, Bios, and Theos. La Salle, IL, Open Court, p. 52.
8. **Maria Mitchell quote**: "Maria Mitchell Quotes," Brainy Media, Inc, BrainyQuote, accessed July 2024, www.brainyquote.com/quotes/maria_mitchell_317047.
9. **Dr. Brian Stone quote**: Don Batten and Carl Wieland, "Prize-winning Professor rejects evolution, Interview with Prof. Brian Stone," Creation.com, Creation Ministries International, accessed July 2024, creation.com/prize-winning-professor-rejects-evolution. This article is from Creation 20(4):52–53, September 1998.
10. **Dr. Ker C. Thomson quote**: Ker C. Thomson, "Chapter 23, Ker C. Thomson, Geophysics," Answers in Genesis, accessed July 2024, answersingenesis.org/answers/books/in-six-days/ker-c-thomson-geophysics/. "The book In Six Days has been graciously provided at no charge to Answers in Genesis by Master Books, a division of New Leaf Press (Green Forest, Arkansas)."
11. **Dr. George Ellis quote**: "Teleological Argument and Scientists," Discover All about Philosophy, accessed July 2024, www.allaboutphilosophy.org/teleological-argument-and-scientists-faq.htm. Quoted from Henry F. Schaefer, Science and Christianity, 71.
12. **Dr. Jim Mason quote**: "Nuclear physicist embraces biblical creation, Jonathan Sarfati chats with Dr Jim Mason," Creation.com, Creation Ministries International, accessed July 2024, /creation.com/jim-mason-nuclear-physicist. This article is from Creation 33(3):47–49, July 2011.

CHAPTER 14

1. **Sir Fred Hoyle quote**: Otis Chandler, "Fred Hoyle > Quotes," GoodReads, accessed July 2024, /www.goodreads.com/author/quotes/199992.Fred_Hoyle.
2. **Dr. Paula M. Dirac quote**: Daniel C. Peterson, "God is a mathematician of a very high order," Sic et Non, Patheos, September 26, 2020, www.patheos.com/blogs/danpeterson/2020/09/god-is-a-mathematician-of-a-very-high-order.html
3. **John Ray quote**: "John Ray Quotes," AZ Quotes, accessed July 2024, www.azquotes.com/quote/818004.
4. **Dr. William D. Phillips quote**: Tihomir Dimitrov, 50 Nobel Laureates and Other Great Scientists Who Believe in God, 1995-2008, accessed July 2024, www.academia.edu/33919294/50_nobel_laureates_and_other_great_scientists_who_believe_in_god.
5. **Blaise Pascal quote**: "Section XII: Proofs of Jesus Christ," Christian Classics Ethereal Library, "Blaise Pascal," Pensees, accessed July 2024, ccel.org/ccel/pascal/pensees/pensees.viii.html.

6. **James Dwight Dana quote**: "James Dwight Dana Quotes," AZ Quotes, accessed July 2024, www.azquotes.com/author/26521-James_Dwight_Dana as quoted from James Dwight Dana (1875). "Manual of Geology: Treating of the Principles of the Science with Special Reference to American Geological History", 770.
7. **Dr. Arno Penzias quote**: Scott Youngren, "Quotes about God to consider…if you think science leads to atheism," God Evidence, August 12, 2010, godevidence.com/2010/08/quotes-about-god-atheism/as quoted from Joshua O. Haberman editor, 'The God I Believe in,' New York, Maxwell Macmillan International, 1994, 184.
8. **Isaac Newton quote**: "Isaac Newton Quotes," Brainy Media, Inc, BrainyQuotes, accessed July 2024, www.brainyquote.com/quotes/isaac_newton_753054.
9. **Dr. George Ellis quote**: "George F. R. Ellis Quotes," AZ Quotes, accessed July 2024, www.azquotes.com/quote/587236.
10. **Blaise Pascal quote**: Otis Chandler, "Blaise Pascal > Quotes > Quotable Quote," GoodReads, accessed July 2024, www.goodreads.com/quotes/801132-there-is-a-god-shaped-vacuum-in-the-heart-of-each.
11. **Dr. Arno Penzias quote**: Tihomir Dimitrov, "Nobel Laureate Arno Penzias on God & Big Bang," Scientific God Inc., 12/13/2011, 2012daily.com/?q=node/92.
12. **Dr. Thomas Sydenham quote**: Stephen Browne, "Sydenham the Physician," CMF: Christian Medical Fellowship, Spring 1996, www.cmf.org.uk/resources/publications/content/?context=article&id=529.
13. **Charles Babbage quote**: Paul James-Griffiths, "Eighteenth and Nineteenth Centuries," Christian Heritage Edinburgh, Aug 23, 2016, www.christianheritageedinburgh.org.uk/category/science/quotes-from-famous-christians-in-science-and-medicine/ as quoted from *Babbage's Ninth Bridgewater Treatise: A Fragment*.

CHAPTER 15

1. **Dr. Roger Penrose quote**: Scott Youngren, "Quotes about God to consider…if you think science leads to atheism," God Evidence, August 12, 2010, godevidence.com/2010/08/quotes-about-god-atheism/as quoted in as cited in Roger Penrose, The Emperor's New Mind.
2. **Dr. Robert Millikan quote**: Joe Spears MS, "What Scientists REALLY Say about God, Creation and Evolution," Triangle Association for the Science of Creation, February 01, 2013, /tasc-creationscience.org/article/what-scientists-really-say-about-god-creation-and-evolution.
3. **Nicolaus Copernicus quote**: "Quotes › Authors › N › Nicolaus Copernicus," AZ Quotes, accessed July 2024, www.azquotes.com/quote/575512 as quoted from Louis E. Van Norman, "Poland: The Knight Among Nations," 1907, 290.
4. **Dr. Tony Rothman quote**: "Tony Rothman Quote," LibQuotes, accessed July 2024, libquotes.com/tony-rothman/quote/lbz8u1m.

5. **Isaac Newton quote**: Otis Chandler, "Isaac Newton > Quotes > Quotable Quote," GoodReads, accessed July 2024, www.goodreads.com/quotes/523034-how-came-the-bodies-of-animals-to-be-contrived-with.
6. **Stephan Hales quote**: "Stephen Hales Quotes," AZ Quotes, accessed July 2024, www.azquotes.com/author/50047-Stephen_Hales.
7. **Dr. Robert Carter quote**: Robert W. Carter, "Skin colour surprises," Creation.com, Creation Ministries International, 23 November 2017, creation.com/skin-colour-surprises.
8. **Dr. George F.R. Ellis quote**: "Quotes › Authors › G › George F. R. Ellis," AZ Quotes, accessed July 2024, www.azquotes.com/quote/784181 as quoted from "The Anthropic Principle: Laws and Environments". Article in "The Anthropic Principle: Proceedings of the Second Venice Conference on Cosmology and Philosophy", book edited by F. Bertola and U. Curi, Cambridge University Press, 1988, 30.
9. **Sir Ernest Chain quote**: Graeme, "Evidence for and against God (quotes)," Wisdom Trove, accessed July 2024, wisdomtrove.com/evidence-for-and-against-god-quotes/.
10. **C.S. Lewis quote**: Otis Chandler, "C.S. Lewis > Quotes > Quotable Quote," GoodReads, accessed July 2024, www.goodreads.com/quotes/7989365-if-god-annihilates-or-deflects-or-creates-a-unit-of.
11. **Dr. Antony Hewish quote**: Tihomir Dimitrov, 50 Nobel Laureates and Other Great Scientists Who Believe in God, 1995-2008, accessed July 2024, www.academia.edu/33919294/50_nobel_laureates_and_other_great_scientists_who_believe_in_god.
12. **Dr. Nevill Mott quote**: Tihomir Dimitrov, "50 Nobelists on God," issue, Aug 31, 2017, issuu.com/tihomirdimitrov1/docs/50_nobelists_on_god.
13. **Dr. Christian Anfinsen quote**: Otis Chandler, "Christian B. Anfinsen > Quotes > Quotable Quote," GoodReads, accessed July 2024, www.goodreads.com/quotes/1280285-i-think-only-an-idiot-can-be-an-atheist-we.
14. **Stephen Taylor quote**: Dr. John Ashton and Prof. Steve Taylor, "Chapter 35, Stephen Taylor, Electrical Engineering," Answers in Genesis, accessed July 2024, answersingenesis.org/answers/books/in-six-days/stephen-taylor-electrical-engineering/. "The book In Six Days has been graciously provided at no charge to Answers in Genesis by Master Books, a division of New Leaf Press (Green Forest, Arkansas)."

CHAPTER 16

1. **James Clerk Maxwell quote**: Ian Hutchinson, "James Clerk Maxwell and the Christian Proposition," MIT IAP Seminar: The Faith of Great Scientists, Jan 1998, 2006, James Clerk Maxwell and the Christian Proposition (mit.edu), silas.pfsc.mit.edu/Maxwell/.
2. **Charles Babbage quote**: Scott Youngren, "Quotes about God to consider… if you think science leads to atheism," God Evidence, August 12, 2010,

godevidence.com/2010/08/quotes-about-god-atheism/ as quoted in Charles Babbage, Passages from the Life of a Philosopher,1864, pp. 396–402.

3. **Dr. William D. Phillips quote**: Daniel C. Peterson, "The religious faith of two Nobel laureate physicists," Sic et Non, Patheos, November 18, 2019, /www.patheos.com/blogs/danpeterson/2019/11/the-religious-faith-of-two-nobel-laureate-physicists.html.
4. **Dr. Alexander Fleming quote**: Scott Youngren, "Quotes about God to consider…if you think science leads to atheism." God Evidence, August 12, 2010, https://godevidence.com/2010/08/quotes-about-god-atheism/.
5. **Michael Faraday quote**: The Doc, "Michael Faraday," Famous Scientists, accessed July 2024, /www.famousscientists.org/michael-faraday/.
6. **Ken Ham quote**: Otis Chandler, "A Flood of Evidence Quotes," GoodReads, accessed July 2024, www.goodreads.com/work/quotes/50936663-a-flood-of-evidence-40-reasons-noah-and-the-ark-still-matter?page=3.
7. **Sir Joseph J. Thomson quote**: Scott Youngren, "Quotes about God to consider…if you think science leads to atheism," God Evidence, August 12, 2010, godevidence.com/2010/08/quotes-about-god-atheism/ as quoted in Presidential Address to the British Association, as quoted in Arthur L. Foley, 'Recent Developments in Physical Science, The Popular Science Monthly (1910), 456.
8. **Dr. Ker C. Thomson quote**: Ker C. Thomson, "Chapter 23, Ker C. Thomson, Geophysics," Answers in Genesis, accessed July 2024, answersingenesis.org/answers/books/in-six-days/ker-c-thomson-geophysics/. "The book In Six Days has been graciously provided at no charge to Answers in Genesis by Master Books, a division of New Leaf Press (Green Forest, Arkansas)."
9. **Ken Ham quote**: Otis Chandler, "A Flood of Evidence Quotes," GoodReads, accessed July 2024, www.goodreads.com/work/quotes/50936663-a-flood-of-evidence-40-reasons-noah-and-the-ark-still-matter.
10. **Dr. Robert Carter quote**: Dr. Robert Carter, "Are Neandertals pre-Flood people?," Creation.com, Creation Ministries International, February 2, 2019, creation.com/neanderthals-pre-flood.
11. **Ken Ham quote**: Otis Chandler, "A Flood of Evidence Quotes," GoodReads, accessed July 2024, www.goodreads.com/work/quotes/50936663-a-flood-of-evidence-40-reasons-noah-and-the-ark-still-matter?page=3.
12. **Gottfried Wilhelm Leibniz quote**: "Gottfried Wilhelm Leibniz Quotes on God (7 quotes)," Today in Science History, accessed July 2024, todayinsci.com/L/Leibniz_Gottfried/LeibnizGottfried-God-Quotations.htm.
13. **Dr. Jim Mason quote**: "Nuclear physicist embraces biblical creation, Jonathan Sarfati chats with Dr Jim Mason," Creation.com, Creation Ministries International, accessed July 2024, creation.com/jim-mason-nuclear-physicist. This article is from Creation 33(3):47–49, July 2011.

14. **Dr. John K.G. Kramer quote**: John K.G. Kramer, "Chapter 3, John K.G. Kramer, Biochemistry," Answers in Genesis, accessed July 2024, answersingenesis.org/answers/books/in-six-days/john-kg-kramer-biochemistry/. "The book In Six Days has been graciously provided at no charge to Answers in Genesis by Master Books, a division of New Leaf Press (Green Forest, Arkansas)."
15. **Dr. Timothy Clarey quote**: "How the rocks fit the Flood, Shaun Doyle talks to Tim Clarey about his research career in Flood geology," Creation.com, Creation Ministries International, accessed July 2024, creation.com/tim-clarey-interview. This article is from Creation 41(4):20–23, October 2019.
16. **Dr. D. Russell Humphreys quote**: D. Russell Humphreys, "Not too old to be Expelled," Creation.com, Creation Ministries International, August 29, 2008, creation.com/not-too-old-to-be-expelled.
17. **Carl Friedrich Gauss quote**: Javier Ordovas, "25 Famous Scientists on God," June 26, 2016, www.catholiceducation.org/en/science/faith-and-science/25-famous-scientists-on-god.html.
18. **Dr. William Henry Bragg quote**: "William Henry Bragg Quotes," AZ Quotes, accessed July 2024, www.azquotes.com/author/21422-William_Henry_Bragg as quoted from Sir William Henry Bragg, "The World of Sound: Six Lectures Delivered at the Royal Institution," 1942.
19. **Dr. Brian Thomas quote**: "How long can proteins last? Jonathan Sarfati chats with expert in protein and ancient fossil chemistry, Dr Brian Thomas," Creation.com, Creation Ministries International, January 25, 2021, creation.com/brian-thomas-interview. This article is from Creation 42(1):38–41, January 2020.

CHAPTER 17

1. **Dr. Brian Thomas quote**: "How long can proteins last? Jonathan Sarfati chats with expert in protein and ancient fossil chemistry, Dr Brian Thomas," Creation.com, Creation Ministries International, January 25, 2021, creation.com/brian-thomas-interview. This article is from Creation 42(1):38–41, January 2020.
2. **Dr. Max Planck quote**: Paul Luminari, "Does Matter Exist Or Is It All Just An Illusion?," SHIFT>, April 4, 2013, www.shift.is/does-matter-exist-or-is-it-all-just-an-illusion/ as quoted from Wolfgang Pauli. "The Influence of Archetypical Presentations on the Development of the Natural Science Theory by Keppler" in Natural Explanation and Psyche, Zuerich 1952, p. 163.
3. **Dr. John Lennox quote**: John Lennox, "Lent Talk on science and religion (transcript)," JohnLennox.org, March 14, 2012, www.johnlennox.org/resources/27/lent-talk-on-science-and. This is a transcript of a 'Lent Talk' given by John Lennox on (BBC) Radio 4 on 14 March 2012.

4. **Dr. Fred Hoyle quote**: Trevor Nace, "20 Inspirational Quotes From Famous Scientists," Forbes Magazine, March 6, 2016, www.forbes.com/sites/trevornace/2016/03/06/20-inspirational-quotes-famous-scientists/.
5. **Dr. John Lennox quote**: "John Lennox Quotes: 100+ Powerful Quotes to Shift Your Perspective," StillFaith.com, accessed July 2024, stillfaith.com/christian-quotes/john-lennox-quotes/.
6. **George Washington Carver quote**: Paul James-Griffiths, "Nineteenth and Twentieth Centuries," Christian Heritage Edinburgh, Aug 23, 2016, www.christianheritageedinburgh.org.uk/category/science/quotes-from-famous-christians-in-science-and-medicine/ as quoted from Ann Lamont, 21 Great Scientists who believed the Bible, Creation Science Foundation, © 1995, p.228.

CHAPTER 18

1. **Galileo Galilei quote**: "Galileo Galilei Quotes," AZ Quotes, accessed July 2024, www.azquotes.com/author/5284-Galileo_Galilei as quoted from Stillman Drake, "Discoveries and Opinions of Galileo," 1957, p. 134-135.
2. **Dr. John G. Leslie quote**: John G. Leslie, Ph.D. and Charles K. Pallaghy, Ph.D., "The Religious Nature of Evolution Theory and its Attack on Christianity," Creation.com, Creation Ministries International, accessed July 2024, creation.com/evolution-religious. This article is from Creation 7(4):42–48, June 1986.
3. **Dr. Neil Huber quote**: Carl Wieland and Don Batten, "Evolution abandoned, A talk with physical anthropologist Dr Neil Huber," Creation.com, Creation Ministries International, January 4, 2017, creation.com/evolution-abandoned. This article is from Creation 24(2):44–45, March 2002.
4. **Dr. John Lennox quote**: "John Lennox Quotes: 100+ Powerful Quotes to Shift Your Perspective," StillFaith.com, accessed July 2024, stillfaith.com/christian-quotes/john-lennox-quotes/.
5. **Carl Friedrich Gauss quote**: "Carl Friedrich Gauss Quotes," AZ Quotes, accessed July 2024, www.azquotes.com/author/5394-Carl_Friedrich_Gauss as quoted from J. R. Newman, "The World of Mathematics," 1956.
6. **Dr. Linus Pauling quote**: "Linus Pauling Quotes," AZ Quotes, accessed July 2024, www.azquotes.com/author/11421-Linus_Pauling.
7. **Dr. Erwin Chargaff quote**: "Erwin Chargaff," wikipedia, March 15, 2024, en.wikipedia.org/wiki/Erwin_Chargaff as found in Weintraub (2002).
8. **Dr. Erwin Schrödinger quote**: Scott Youngren, "Quotes about God to consider...if you think science leads to atheism," God Evidence, August 12, 2010, godevidence.com/2010/08/quotes-about-god-atheism/.
9. **Dr. Albert Einstein quote**: "Albert Einstein," All Great Quotes, accessed July 2024, www.allgreatquotes.com/quote-10984/.
10. **Dr. Markus Blietz quote**: "Galaxies, Black Holes, and Creation, Jonathan Sarfati chats with astrophysicist Dr Markus Blietz," Creation.com, Creation Ministries International, January 11, 2023, creation.com/

markus-blietz-astrophysicist-interview. This article is from Creation 35(2):34–36, April 2013.

11. **Dr. John Ambrose Fleming quote**: "Quotes by John Ambrose Fleming," DictionaryBook, accessed July 2024, www.dictionarybook.org/quotes/john-ambrose-fleming.html.
12. **Werner Heisenberg quote**: Otis Chandler, "Werner Heisenberg > Quotes > Quotable Quote," GoodReads, accessed July 2024, www.goodreads.com/quotes/9627737-the-first-gulp-from-the-glass-of-natural-sciences-will.
13. **Dr. David A. Dewitt quote**: David A. DeWitt, Ph.D., "Why I rejected 'theistic evolution,'" Creation.com, Creation Ministries International, January 12, 2004, creation.com/why-i-rejected-theistic-evolution.
14. **Blaise Pascal quote**: "Quotations, Blaise Pascal," MacTutor, School of Mathematics and Statistics, University of St Andrews, Scotland, accessed July 2024, mathshistory.st-andrews.ac.uk/Biographies/Pascal/quotations/ as quoted in Des MacHale, Wisdom, (London, 2002).
15. **Sir Ghillean T. Prance quote**: Aaron D. Davis, "What Does Science Say About God?," TattooPreacher.com, accessed July 2024, tattoopreacher.com/science-god/.
16. **Dr. Saami Shaibani quote**: Carl Wieland and Don Batten, "Physics, faith and the FBI, A talk with physicist Saami Shaibani," Creation.com, Creation Ministries International, accessed July 2024, creation.com/physics-faith-and-the-fbi. This article is from Creation 23(1):18–21, December 2000.
17. **Dr. Dudley Eirich quote**: Don Batten, "Separation of faith and reality, Editorial," Creation.com, Creation Ministries International, accessed July 2024, creation.com/separation-of-faith-and-reality. This article is from Creation 27(1):6, December 2004.
18. **Robert Millikan quote**: Joe Spears MS, "What Scientists REALLY Say about God, Creation and Evolution," Triangle Association for the Science of Creation, February 01, 2013, tasc-creationscience.org/article/what-scientists-really-say-about-god-creation-and-evolution as quoted from RA Millikan, The Autobiography of Robert A. Millikan, Prentice-Hall, Inc., New York, 1950, 288-287.
19. **Dr. Erwin Chargaff quote**: "Erwin Chargaff Quotes," AZ Quotes, accessed July 2024, www.azquotes.com/quote/586646 as quoted from Erwin Chargaff, "Voices in the labyrinth: nature, man, and science," Harper, San Francisco, 1977.
20. **Dr. Keith Wanser quote**: Carl Wieland and Jonathan Sarfati, "God and the electron, A talk with physicist Keith Wanser," Creation.com, Creation Ministries International, accessed July 2024, creation.com/god-and-the-electron. This article is from Creation 21(4):38–41, September 1999.

21. **Dr. Brian Schmidt quote**: Rajika Jayatilake, "Brian Schmidt: Neliti Face of Science 008," Breakthrough by neliti, August 16, 2023, breakthrough.neliti.com/brian-schmidt/.
22. **James A. Garfield quote**: "Astrology Quotes (46 quotes)," Today in Science History, accessed July 2024, todayinsci.com/QuotationsCategories/A_Cat/Astrology-Quotations.htm as quoted from "Speech (16 Dec 1867), given while a member of the U.S. House of Representatives, introducing resolution for the appointment of a committee to examine the necessities for legislation upon the subject of the ninth census to be taken the following year. Quoted in John Clark Ridpath, The Life and Work of James A. Garfield, 1821, 216."

CHAPTER 19

1. **Dr. Albert Einstein quote**: Tihomir Dimitrov, "50 Nobelists on God," issuu, Aug 31, 2017, issuu.com/tihomirdimitrov1/docs/50_nobelists_on_god.
2. **Abraham Lincoln quote**: "Science Quotes by Abraham Lincoln (9 quotes)," Today in Science History, accessed July 2024, todayinsci.com/L/Lincoln_Abraham/LincolnAbraham-Quotations.htm as cited, "Opening sentences of lecture 'Discoveries and inventions,' (1860) in Discoveries and Inventions (1915)."
3. **Dr. John Lennox quote**: "Top 40 John C. Lennox Quotes (2024 Update)," quotefancy, accessed July 2024, quotefancy.com/john-c-lennox-quotes.
4. **Joseph Henry quote**: "Joseph Henry Quotations," Smithsonian Institute Archives, accessed July 2024, siarchives.si.edu/sites/default/files/pdfs/jhpp/JHP_Joseph_Henry_Quotations.pdf as quoted first in the Smithsonian Annual Report for 1852, p. 15.
5. **Blaise Pascal quote**: "Blaise Pascal Quotes About Truth," AZ Quotes, accessed July 2024, www.azquotes.com/author/11361-Blaise_Pascal/tag/truth.
6. **James Clerk Maxwell quote**: Paul James-Griffiths, "Eighteenth and Nineteenth Centuries," Christian Heritage Edinburgh, Aug 23, 2016, www.christianheritageedinburgh.org.uk/category/science/quotes-from-famous-christians-in-science-and-medicine/. Quoted from Ann Lamont, 21 Great Scientists who believed the Bible, Creation Science Foundation, © 1995, p.228.

Chapter title page and section photographs from Pexels.com.

TOPIC & PERSON INDEX

SCRIPTURE INDEX

GENESIS

1 KINGS

2 KINGS

1 CHRONICLES

2 CHRONICLES

JOB

PSALMS

1 CORINTHIANS

2 CORINTHIANS

GALATIANS

EPHESIANS

PHILIPPIANS

COLOSSIANS

1 TIMOTHY

2 TIMOTHY

TITUS

Made in the USA
Las Vegas, NV
08 December 2024

13627660R00312